Daniels and Worthingham's

MUSCLE TESTING

Techniques of Manual Examination

Daniels and Worthingham's
MUSCLE TESTING

6th Edition

Techniques of Manual Examination

Helen J. Hislop, Ph.D., Sc.D., FAPTA

Professor and Chair, Department of Biokinesiology and Physical Therapy
University of Southern California, Los Angeles
Los Angeles, California

Jacqueline Montgomery, M.A., P.T.

Director of Physical Therapy
Rancho Los Amigos Medical Center, Downey, California
Clinical Professor, Department of Biokinesiology and Physical Therapy
University of Southern California, Los Angeles
Los Angeles, California

Contributor:

Barbara Connelly, Ed.D., P.T.

Chairman, Department of Rehabilitation Sciences
University of Tennessee
Memphis, Tennessee

W.B. SAUNDERS COMPANY
A Division of Harcourt Brace & Company
Philadelphia • London • Toronto • Montreal • Sydney • Tokyo

W.B. Saunders Company
A Division of Harcourt Brace & Company

The Curtis Center
Independence Square West
Philadelphia, Pennsylvania 19106

Library of Congress Cataloging-in-Publication Data

Hislop, Helen J.
    Daniels and Worthingham's muscle testing: techniques of manual examination.
— 6th ed. / Helen J. Hislop, Jacqueline Montgomery; contributor, Barbara
Connelly.
        p. cm.
    Rev. ed. of: Muscle testing / Lucille Daniels, Catherine Worthingham. 5th ed.
1986.
    Includes bibliographical references and index.
    ISBN 0–7216–4305–1
    1. Muscles—Examination.   I. Montgomery, Jacqueline.
II. Connelly, Barbara.   III. Daniels, Lucille.   Muscle testing.
IV. Title.   V. Title: Muscle testing.
    [DNLM: 1. Muscles—physiology.   2. Physical Examination—methods.
WE 500 H673d 1995]
RC925.7.D36   1995   616.7′40754--dc20
DNLM/DLC   95-13454

DANIELS AND WORTHINGHAM'S MUSCLE TESTING:
TECHNIQUES OF MANUAL EXAMINATION, Sixth Edition   ISBN 0–7216–4305–1
                                        International Edition: ISBN 0–7216–6774–0

Printed in the United States of America.

Last digit is the print number: 9   8   7   6   5   4   3   2   1

To two of the most illustrious and worthy physical therapists of their day and any other, in grateful appreciation for their majestic contributions to the profession of physical therapy

Catherine A. Worthingham, P.T., Ph.D.

Jacquelin Perry, P.T., M.D.

Preface

This sixth edition of *Daniels and Worthingham's Muscle Testing* represents a major departure from the earlier editions. The authors are new. The content has new sections on testing infants, patients with upper motor neuron diseases, and the respiratory muscles, and expanded sections on neck and bulbar testing. The text also includes a synopsis of muscle anatomy and muscle innervation so that readers, particularly students, can readily refresh their memories about the details of muscle topography and function. Also to assist the reader, each muscle has been given a constant reference number to speed cross-referencing and to locate details of any given muscle quickly in Chapter 9, Ready Reference.

The current authors gratefully acknowledge the basic contributions to this text of the original authors, Lucille Daniels and Catherine Worthingham. They established the format still followed, and they detailed the muscle testing techniques, modified slightly in this edition.

This book is a handbook of manual evaluation of muscular strength and is not intended for use as a comprehensive text for rationale and variations on such testing. Nor does its scope include instrumented methods of evaluation of force, work, and power.

The ultimate message of this book is that here are tried and true methods for assessing and grading skeletal muscle function. The techniques are not such that skill is achieved quickly despite the considerable detail used to describe them. The only way to acquire mastery of clinical evaluation procedures, including manual muscle testing, is to practice over and over again. As experience with patients matures over time, the nuances that can never be fully described for the wide variety of patients encountered by the clinician will become as much intuition as science. The master clinician will include muscle testing as part and parcel of every patient evaluation, whether a detailed form is completed or whether a prelude to program planning. It is, indeed, among the most fundamental skills of the physical therapist as well as the occupational therapist and other practitioners. Learn well. Develop a high level of skill, and maintain high standards. The patient will be the beneficiary of how well you achieve high proficiency.

We wish to acknowledge very substantial contributions from a number of persons who assisted with content review, editing, and suggestions. These include Barbara Connelly, for her chapter on manual testing in the infant; Maureen Rodgers, who reviewed every single test sequence and was the tester for the new illustrations used in this edition and is certainly the most gifted muscle tester we know; Jacquelin Perry, for her incredible knowledge of functional anatomy; Arthur Hsu, who let not even the slightest detail of anatomy go unchallenged; Linda Wood, for her insight into editing, which let the authors speak clearly to the reader; to our illustrators, Larry Ward, who did all of the clinical testing sequences, and Walter Stuart, for wonderful innervation figures and the new anatomy figures, including the three plates; and to Hazel Adkins and the memory of Viola Robins, for their pivotal work on respiratory and bulbar muscle testing in the days of polio. Finally, we wish to acknowledge Florence Kendall, not only for her pioneering work in muscle testing and her constant demand for high standards of patient care, but also for her friendship and discussion of testing procedures included here. We are deeply grateful to them all.

HELEN J. HISLOP, PH.D., P.T.
JACQUELINE MONTGOMERY, M.A., P.T.

Introduction

This book presents an approach to the assessment of muscular strength and function as fundamental components of movement and performance. Classic muscle testing involves manual methods of evaluation and draws upon the work and experience of a number of clinical scientists, some of whose work is corroborated by formal research. The majority of manual muscle testing procedures are just coming under scientific scrutiny, but almost a century of clinical use has provided a wealth of clinical corroboration for the empirical validity of such tests.

Use of manual muscle testing is valid in normal persons and those with weakness or paralysis secondary to motor unit disorders (lower motor neuron lesions and muscle disorders). Its use in persons with disturbances of the higher neural centers is flawed because of interference by abnormal sensation, or disturbed tone or motor control.

Nevertheless, muscle function must be assessed in such patients, although the procedures used may be quite different. One approach to overall movement analysis that can be used in patients with upper motor neuron disturbances is included in this book. Additional tests for these people remain to be codified, and other procedures, which probably will require the use of extensive technology, may be available for routine clinical use by the turn of the century.

This book, as in previous editions, directs its focus on manual procedures. Its organization is based on joint motions (e.g., hip flexion) rather than on individual muscles (e.g., iliopsoas). The reason for this approach is that each motion generally is the result of activity by more than one muscle, and although so-called prime movers can be identified, the importance of secondary or accessory movers should never be discounted. Rarely is a prime mover the only active muscle, and rarely is it used under isolated control for a given movement. For example, knee extension is the prerogative of the five muscles of the Quadriceps femoris, yet none of the five extends the knee in isolation from its synergists. Regardless, definitive activity of any muscle in a given movement can be precisely detected only by kinesiologic electromyography, and such studies, although numerous, remain incomplete.

There are examples of manual testing in which an examiner pre-positions a limb with the intent of ruling out a particular muscle from acting in a given movement. Newer work reporting on electromyographic recordings of muscles participating in manual tests, however, will shed light on the actual contributions of participating muscles in specific motions. One example of this is the test used to isolate the Soleus. The Gastrocnemius never turns off in any plantar flexion motion; therefore, it will contaminate any test that purports to isolate the Soleus. The Gastrocnemius does diminish its activity with the knee flexed, most notably when the knee is flexed beyond 45 degrees. The Gastrocnemius still contributes to plantar flexion in that posture, however, so the Soleus is not, in actual fact, totally "isolated." The reader is referred to the tests on plantar flexion for further details.

Range of motion in this book is presented only as information the physical therapist requires to test muscles correctly. A consensus of typical ranges is presented with each test, but the techniques of measurement used are not within the scope of this text.

Brief History of Muscle Testing

Wilhelmine Wright and Robert W. Lovett, M.D., Professor of Orthopedic Surgery at Harvard University Medical School, were the originators of the muscle testing system that incorporated the effect of gravity. Janet Merrill, P.T., Director of Physical Therapeutics at Children's Hospital and the Harvard Infantile Paralysis Commission in Boston, an early colleague of Dr. Lovett, stated that the tests were used first by Wright in Lovett's office gymnasium in 1912.[1] The seminal description of the tests used largely today was written by Wright and published in 1912[2]; this was followed by an article by Lovett and Martin in 1916[3] and by Wright's book in 1928.[4] Miss Wright was a precursor of the physical therapist of today, there being no educational programs in physical therapy in her time, but she headed Lovett's physical therapeutic clinic. Lovett credits her fully in his 1917 book, *Treatment of Infantile Paralysis,*[5] with developing the testing for polio (see Sidebar). In this book, muscles are tested using a resistance-gravity system and graded on a scale of 0 to 6. Another early numerical scale in

H.S. Stewart, a physician, published a description of muscle testing in 1925 that was very brief and was not anatomically or procedurally consistent with what is done today.[8] His descriptions included a resistance-based grading system not substantially different from that in use today: maximal resistance for a normal muscle; completion of the motion against gravity with no other resistance for a grade of Fair, and so forth.

Among the earliest clinicians to organize muscle testing and support such testing with sound and documented kinesiologic procedures in the way they are used today were Henry and Florence Kendall. Their earliest published documents on comprehensive manual muscle testing became available in 1936 and 1938.[9,10] The 1938 monograph on muscle testing was published and distributed to all Army hospitals in the United States by the U.S. Public Health Service. Another early contribution came from Signe Brunnstrom and Marjorie Dennen in 1931; their syllabus described a system of grading movement rather than individual muscles as a modification of Lovett's work with gravity and resistance.[11]

In this same time period, Elizabeth Kenny came to the United States from her unique experiences of treating polio victims in the Australian back country. Kenny made no contributions to muscle testing, and in her own book and speeches she was clearly against such an evaluative procedure, which she deemed to be harmful.[12] Her one contribution was to heighten the awareness of organized medicine to the dangers of prolonged and injudicious immobilization of the polio patient, something that physical therapists in this country had been saying for some time but were not widely heeded at the time.[12,13] Kenny also advocated the early use of "hot fomentations" (hot packs) in the acute phase of the disease.[12] In fact, Kenny vociferously maintained that poliomyelitis was not a central nervous system disease resulting in flaccid paresis or paralysis but rather "mental alienation" of muscles from the brain.[12,14] In her system "deformities never occurred,"[12] but neither did she ever present data on muscular strength or imbalance in her patients at any point in the course of their disease.[13,14]

Another muscle testing book was published by muscle testing was described by Charles L. Lowman, M.D., founder and medical director of Orthopedic Hospital, Los Angeles. Lowman's system (1927) covered the effects of gravity and the full range of movement on all joints and was particularly helpful for assessing extreme weakness.[6] Lowman further described muscle testing procedures in the Physical Therapy Review in 1940.[7]

A.T. Legg and Janet Merrill in 1932 and used extensively in schools during the early 1940s.[15] This book offered a comprehensive system of muscle testing; muscles were graded on a scale of 0 to 5, and a plus or minus designation was added to all grades except 1 and zero.

The first comprehensive text on muscle testing still in print is the predecessor of this sixth edition of manual muscle testing; it was written by Lucille Daniels, M.A., P.T., Marian Williams, Ph.D., P.T., and Catherine Worthingham, Ph.D., P.T., and published in 1946.[16] These three authors prepared a comprehensive handbook on the subject of manual testing procedures that was concise and easy to use. It remains one of the most used texts the world over at the present time. Williams died in 1962, and her two coauthors revised the 5th edition alone.

The Kendalls (together and then Florence alone after Henry's death in 1979) developed and published work on muscle testing and related subjects for more than six decades, certainly one of the more remarkable sagas in physical therapy or even medical history.[10, 17–19] Their first edition of *Muscles: Testing and Function* appeared in 1949.[17] Earlier, the Kendalls had developed a percentage system ranging from 0 to 100 to express muscle grades as a reflection of normal; they then reduced the emphasis on this scale, only to return to it in the latest edition (1993), in which Florence again advocated the 0 to 10 scale.[19] The contributions of the Kendalls, however, should not be thought to be limited to grading scales. Their integration of muscle function with posture and pain in two separate books[17,18] and then in one book[19] is a unique and extremely valuable contribution to the clinical science of physical therapy.

Muscle testing procedures used in national field trials that examined the use of gamma globulin in the prevention of paralytic poliomyelitis were described by Carmella Gonnella, Georgianna Harmon, and Miriam Jacobs, all physical therapists.[20] The later field trials for the Salk vaccine also used muscle testing procedures.[21] The epidemiology teams at the Centers for Disease Control were charged with assessing the validity and reliability of the vaccine. Because there was no other method of accurately "measuring" the presence or absence of muscular weakness, manual muscle testing techniques were used.

A group from the D.T. Watson School of Physiatrics near Pittsburgh, which included Jesse Wright, M.D., Mary Elizabeth Kolb, P.T., and Miriam Jacobs, P.T., devised a test procedure that eventually was used in the field trials.[22] The test was an abridged version of the complete test procedure but did test key muscles in each functional group and body part. It used numerical values that were assigned grades, and each muscle or muscle group also had an arbitrary assigned factor that corresponded (as closely as possible) to the bulk of the tissue. The bulk factor multiplied by the test grade resulted in an "index of involvement" expressed as a ratio.[21]

Before the trials, Kolb and Jacobs were sent to Atlanta to train physicians to conduct the muscle tests, but it was decided that experienced physical therapists would be preferable to maintain the reliability of the test scores.[22] Lucy Blair, then the Poliomyelitis Consultant in the American Physical Therapy Association, was asked by Catherine Worthingham of the National Foundation for Infantile Paralysis to assemble a team of experienced physical therapists to conduct the muscle tests for the field trials. A group of 67 therapists was trained by Kolb and Jacobs in the use of the abridged muscle test. A partial list of participants was appended to the Lilienfeld paper in the Physical Therapy Review in 1954 (p. 289).[21] This approach and the evaluations by the physical therapists of the presence or absence of weakness and paralysis in the field trial samples eventually resulted in resounding approval of the Salk vaccine.

Since the polio vaccine field trials, sporadic research in manual muscle testing has occurred as well as continued challenges of its worth as a valid clinical assessment tool. Iddings and colleagues noted that inter-tester reliability among practitioners varied by about 4 percent, which compares favorably with the 3 percent variation among the carefully trained therapists who participated in the vaccine field trials.[23]

There is growing interest in establishing norms of muscular strength and function. Early efforts in this direction were begun by Willis Beasley[24] (although his earliest work was presented only at scientific meetings) and continued by Marian Williams[25] and Helen J. Hislop,[26,27] which set the stage for objective measures by Bohannon[28] and others. The literature on objective measurement increases yearly, an effort that is long overdue. The data from these studies must be applied to manual testing so that correlations between instrumented muscle assessment and manual assessment can ensue.

In the meantime, until instrumented methods become affordable for every clinic, manual techniques of muscle testing will remain in use. The skill of manual muscle testing is a critical clinical tool that every physical therapist must not only learn but must master. A physical therapist who aspires to recognition as a master clinician will never achieve that status without acquiring exquisite skills in manual muscle testing and precise assessment of muscle performance.

How to Use This Book

The general principles that govern manual muscle testing are described in Chapter 1. Chapters 2 through 8 present the techniques for testing motions of skeletal muscle groups in the body region covered by that chapter. Each muscle test is described in sequential detail and is accompanied by illustrations that help the user perform the test.

For instant access to anatomical information without carrying a large anatomy text to a muscle testing session, a Ready Reference Anatomy section is given in Chapter 9. This chapter is a synopsis of muscle anatomy, muscles as part of motions, muscle innervation, and myotomes.

To assist readers, each muscle has been assigned a reference number based on a regional sequence beginning with the head and face and proceeding through the neck, thorax, abdomen, perineum, upper extremity, and lower extremity. This reference number is retained throughout the text for cross-reference purposes. For example, the Multifidi are referenced as muscle number 94; the Flexor digiti minimi brevis in the hand is number 160; and the muscle of the same name in the foot is number 216. The purpose of these reference numbers is to allow the reader to refer quickly from a muscle listed on the testing page to a more detailed description of its anatomy and innervation in the Ready Reference Anatomy section.

Two lists of muscles with their reference numbers are presented: one alphabetical and one by region to assist readers in finding muscles in the Ready Reference section.

NAMES OF THE MUSCLES

Muscle names have conventions of usage. The most formal usage (and the correct form for many journal manuscripts) is the terminology established by the International Anatomical Nomenclature Committee and approved or revised in 1955, 1960, and 1965.[29] Common usage, however, often neglects these prescribed names in favor of shorter or more readily pronounced names. The authors of this text make no apologies for not keeping strictly to formal usage. The majority of the muscles cited do follow the Nomina Anatomica. Others are listed by the names in most common use. The alphabetical list of muscles (see page 328) gives the name used in this text and the correct Nomina Anatomica term, when it differs, in parentheses.

THE CONVENTION OF ARROWS IN THE TEXT

Open arrows in the text denote the direction of movement of a body part, either actively by the patient or passively by the examiner. The length and direction of the arrow indicates the relative excursion of the part.

Examples:

Closed arrows in the text denote resistance by the examiner. The arrow indicates distance, and the width gives some relative idea of whether resistance is large or small.

Examples:

REFERENCES

1. Merrill J. Personal letter dated January 5, 1945 to Lucille Daniels.
2. Wright WG. Muscle training in the treatment of infantile paralysis. M. & S. J. 167:567–574, 1912.
3. Lovett RW, Martin EG. Certain aspects of infantile paralysis and a description of a method of muscle testing. JAMA 66:729–733, 1916.
4. Wright WG. *Muscle Function.* New York: Paul B. Hoeber, 1928.
5. Lovett RW. *Treatment of Infantile Paralysis,* 2nd ed. Philadelphia: Blakiston's Son & Co., 1917.
6. Lowman CL. A method of recording muscle tests. Am J Surg 3:586–591, 1927.
7. Lowman CL. Muscle strength testing. Physiother Rev 20:69–71, 1940.
8. Stewart HS. *Physiotherapy: Theory and Clinical Application.* New York: Paul B. Hoeber, 1925.
9. Kendall HO. Some interesting observations about the after care of infantile paralysis patients. J Excep Child 3:107, 1936.
10. Kendall HO, Kendall FP. Care during the recovery period of paralytic poliomyelitis. U.S. Public Health Bulletin No. 242. Washington D.C.: U.S. Government Printing Office, 1938.
11. Brunnstrom S, Dennen M. Round table on muscle testing. New York: Annual Conference of the American Physical Therapy Association, Federation of Crippled and Disabled, Inc. (mimeographed), 1931.
12. Kenny E. Paper read at Northwestern Pediatric Conference at St. Paul University Club; November 14, 1940.
13. Plastridge AL. Personal report to the National Foundation for Infantile Paralysis after a trip to observe work of Sister Kenny, 1941.
14. Kendall HO, Kendall FP. Report on the Sister Kenny Method of Treatment in Anterior Poliomyelitis made to the National Foundation for Infantile Paralysis. New York, March 10, 1941.
15. Legg AT, Merrill J. Physical therapy in infantile paralysis. In: Mock. *Principles and Practice of Physical Therapy,* Vol. 2. Hagerstown, MD: W.F. Prior, 1932.
16. Daniels L, Williams M, Worthingham CA. *Muscle Testing: Techniques of Manual Examination.* Philadelphia: W.B. Saunders, 1946.
17. Kendall HO, Kendall FP. *Muscles: Testing and Function.* Baltimore: Williams & Wilkins, 1949.
18. Kendall HO, Kendall FP. *Posture and Pain.* Baltimore, Williams & Wilkins, 1952.
19. Kendall FP, McCreary EK, Provance PG. *Muscles: Testing and Function,* 4th ed. Baltimore: Williams & Wilkins, 1993.
20. Gonnella C, Harmon G, Jacobs M. The role of the physical therapist in the gamma globulin poliomyelitis prevention study. Phys Ther Rev 33:337–345, 1953.
21. Lilienfeld AM, Jacobs M, Willis M. Study of the reproducibility of muscle testing and certain other aspects of muscle scoring. Phys Ther Rev 34:279–289, 1954.
22. Kolb ME. Personal communication, October 1993.
23. Iddings DM, Smith LK, Spencer WA. Muscle testing. Part 2: Reliability in clinical use. Phys Ther Rev 41:249–256, 1961.
24. Beasley W. Quantitative muscle testing: Principles and applications to research and clinical services. Arch Phys Med Rehabil 42:398–425, 1961.

25. Williams M, Stutzman L. Strength variation through the range of joint motion. Phys Ther Rev 39:145–152, 1959.
26. Hislop HJ. Quantitative changes in human muscular strength during isometric exercise. Phys Ther 43:21–36, 1963.
27. Hislop HJ, Perrine JJ. Isokinetic concept of exercise. Phys Ther 47:114–117, 1967.
28. Bohannon RW. Manual muscle test scores and dynamometer test scores of knee extension strength. Arch Phys Med Rehabil 67:204, 1986.
29. International Anatomical Nomenclature Committee. *Nomina Anatomica*. Amsterdam: Excerpta Medica Foundation, 1965.

OTHER READINGS

Bailey JC. Manual muscle testing in industry. Phys Ther Rev 41:165–169, 1961.

Bennett RL. Muscle testing: A discussion of the importance of accurate muscle testing. Phys Ther Rev 27:242–243, 1947.

Borden R, Colachis S. Quantitative measurement of the Good and Normal ranges in muscle testing. Phys Ther 48:839–843, 1968.

Brunnstrom S. Muscle group testing. Physiother Rev 21:3–21, 1941.

Currier DP. Maximal isometric tension of the elbow extensors at varied positions. Phys Ther 52:52, 1972.

Downer AH. Strength of the elbow flexor muscles. Phys Ther Rev 33:68–70, 1953.

Fisher FJ, Houtz SJ. Evaluation of the function of the gluteus maximus muscle. Am J Phys Med 47:182–191, 1968.

Frese E, Brown M, Norton BJ. Clinical reliability of manual muscle testing: Middle trapezius and gluteus medius muscles. Phys Ther 67:1072–1076, 1987.

Gonnella C. The manual muscle test in the patient's evaluation and program for treatment. Phys Ther Rev 34:16–18, 1954.

Granger CV. The clinical discernment of muscle weakness. Arch Phys Med 44:430–438, 1963.

Hoppenfeld S. *Physical Examination of the Spine and Extremities*. New York: Appleton-Century-Crofts, 1976.

Janda V. *Muscle Function Testing*. Boston: Butterworths, 1983.

Jarvis DK. Relative strength of hip rotator muscle groups. Phys Ther Rev 32:500–503, 1952.

Kendall FP. Testing the muscles of the abdomen. Phys Ther Rev 21:22–24, 1941.

Palmer ML, Epler ME. *Clinical Assessment Procedures in Physical Therapy*. Philadelphia: J.B. Lippincott, 1990.

Salter N, Darcus HD. Effect of the degree of elbow flexion on the maximum torque developed in pronation and supination of the right hand. J Anat 86-B:197, 1952.

Smidt GL, Rogers MW. Factors contributing to the regulation and clinical assessment of muscular strength. Phys Ther 62:1283–1289, 1982.

Wadsworth CT, Krishnan R, Sear M, et al. Intrarater reliability of manual muscle testing and hand held dynametric testing. Phys Ther 67:1342–1347, 1987.

Wintz M. Variations in current muscle testing. Phys Ther Rev 39:466–475, 1959.

Zimny N, Kirk C. Comparison of methods of manual muscle testing. Clin Manag 7:6–11, 1987.

Contents

Chapter 1

Principles of Manual Muscle Testing 1
The Grading System, 2
Criteria for Assigning a Muscle Test Grade, 4
Screening Tests, 6
Preparing for the Muscle Test, 6

Chapter 2

Testing the Muscles of the Neck 11
Capital Extension, 12
Cervical Extension, 16
Combined Neck Extension (Capital plus cervical), 19
Capital Flexion, 21
Cervical Flexion, 24
Combined Cervical Flexion (Capital plus cervical), 28
Combined Flexion to Isolate a Single Sternocleidomastoid, 30
Cervical Rotation, 31

Chapter 3

Testing the Muscles of the Trunk 33
Trunk Extension, 34
Elevation of the Pelvis, 38
Trunk Flexion, 41
Trunk Rotation, 45
Inspiration (quiet), 50
Forced Expiration, 54

Chapter 4

Testing the Muscles of the Upper Extremity 57
Scapular Abduction and Upward Rotation (Serratus anterior), 58
Scapular Elevation (Trapezius, upper fibers), 65

Scapular Adduction (Trapezius, middle fibers), 69
Scapular Depression and Adduction (Trapezius, lower fibers), 73
Scapular Adduction and Downward Rotation (Rhomboids), 76
Shoulder Flexion (Anterior Deltoid, Supraspinatus, and Coracobrachialis), 81
Shoulder Extension (Latissimus dorsi, Teres major, Posterior Deltoid), 84
Shoulder Scaption (Deltoid and Supraspinatus), 88
Shoulder Abduction (Middle Deltoid and Supraspinatus), 90
Shoulder Horizontal Abduction (Posterior Deltoid), 94
Shoulder Horizontal Adduction (Pectoralis major), 97
Shoulder External Rotation (Infraspinatus and Teres minor), 102
Shoulder Internal Rotation (Subscapularis), 105
Elbow Flexion (Biceps, Brachialis, and Brachioradialis), 108
Elbow Extension (Triceps brachii), 114
Forearm Supination (Supinator and Biceps brachii), 118
Forearm Pronation (Pronator teres and Pronator quadratus), 121
Wrist Flexion (Flexor carpi radialis and Flexor carpi ulnaris), 124
Wrist Extension (Extensor carpi radialis longus, Extensor carpi radialis brevis, and Extensor carpi ulnaris), 128
Finger MP Flexion (Lumbricales and Interossei), 132
Finger PIP and DIP Flexion (Flexor digitorum superficialis and Flexor digitorum profundus), 135
PIP Tests, 136
DIP Tests, 138
Finger MP Extension (Extensor digitorum, Extensor indicis, Extensor digiti minimi), 139
Finger Abduction (Dorsal Interossei), 142
Finger Adduction (Palmar Interossei), 146
Thumb MP and IP Flexion (Flexor pollicis brevis and Flexor pollicis longus), 148
Thumb MP Flexion Tests (Flexor pollicis brevis), 149
Thumb IP Flexion Tests (Flexor pollicis longus), 150
Thumb MP and IP Extension (Extensor pollicis brevis and Extensor pollicis longus), 152

Thumb MP Extension Tests (Extensor pollicis brevis), 153

Thumb IP Extension Tests (Extensor pollicis longus), 154

Thumb Abduction (Abductor pollicis longus and Abductor pollicis brevis), 156

Abductor Pollicis Longus Test, 157

Abductor Pollicis Brevis Test, 158

Thumb Adduction (Adductor pollicis), 160

Opposition (Thumb to Little Finger) (Opponens pollicis and Opponens digiti minimi), 163

Chapter **5**

Testing the Muscles of the Lower Extremity 167

Introduction to Testing the Hip, 168

Hip Flexion (Psoas major and Iliacus), 169

Hip Flexion, Abduction, and External Rotation with Knee Position (Sartorius), 173

Hip Extension (Gluteus maximus and Hamstrings), 176

Hip Extension Tests Modified for Hip Flexion Tightness, 181

Hip Abduction (Glutei medius and minimus), 182

Hip Abduction from Flexed Position (Tensor fasciae latae), 186

Hip Adduction (Adductors magnus, brevis and longus; Pectineus and Gracilis), 190

Hip External Rotation (Obturators internus and externus, Gemellae superior and inferior. Piriformis, Quadratus femoris, Gluteus maximus {posterior}), 194

Hip Internal Rotation (Glutei minimus and medius; Tensor fasciae latae), 198

Knee Flexion (All hamstring muscles), 202

Knee Extension (Quadriceps femoris), 207

Ankle Plantar Flexion (Gastrocnemius and Soleus), 211

Gastrocnemius and Soleus Test, 212

Plantar Flexion, Soleus Only, 215

Foot Dorsiflexion and Inversion (Anterior tibialis), 218

Foot Inversion (Posterior tibialis), 221

Foot Eversion with Plantar Flexion or Dorsiflexion (Peronei longus and brevis), 224

Hallux and Toe MP Flexion (Lumbricales and Flexor hallucis brevis), 227

Hallux MP Flexion (Flexor hallucis brevis), 228

Toe MP Flexion (Lumbricales), 229

Hallux and Toe DIP and PIP Flexion (Flexor digitorum longus, Flexor digitorum brevis, Flexor hallucis longus), 230

Hallux and Toe MP and IP Extension (Extensor digitorum longus and brevis; Extensor hallucis longus), 232

Chapter **6**

Testing in Infants and Children 235
Barbara Connolly, Ed. D., P.T.

Chapter **7**

Assessment of Muscles Innervated by Cranial Nerves 261

Introduction to Testing and Grading, 262

Extraocular Muscles of the Eye, 263

The Muscles of the Face, 267

Muscles of the Eyelids, Eyebrows, and Forehead, 268

The Nose Muscles, 276

Muscles of the Mouth, 278

Muscles of Mastication, 284

Muscles of the Tongue, 290

Muscles of the Palate, 298

Muscles of the Pharynx, 304

Muscles of the Larynx, 308

Swallowing, 314

Muscle Actions in Swallowing, 314

Testing Swallowing, 315

Chapter **8**

Upright Motor Control 319

The Test for Upright Control, 320

Flexion Control Test, 320

Extension Control Test in Parts 4, 5, and 6, 321

Chapter **9**

Ready Reference Anatomy 327

Using This Ready Reference Section, 328

PART I. ALPHABETICAL LIST OF MUSCLES ▪ 328

PART II. LIST OF MUSCLES BY REGION ▪ 331

PART III. SKELETAL MUSCLES OF THE HUMAN BODY ▪ 334

PART IV. MOTIONS AND THEIR PARTICIPATING MUSCLES (MOTIONS OF THE NECK, TRUNK, AND LIMBS) ▪ 401

PART V. CRANIAL AND PERIPHERAL NERVES AND THE MUSCLES THEY INNERVATE ▪ 410

PART VI. MYOTOMES; THE MOTOR NERVE ROOTS AND THE MUSCLES THEY INNERVATE ▪ 417

Bibliography 427
Cited References 427
Index 429

Principles of Manual Muscle Testing

The Grading System
    The Break Test
    Active Resistance Test
    Application of Resistance
    The Examiner and the Value of the Muscle Test
    Influence of the Patient on the Test
Criteria for Assigning a Muscle Test Grade
    The Grade 5 (Normal) Muscle
    The Grade 4 (Good) Muscle
    The Grade 3 (Fair) Muscle
    The Grade 2 (Poor) Muscle
    The Grade 1 (Trace) Muscle
    The Grade 0 (Zero) Muscle
    Plus (+) and Minus (−) Grades
    Available Range of Motion
Screening Tests
Preparing for the Muscle Test
Summary

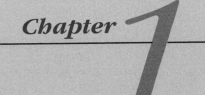

Chapter 1

The Grading System

Grades for a manual muscle test are recorded as numerical scores ranging from zero (0), which represents no activity, to five (5), which represents a normal response to the test, or as normal a response as can be evaluated by a manual muscle test. Because this text is based on tests of motions rather than tests of individual muscles, the grade represents the performance of all muscles in that motion. The 5 to 0 system is the most commonly used convention.

Each numerical grade can be paired with a word that describes the test performance in qualitative terms. These qualitative terms, when written, are capitalized to indicate that they too represent a score.

Numerical Score	Qualitative Score
5	Normal (N)
4	Good (G)
3	Fair (F)
2	Poor (P)
1	Trace activity (T)
0	Zero (no activity) (0)

These grades are based on several factors of testing and response.

THE BREAK TEST

Manual resistance is applied to a limb or other body part after it has completed its range of movement or after it has been placed at end range by the examiner. The term resistance is always used to denote a force that acts in opposition to a contracting muscle. Manual resistance should always be applied in the direction of the "line of pull" of the participating muscle or muscles. At the end of the available range, or at a point in the range where the muscle is most challenged, the patient is asked to hold the part at that point and not allow the examiner to "break" the hold with manual resistance. For example, a seated subject is asked to flex the elbow to its end range; when that position is reached, the examiner applies resistance at the wrist, trying to force the elbow to "break" its hold and move downward into extension. This is called a break test, and it is the procedure most commonly used in manual muscle testing today.

As a recommended alternative procedure, the examiner may choose to place the muscle group to be tested in the end or test position rather than have the patient actively move it there. In this procedure the examiner ensures correct positioning and stabilization for the test.

ACTIVE RESISTANCE TEST

An alternative to the break test is the application of manual resistance against an actively contracting muscle or muscle group (i.e., against the direction of the movement as if to prevent that movement). This may be called an active resistance test. During the motion, the examiner gradually increases the amount of manual resistance until it reaches the maximal level the subject can tolerate and motion ceases. This kind of manual muscle test requires considerable skill and experience to perform and is so often equivocal that its use is not recommended.

APPLICATION OF RESISTANCE

The principles of manual muscle testing presented here and in all published sources since 1921 follow the basic tenets of muscle length–tension relationships as well as those of joint mechanics.[1,2] In the case of the Biceps brachii, for example, when the elbow is straight, the Biceps lever is short; leverage increases as the elbow flexes and becomes maximal (most efficient) at 90 degrees, but as flexion continues beyond that point, the lever arm again decreases in length and efficiency.

In manual muscle testing, the application of external force (resistance) at the end of the range in one-joint muscles allows consistency of procedure rather than an attempt to select the estimated midrange position. In two-joint muscles the point of maximum resistance is generally at or near midrange (e.g., the medial or lateral hamstring muscles).

The point on an extremity or part where the examiner should apply resistance is near the distal end of the segment to which the muscle attaches. There are two common exceptions to this rule: the hip abductors and the scapular muscles. In testing the hip abductor muscles, resistance would be applied at the distal end of the femur just above the knee. The abductor muscles are so strong, however, that most examiners, in testing a patient with normal knee strength and joint integrity, will choose to apply resistance at the ankle. The longer lever provided by resistance at the ankle is a greater challenge for the abductors and is more indicative of the functional demands required in gait. In the patient who has a weak knee, resistance to the abductors should be applied at the distal femur just above the knee. When using the short lever, abductor strength must be graded no better than Grade 4 (Good) even when the muscle takes maximum resistance.

An example of testing with a short lever occurs in the patient with an above-knee amputation, where the grade awarded, even when the patient can hold against maximal resistance, is Grade 4 (Good). This is done because of the loss of the weight of the leg and is particularly important when the examiner is evaluating the patient for a prosthesis. The muscular force available should not be overestimated in predicting the patient's ability to use the prosthesis.

In testing the vertebroscapular muscles (e.g., Rhomboids), the preferred point of resistance is on the arm rather than on the scapula where these muscles insert. The longer lever more closely reflects the functional demands that incorporate the weight of the arm. Other exceptions to the general rule of applying distal resistance include contraindications such as a painful condition or a healing wound in a place where resistance might otherwise be given.

The application of manual resistance to a part should never be sudden or uneven (jerky). The examiner should apply resistance somewhat slowly and gradually, allowing it to build to the maximum tolerable intensity.

THE EXAMINER AND THE VALUE OF THE MUSCLE TEST

The knowledge and skill of the examiner determine the accuracy and defensibility of a manual muscle test. Specific aspects of these qualities include the following:

- Knowledge of the location and anatomic features of the muscles in a test. In addition to knowing the muscle attachments, the examiner should be able to visualize the location of the tendon and its relationship to other tendons and structures in the same area (e.g., the tendon of the extensor carpi radialis longus lies on the radial side of the tendon of the extensor carpi radialis brevis at the wrist).
- Knowledge of the direction of muscle fibers and their "line of pull" in each muscle.
- Knowledge of the function of the participating muscles (e.g., synergists, prime movers, accessories).
- Familiarity with the positioning and stabilization needed for each test procedure.
- Ability to identify patterns of substitution in a given test and how they can be spotted based on a knowledge of which other muscles can be substituted for the one(s) being tested.
- Ability to detect contractile activity during both contraction and relaxation, especially in minimally active muscle.
- Sensitivity to differences in contour and bulk of the muscles being tested in contrast to the con-

tralateral side or to normal expectations based on body size, occupation, and so on.
- Awareness of any deviation from normal values for range of motion and the presence of any joint laxity or deformity.
- Understanding that the muscle belly must not be grasped at any time during a manual muscle test except specifically to assess tenderness or pain and muscle mass.
- Ability to identify muscles with the same innervation, which will ensure a comprehensive muscle evaluation and accurate interpretation of test results (because weakness of one muscle in a myotome should require examination of all).
- Knowledge of the relationship of the diagnosis to the sequence and extent of the test (e.g., the C7 complete quadriplegic client will require definitive muscle testing of the upper extremity but only confirmatory tests in the lower extremities).
- Ability to modify test procedures when necessary while not compromising the test result and understanding of the influence of the modification on the result.
- Knowledge of the effect of fatigue on the test results, especially muscles tested late in a long testing session, and a sensitivity to fatigue in certain diagnostic conditions such as myasthenia gravis or Eaton-Lambert syndrome.
- Understanding of the effect of sensory loss on movement.

The examiner also may inadvertently *influence* the test results and should be especially alert when testing in the following situations:

- The patient with open wounds or other conditions requiring gloves, which may blunt palpation skills.
- The patient who must be evaluated under difficult conditions: such as the patient in an intensive care unit with multiple tubes and monitors, the patient in traction, the patient in whom turning is contraindicated, the patient on a ventilator, and the patient in restraints.

The novice muscle tester must avoid the temptation to use short cuts or "tricks of the trade" before mastering the basic procedures lest such short cuts become an inexact personal standard. One such pitfall for the novice tester is to inaccurately assign a muscle grade from one test position that the patient could not perform successfully to a lower grade without actually testing in the position required for the lower grade.

For example, when testing trunk flexion, a patient partially clears the scapula from the surface with the hands clasped behind the head (the position for the Grade 5 test). The temptation may exist to assign a grade of 4 to this test, but this may "overrate" the true strength of trunk flexion unless the patient is actually tested with the arms across the chest to confirm Grade 4.

EARLY KENDALL

Accuracy in giving examinations depends primarily on the examiner's knowledge of the isolated and combined actions of muscles in individuals with normal, as well as those with weak or paralyzed, muscles.

The fact that muscles act in combination permits substitution of a strong muscle for a weaker one. For accurate muscle examinations, no substitutions should be permitted; that is, the movement described as a test movement should be done without shifting the body or turning the part to allow other muscles to perform the movement for the weak or paralyzed group. The only way to recognize substitution is to know normal function, and realize the ease with which a normal muscle performs the exact test movement.

KENDALL HO, KENDALL FP

From *Care During the Recovery Period in Paralytic Poliomyelitis.* Public Health Bulletin No. 242. Washington, D.C., U.S. Government Printing Office, 1937, 1939, p. 26.

The good clinician never ignores a patient's comments and must be a good listener, not just to questions but also to the words the patient uses and their meaning. This quality is the first essential of good communication and the means of encouraging understanding and respect between therapist and patient. The patient is the best guide to a successful muscle test.

INFLUENCE OF THE PATIENT ON THE TEST

The intrusion of a living, breathing, feeling person into the neat test package may distort scoring for the unwary examiner. The following circumstances should be recognized:

- There may be variation in the assessment of the true effort expended by a patient in a given test (reflecting the patient's desire to do well or to seem more impaired than is actually the case).
- The patient's willingness to endure discomfort or pain may vary (e.g., in the stoic, the whiner, the high competitor).
- The patient's ability to understand the test requirements may be limited in some cases because of comprehension and language barriers.
- The motor skills for the test may be beyond some patients (e.g., in the clumsy or inept patient who just cannot perform as requested).
- Lassitude and depression may cause the patient to be indifferent to the test and the examiner.
- Cultural, social, and gender issues may be associated with palpation and exposure of a body part for testing.

PRINCIPLES OF TESTING (1925)

The following points are applicable to nearly every case requiring muscle [testing] and are of the utmost importance for successful work:

1. *Determine just what muscles are involved by careful testing and chart the degree of power in each muscle or group to be treated.*

2. *Insist on such privacy and discipline as will gain the patient's cooperation and undivided attention . . . .*

3. *Use some method of preliminary warming up of the muscles . . . doubly essential in the cold, cyanotic and weakened muscles . . . .*

4. *Have the entire part free from covering and so supported as not to bring strain . . . from gravity . . . or antagonists.*

HARRY EATON STEWART, M.D.

From *Physiotherapy: Theory and Clinical Application.* New York: Paul B. Hoeber, 1925.

Criteria for Assigning a Muscle Test Grade

The grade given on a manual muscle test comprises both subjective and objective factors. Subjective factors include the examiner's impression of the amount of resistance to give before the actual test and then the amount of resistance the patient actually tolerates during the test. Objective factors include the ability of the patient to complete a full range of motion, or to hold the position once placed there, and to move the part against gravity or an inability to move it at all. All these factors require clinical judgment, which makes manual muscle testing an exquisite skill that requires considerable experience to master. An accurate test grade is important not only to establish a functional diagnosis but also to assess the patient's longitudinal progress during the period of recovery and treatment.

THE GRADE 5 (NORMAL) MUSCLE

The wide range of "normal" muscle performance leads to a considerable underestimation of a muscle's capability. If the examiner has no experience in examining persons who are free of disease or injury, it is unlikely that there will be any realistic judgment of what is Normal and how much normality can vary. Generally, a physical therapy student learns manual muscle testing by practicing on classmates, but this provides only minimal experience compared to what is needed to master the skill. It should be recognized, for example, that the average physical therapist cannot "break" knee extension in a reasonably fit young man, even by doing a hand stand on his leg! This and similar observations were derived by objective comparisons of movement performance acquired by assessing the amount of resistance given and then testing the muscle group's maximal capacity on an isokinetic dynamometer.[3–6]

The examiner should test normal muscles at every opportunity, especially when testing the contralateral limb in a patient with a unilateral problem. In almost every instance when the examiner cannot break the patient's hold position, a grade of 5 (Normal) is assigned. This value must be accompanied by the ability to complete a full range of motion or maintain end-point range against maximum resistance.

THE GRADE 4 (GOOD) MUSCLE

The grade of 4 (Good) represents the true weakness in manual muscle testing procedures (pun intended). Sharrard counted alpha motor neurons in the spinal cords of poliomyelitis victims at the time of autopsy.[7] He correlated the manual muscle test grades in the patient's chart with the number of motor neurons remaining in the anterior horns. His data revealed that more than 50 percent of the pool of motor neurons to a muscle group were gone when the muscle test result had been recorded as Grade 4 (Good). Thus, when the muscle can withstand considerable but less than "normal" resistance, it has already been deprived of at least half of its innervation.

Grade 4 is used to designate a muscle group that is able to complete a full range of motion against gravity and can tolerate strong resistance without breaking the test position. The Grade 4 muscle "gives" or "yields" to some extent at the end of its range with maximum resistance. When maximum resistance clearly results in a break, the muscle is assigned a grade of 4 (Good).

THE GRADE 3 (FAIR) MUSCLE

The Grade 3 muscle test is based on an objective measure. The muscle or muscle group can complete a full range of motion against only the resistance of gravity. If a tested muscle can move through the full range against gravity but additional resistance, however mild, causes the motion to break, the muscle is assigned a grade of 3 (Fair).

Sharrard cited a residual autopsy motor neuron count of 15 percent in polio-paretic muscles that had been assessed as Grade 3, meaning that 85 percent of the innervating neurons had been destroyed.[7]

Direct force measurements have demonstrated that the force level of the Grade 3 muscle usually is low, so that a much greater span of functional loss

exists between Grades 3 and 5 than between Grades 3 and 1. A grade of 3 (Fair) may be said to represent a definite *functional threshold* for each movement tested, indicating that the muscle or muscles can achieve the minimum task of moving the part upward against gravity through its range of motion. Although this ability is significant for the upper extremity, it falls far short of the functional requirements of many lower extremity muscles used in walking, particularly such groups as the hip abductors and the plantar flexors. The examiner must be sure that muscles given a grade of 3 are not in the joint "locked" position during the test (e.g., locked knee when testing knee extension).

THE GRADE 2 (POOR) MUSCLE

The Grade 2 (Poor) muscle is one that can complete the full range of motion in a position that minimizes the force of gravity. This "gravity-eliminated" position often is described as the horizontal plane of motion.

THE GRADE 1 (TRACE) MUSCLE

The Grade 1 (Trace) muscle means that the examiner can detect visually or by palpation some contractile activity in one or more of the muscles that participate in the movement being tested (provided that the muscle is superficial enough to be palpated). The examiner also may be able to see or feel a tendon pop up or tense as the patient tries to perform the movement. However, there is no movement of the part as a result of this minimal contractile activity.

A Grade 1 muscle can be detected with the patient in almost any position. When a Grade 1 muscle is suspected, the examiner should passively move the part into the test position and ask the patient to hold the position and then relax; this will enable the examiner to palpate the muscle or tendon, or both, during the patient's attempts to contract the muscle and also during relaxation.

THE GRADE 0 (ZERO) MUSCLE

The Grade 0 (Zero) muscle is completely quiescent on palpation or visual inspection.

PLUS (+) AND MINUS (−) GRADES

Use of a + or − addition to a manual muscle test grade is discouraged except in two instances—Fair+ and Poor−. Scalable gradations in other instances can be described in documentation as improved or deteriorated within a given test grade (such as Grade 4) without resorting to the use of + or − labels. The purpose of avoiding the use of + or − signs is to restrict the variety of manual muscle test grades to those that are meaningful and defendable.

The Grade 3+ (Fair+) Muscle

The Grade 3+ muscle can complete a full range of motion against gravity, and the patient can hold the end position against mild resistance. There are functional implications associated with this grade.

For example, the patient with weak wrist extensors at Grade 3 cannot use a wrist-hand orthosis (WHO) effectively, but a patient with a Grade 3+ muscle can use such a device. Likewise, the patient with only Grade 3 ankle dorsiflexion cannot use a shoe-insert type of ankle-foot orthosis functionally. The patient with Grade 3+ dorsiflexors can tolerate the added weight of the brace, which is comparable to the mild resistance used in the test.

The + addition to Grade 3 is considered by many clinicians to represent not just strength but the additional endurance that is lacking in a simple Grade 3 muscle.

The Grade 2− (Poor−) Muscle

The Grade 2− (Poor−) muscle can complete partial range of motion in the horizontal plane, the gravity-minimized position. The difference between Grade 2 and Grade 1 muscles represents such a broad functional difference that a minus sign is important in assessing even minor improvements in return of function.

For example, the patient with infectious neuronitis (Landry-Guillain-Barré syndrome) who moves from muscle Grade 1 to Grade 2− demonstrates a quantum leap forward in terms of recovery and prognosis.

AVAILABLE RANGE OF MOTION

When any condition limits joint range of motion, the patient can perform only within the range available. In this circumstance, the *available range* is the full range of motion for that patient at that time, even though it is not "normal." This is the range used to assign a muscle testing grade.

For example, the normal knee extension range is 135 to 0 degrees. A patient with a 20-degree knee flexion contracture is tested for knee extension strength. This patient's maximum range into extension is −20 degrees. If this range (in sitting) can be completed with maximal resistance, the grade assigned would be a 5 (Normal). If the patient cannot complete that range, the grade assigned MUST be less than 3 (Fair). The patient then should be repositioned in the side-lying position to ascertain the correct grade.

Screening Tests

In the interests of time and cost-efficient care, it is rarely necessary to perform a muscle test for the entire body. Two exceptions among several are patients with Landry-Guillain-Barré syndrome and those with incomplete spinal cord injuries. To screen for areas that need definitive testing, the examiner can use a number of maneuvers to rule out parts that do not need testing. Covert observation of the patient before the examination will provide valuable clues to muscular weakness and performance deficits. For example, the examiner can

- Watch the patient as he or she enters the treatment area to detect gross abnormalities of gait.
- Watch the patient sit and rise from a chair, fill out admission or history forms, or remove street clothing.
- Ask the seemingly normal patient to walk on the toes and then on the heels.
- Ask the patient to grip the examiner's hand.
- Perform gross checks of bilateral muscle groups.

Preparing for the Muscle Test

The examiner and the patient must work in harmony if the test session is to be successful. This means that some basic principles and inviolable procedures should be second nature to the examiner.

1. The patient should be as free as possible from discomfort or pain for the duration of each test. It may be necessary to allow some patients to move or be positioned differently between tests.
2. The environment for testing should be quiet and nondistracting. The temperature should be comfortable for the partially disrobed subject.
3. The plinth or mat table for testing must be firm. The ideal is a hard surface, minimally padded or not padded at all. The hard surface will not allow the trunk or limbs to "sink in." Friction of the surface material should be kept to a minimum. When the patient is reasonably mobile a plinth is fine, but its width should not be so narrow that the patient is terrified of falling or sliding off. When the patient is severely paretic, a mat table is the more practical choice. The height of the table should be adjustable to allow the examiner to use proper leverage and body mechanics.
4. Patient position should be carefully organized so that position changes in a test sequence are minimized. The patient's position must permit adequate stabilization of the part or parts being tested by virtue of body weight or with help provided by the examiner.

DOCUMENTATION OF MUSCLE EXAMINATION

LEFT					RIGHT		
3	2	1	Date of Examination Examiner's Name		1	2	3
			NECK				
			Capital extension				
			Cervical extension				
			Combined extension (capital plus cervical)				
			Capital flexion				
			Cervical flexion				
			Combined flexion (capital plus cervical)				
			Combined flexion and rotation (Sternocleidomastoid)				
			Cervical rotation				
			TRUNK				
			Extension—Lumbar				
			Extension—Thoracic				
			Pelvic elevation				
			Flexion				
			Rotation				
			Diaphragm strength				
			Maximal inspiration less full expiration (indirect intercostal test)(inches)				
			Cough (indirect forced expiration) (F, WF, NF, O)				
			UPPER EXTREMITY				
			Scapular abduction and upward rotation				
			Scapular elevation				
			Scapular adduction				
			Scapular adduction and downward rotation				
			Shoulder flexion				
			Shoulder extension				
			Shoulder scaption				
			Shoulder abduction				
			Shoulder horizontal abduction				
			Shoulder horizontal adduction				
			Shoulder external rotation				
			Shoulder internal rotation				
			Elbow flexion				
			Elbow extension				
			Forearm supination				
			Forearm pronation				
			Wrist flexion				
			Wrist extension				
			Finger metacarpophalangeal flexion				
			Finger proximal interphalangeal flexion				
			Finger distal interphalangeal flexion				
			Finger metacarpophalangeal extension				
			Finger abduction				
			Finger adduction				
			Thumb metacarpophalangeal flexion				
			Thumb interphalangeal flexion				

*After Hislop and Montgomery

Figure 1–1

Illustration continued on following page

LEFT			MUSCLE EXAMINATION - Page 2	RIGHT		
3	2	1		1	2	3
			Thumb metacarpophalangeal extension (motion superior to plane of metacarpals)			
			Thumb interphalangeal extension			
			Thumb carpometacarpal abduction (motion perpendicular to plane of palm)			
			Thumb carpometacarpal abduction and extension (motion parallel to plane of palm)			
			Thumb adduction			
			Thumb opposition			
			Little finger opposition			
			LOWER EXTREMITY			
			Hip flexion			
			Hip flexion, abduction, and external rotation with knee flexion (Sartorius)			
			Hip extension			
			Hip extension (Gluteus maximus)			
			Hip abduction			
			Hip abduction and flexion			
			Hip adduction			
			Hip external rotation			
			Hip internal rotation			
			Knee flexion			
			Knee flexion with leg external rotation			
			Knee flexion with leg internal rotation			
			Knee extension			
			Ankle plantar flexion			
			Ankle plantar flexion (soleus)			
			Foot dorsiflexion and inversion			
			Foot inversion			
			Foot eversion with plantar flexion			
			Foot eversion with dorsiflexion			
			Great toe metatarsophalangeal flexion			
			Toe metatarsophalangeal flexion			
			Great toe interphalangeal flexion			
			Toe interphalangeal flexion			
			Great toe metatarsophalangeal extension			
			Toe metatarsophalangeal extension			
			Great toe interphalangeal extension			
			Toe interphalangeal extension			

Comments: _____

_____

_____

_____

Diagnosis _____ Onset _____ Age _____ Birth date _____

Patient Name _____

last   first   middle   ID number

Figure 1–1 *Continued*

5. All materials needed for the test must be at hand. This is particularly important when the patient is anxious for any reason or is too weak to be safely left untended.

Materials needed include:

- Muscle test documentation forms (Fig. 1–1)
- Pen, pencil, or computer terminal
- Pillows, towels, pads, and wedges for positioning
- Sheets or other draping linen
- Goniometer
- Interpreter (if needed)
- Assistance for turning, moving, or stabilizing the patient
- Emergency call system (if no assistant is available)
- Reference material

Summary

From the foregoing, it should be clear that manual muscle testing is an exacting clinical skill. Experience, experience, and more experience are essential to bring such a skill to an acceptable level of clinical proficiency, to say nothing of clinical mastery.

REFERENCES

1. LeVeau B. *Biomechanics of Human Motion,* 2nd ed. Philadelphia: W.B. Saunders, 1977.
2. Soderberg GL. *Kinesiology: Application to Pathological Motion.* Baltimore: Williams & Wilkins, 1986.
3. Beasley WC. Influence of method on estimates of normal knee extensor force among normal and post-polio children. Phys Ther Rev 36:21–41, 1956.
4. Williams M, Stutzman L. Strength variation through the range of motion. Phys Ther Rev 39:145–152, 1959.
5. Bohannon RW. Test retest reliability of hand held dynamometry during single session of strength assessment. Phys Ther 66:206–209, 1986.
6. Bohannon RW. Manual muscle test scores and dynamometer test scores of knee extension strength. Arch Phys Med Rehabil 67:204, 1986.
7. Sharrard WJW. Muscle recovery in poliomyelitis. J Bone Joint Surg 37B:63–69, 1955.

Testing the Muscles of the Neck

Capital Extension
Cervical Extension
Combined Neck Extension (Capital plus cervical)
Capital Flexion
Cervical Flexion
Combined Cervical Flexion (Capital plus cervical)
Combined Flexion to Isolate a Single Sternocleidomastoid
Cervical Rotation

Note: This section of the book on testing the neck muscles is divided into tests for capital and cervical extension and flexion and their combination. This distinction was first described by Perry as a necessary and effective way of managing nuchal weakness or paralysis.[1]

Chapter 2

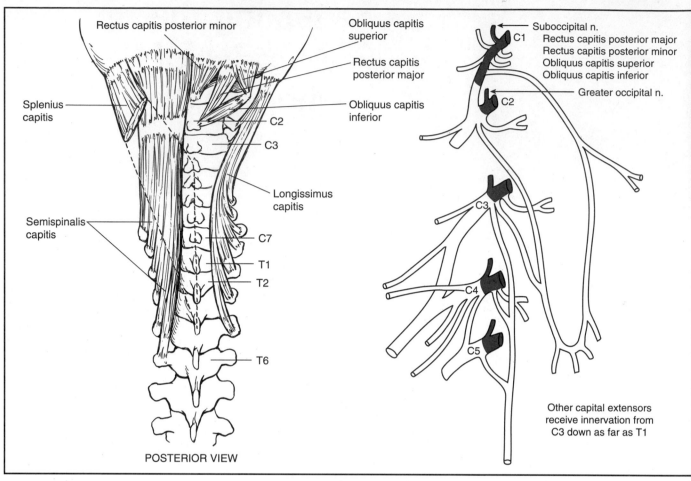

Figure 2–1 Participating muscles

POSTERIOR VIEW

Figure 2–2 Innervation

Other capital extensors receive innervation from C3 down as far as T1

Table 2–1	CAPITAL EXTENSION	
Muscle	Origin	Insertion
56. Rectus capitis post. major	Axis (spinous process)	Occiput (inf. nuchal line)
57. Rectus capitis post. minor	Atlas	Occiput (inf. nuchal line)
60. Longissimus capitis	T1–T5 vertebrae (transverse processes) C4–C7 vertebrae (articular processes)	Temporal bone (mastoid)
58. Obliquus capitis superior	Atlas (transverse process)	Occiput (between sup. and inf. nuchal lines)
59. Obliquus capitis inferior	Axis (spinous process)	Atlas (dorsal transverse process)
61. Splenius capitis	C3–C7 vertebrae (ligamentum nuchae) C7–T4 vertebrae (spinous processes)	Temporal bone (mastoid) Occiput (below nuchal line)
62. Semispinalis capitis	C7–T6 vertebrae (transverse processes) C4–C6 vertebrae (articular processes)	Occiput (between sup. and inf. nuchal lines)

Range of Motion:

0° to 25°

All muscles acting on the head are inserted on the skull. Those muscles that lie behind the coronal midline are termed capital extensors. Motion is centered at the atlanto-occipital and atlantoaxial joints.[2,3]

Grade 5 (Normal) and Grade 4 (Good)

Position of Patient: Prone with head off end of table. Arms at sides.

Position of Therapist: Standing at side of patient next to the head. One hand provides resistance over the occiput (Fig. 2–3). The other hand is placed beneath the overhanging head, prepared to support the head should it give way with the applied resistance.

Test: Patient extends head by tilting chin upward in a nodding motion. (Cervical spine is not extended.)

Instructions to Patient: "Look at the wall. Hold it. Don't let me tilt your head down."

Grading

Grade 5 (Normal): Patient completes available range of motion without substituting cervical extension. Tolerates maximal resistance. (This is a strong muscle group.)

Grade 4 (Good): Patient completes available range of motion without substituting cervical extension. Tolerates strong to moderate resistance.

Grade 3 (Fair)

Position of Patient: Prone with head off end of table and supported by therapist. Arms at sides.

Position of Therapist: Standing at side of patient's head. One hand should remain under the head to catch it should the muscles fail to hold position (Fig. 2–4).

Instructions to Patient: "Look at the wall."

Test: Patient completes available range of motion with no resistance.

Figure 2–3

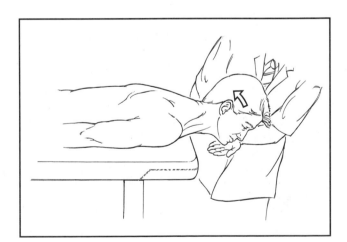

Figure 2–4

Grade 2 (Poor), Grade 1 (Trace), and Grade 0 (Zero)

Position of Patient: Supine with head on table. Arms at sides.

Position of Therapist: Standing at end of table facing patient. Head is supported with two hands under the occiput. Fingers should be placed just at the base of the occiput lateral to the vertebral column to attempt to palpate the capital extensors. Head may be slightly lifted off table to reduce friction (Fig. 2–5).

Test: Patient attempts to look back toward examiner without lifting the head from the table.

Instructions to Patient: "Tilt your chin up," or "Look back at me. Don't lift your head."

Grading

Grade 2 (Poor): Patient completes limited range of motion.

Grade 1 (Trace) and Grade 0 (Zero): Palpation of the capital extensors at the base of the occiput just lateral to the spine may be difficult; the Splenius capitis lies most lateral and the Recti lie just next to the spine.

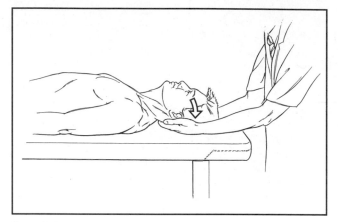

Figure 2–5

HELPFUL HINTS

1. Clinicians are reminded that the head is a very heavy object suspended on thin support. Whenever testing with the patient's head off the table, extreme caution should be used for the patient's safety, especially in the presence of suspected or known neck or trunk weakness. When in doubt, always place a hand under the head to catch it should the muscles give way.

2. Significant weakness of the capital extensor muscles combined with laryngeal and pharyngeal weakness can result in a nonpatent airway. There also may be inability to swallow. Both of these problems occur because the loss of capital extensors leaves the capital flexors unopposed, and the resultant head position favors the chin tucked on the chest, especially in the supine position.[1] This problem is not limited to patients with severe polio paralysis; it is also very much evident in patients with severe rheumatoid arthritis.

PLATE 1

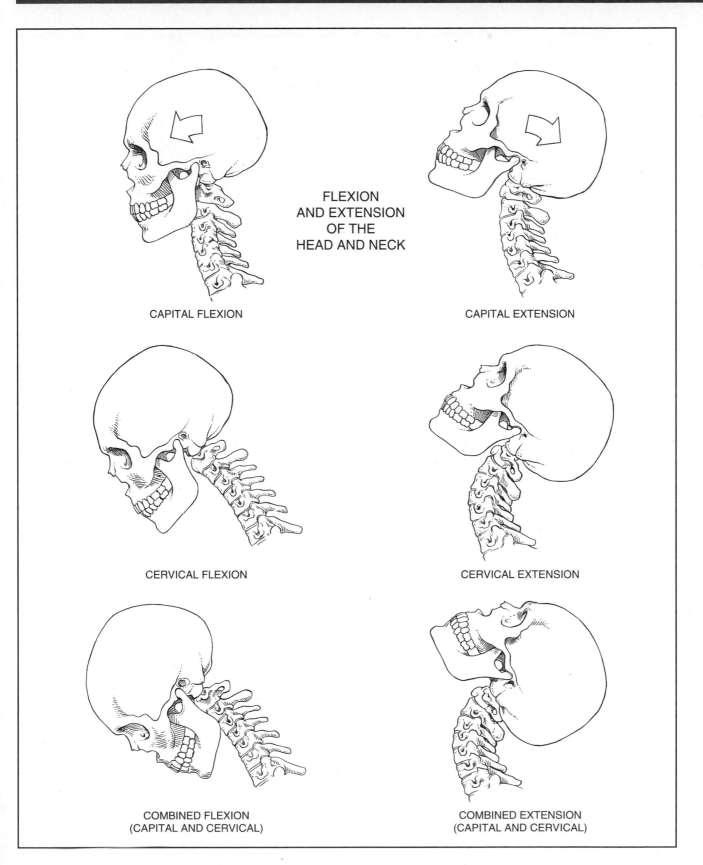

FLEXION
AND EXTENSION
OF THE
HEAD AND NECK

CAPITAL FLEXION

CAPITAL EXTENSION

CERVICAL FLEXION

CERVICAL EXTENSION

COMBINED FLEXION
(CAPITAL AND CERVICAL)

COMBINED EXTENSION
(CAPITAL AND CERVICAL)

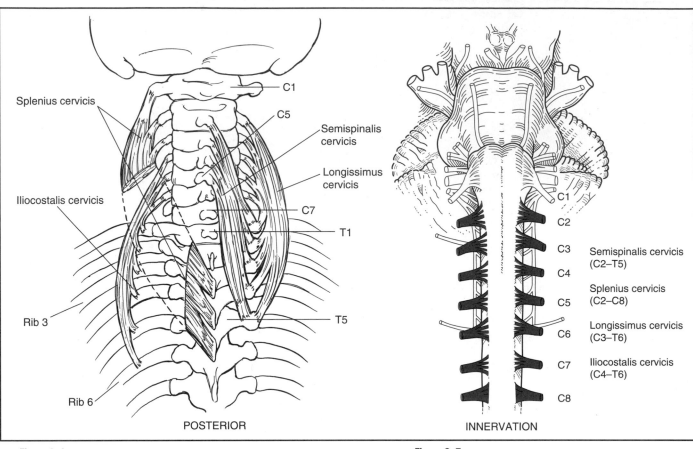

Splenius cervicis
C1
C5
Semispinalis cervicis
Longissimus cervicis
Iliocostalis cervicis
C7
T1
Rib 3
T5
Rib 6
POSTERIOR

C1
C2
C3
C4
C5
C6
C7
C8
Semispinalis cervicis (C2–T5)
Splenius cervicis (C2–C8)
Longissimus cervicis (C3–T6)
Iliocostalis cervicis (C4–T6)
INNERVATION

Figure 2–6

Figure 2–7

Table 2–2 CERVICAL EXTENSION		
Muscle	**Origin**	**Insertion**
64. Longissimus cervicis	T1–T5 vertebrae (transverse processes)	C2–C6 vertebrae (transverse processes)
65. Semispinalis cervicis	T1–T5 vertebrae (transverse processes)	Axis to C5 vertebrae (spinous processes)
66. Iliocostalis cervicis	Ribs 3–6 (angles)	C4–C6 vertebrae (transverse processes)
67. Splenius cervicis	T3–T6 vertebrae (spinous processes)	C1–C3 vertebrae (transverse processes)

Range of Motion:

0° to less than 30°

The cervical extensor muscles are limited to those that act only on the cervical spine with motion centered in the lower cervical spine.[2,3]

Grade 5 (Normal) and Grade 4 (Good)

Position of Patient: Prone with head off end of table. Arms at sides.

Position of Therapist: Standing next to patient's head. One hand is placed over the parieto-occipital area for resistance (Fig. 2–8). The other hand is placed below the chin, ready to catch the head if it gives way suddenly during resistance.

Test: Patient extends neck without tilting chin.

Instructions to Patient: "Push up on my hand but keep looking at the floor. Hold it. Don't let me push it down."

Grading

Grade 5 (Normal): Patient completes full range of motion and holds against maximal resistance. Examiner must use clinical caution because these muscles are not strong, and their maximal effort will not tolerate much resistance.

Grade 4 (Good): Patient completes full range of motion against moderate resistance.

Grade 3 (Fair)

Position of Patient: Prone with head off end of table. Arms at sides.

Position of Therapist: Standing next to patient's head with one hand supporting (or ready to support) the forehead (Fig. 2–9).

Test: Patient extends neck without looking up or tilting chin.

Instructions to Patient: "Lift your forehead from my hand and keep looking at the floor."

Grading

Grade 3 (Fair): Patient completes range of motion but takes no resistance.

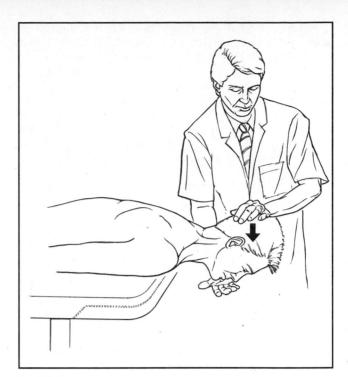

Figure 2–8

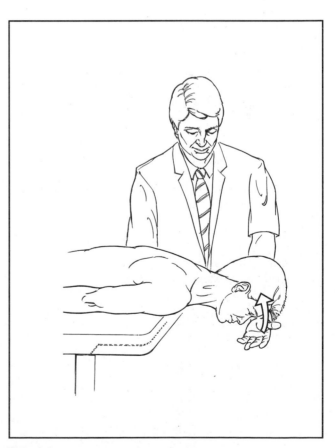

Figure 2–9

Alternate Test for Grade 3: This test should be used if there is known or suspected trunk extensor weakness. The examiner should always have an assistant participate to provide protective guarding under the patient's forehead. This test is identical to the preceding Grade 3 test except that stabilization is provided by the therapist if needed to accommodate trunk weakness. Stabilization is provided to the upper back by the forearm placed over the upper back with the hand cupped over the shoulder (Fig. 2–10).

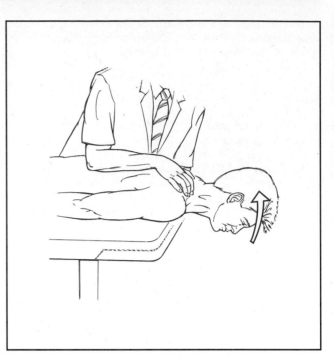

Figure 2–10

Grade 2 (Poor), Grade 1 (Trace), and Grade 0 (Zero)

Position of Patient: Supine with head fully supported by table. Arms at sides.

Position of Therapist: Standing at head end of table facing the patient. Both hands are placed under the head. Fingers are distal to the occiput at the level of the cervical vertebrae for palpation (Fig. 2–11).

Test: Patient attempts to extend neck into table.

Instructions to Patient: "Try to push your head down into my hands."

Grading

Grade 2 (Poor): Patient moves through small range of neck extension by pushing into therapist's hands.

Grade 1 (Trace): Contractile activity palpated in cervical extensors.

Grade 0 (Zero): No palpable muscle activity.

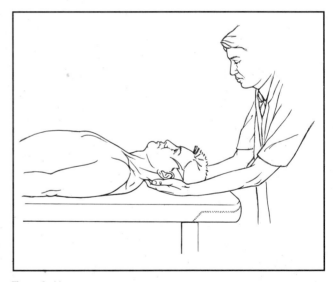

Figure 2–11

Grade 5 (Normal) and Grade 4 (Good)

Position of Patient: Prone with head off end of table. Arms at sides.

Position of Therapist: Standing next to patient's head. One hand is placed over the parieto-occipital area to give resistance, which is directed both down and forward (Fig. 2–12). The other hand is below the chin ready to catch the head if muscles give way during resistance.

Test: Patient extends head and neck through available range of motion by lifting head and looking up.

Instructions to Patient: "Lift your head and look at the ceiling. Hold it. Don't let me push your head down."

Grading

Grade 5 (Normal): Patient completes available range of motion against maximal resistance.

Grade 4 (Good): Patient completes available range of motion against moderate resistance.

Grade 3 (Fair)

Position of Patient: Patient prone with head off end of table. Arms at sides.

Position of Therapist: Standing next to patient's head.

Test: Patient extends head and neck by raising head and looking up (Fig. 2–13).

Instructions to Patient: "Raise your head from my hand and look up to the ceiling."

Range of Motion:
0° to 30°

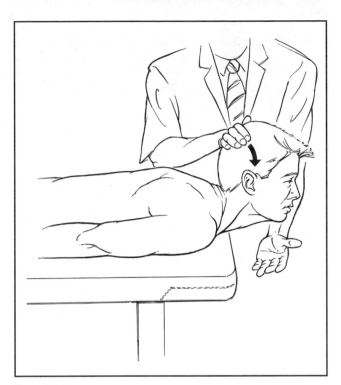

Figure 2–12

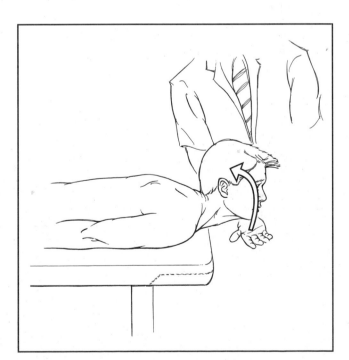

Figure 2–13

Grading

Grade 3 (Fair): Patient completes available range of motion without resistance except that of gravity.

Alternate Test for Grade 3: This test is used when the patient has trunk or hip extensor weakness. The test is identical to the previous test except that stabilization of the upper back is provided by the therapist (Fig. 2–14).

Grade 2 (Poor), Grade 1 (Trace), and Grade 0 (Zero)

Position of Patient: Patient prone with head fully supported on table. Arms at sides.

Position of Therapist: Standing next to patient's upper trunk. Both hands on cervical region and base of occiput for palpation.

Test: Patient attempts to raise head and look up.

Instructions to Patient: "Try to raise your head off the table and look at the ceiling."

Grading

Grade 2 (Poor): Patient moves through partial range of motion.

Grade 1 (Trace): Palpable contractile activity in both capital and cervical extensor muscles, but no movement.

Grade 0 (Zero): No palpable activity in muscles.

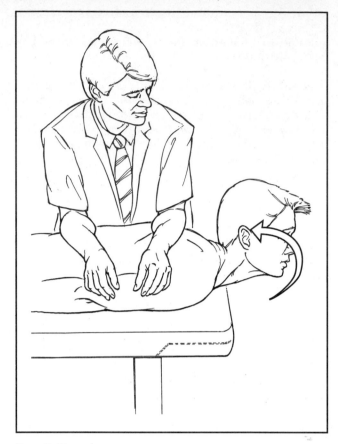

Figure 2–14

> ### HELPFUL HINTS
>
> Extensor muscles on the right (or left) may be tested by having the patient rotate the head to the right (or left) and extend the head and neck.

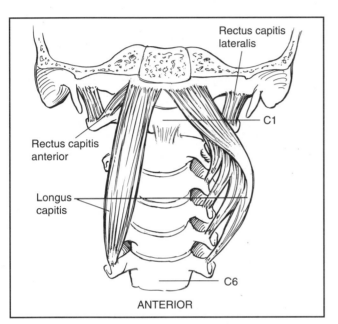

Figure 2–15

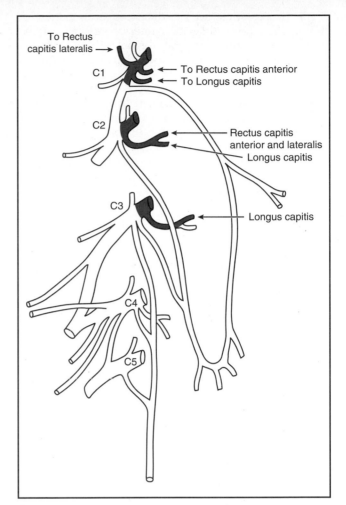

Figure 2–16

Table 2–3	CAPITAL FLEXION	
Muscle	**Origin**	**Insertion**
72. Rectus capitis anterior	Atlas (lateral)	Occiput (inferior surface)
73. Rectus capitis lateralis	Atlas (transverse process)	Occiput (jugular process)
74. Longus capitis	C3–C6 vertebrae (transverse processes)	Occiput (basilar part)

Range of Motion:
0° to 10°–15°

All muscles that act on the head are inserted on the skull. Those that are anterior to the coronal midline are termed capital flexors. Their center of motion is in the atlanto-occipital or atlantoaxial joints.[2,3]

Starting Position of Patient: In all capital, cervical, and combined flexion tests, patient is supine with head supported on table and arms at sides (Fig. 2–17).

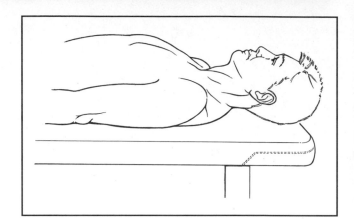

Figure 2–17

Grade 5 (Normal) and Grade 4 (Good)

Position of Patient: Supine with head on table. Arms at sides.

Position of Therapist: Standing at head of table facing patient. Both hands are cupped under the mandible and touching the cheeks to give resistance in an upward and backward direction (Fig. 2–18).

Test: Patient tucks chin into neck without raising head from table. No motion should occur at the cervical spine. This is the downward motion of nodding.

Instructions to Patient: "Tuck your chin. Don't lift your head from the table. Hold it. Don't let me lift up your chin."

Grading

Grade 5 (Normal): Patient completes available range of motion against maximal resistance. These are very strong muscles.

Grade 4 (Good): Patient completes available range of motion against moderate resistance.

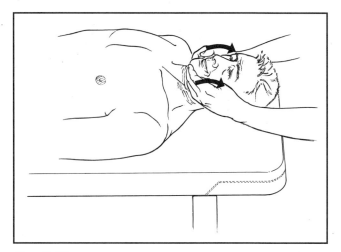

Figure 2–18

Grade 3 (Fair)

Position of Patient: Supine with head supported on table. Arms at sides.

Position of Therapist: Standing at head of table facing patient.

Test: Patient tucks chin without lifting head from table (Fig. 2–19).

Instructions to Patient: "Tuck your chin into your neck. Do not raise your head from the table."

Grading

Grade 3 (Fair): Patient completes available range of motion with no resistance.

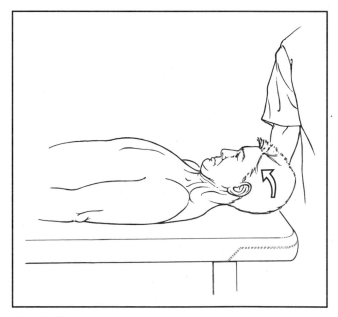

Figure 2–19

Grade 2 (Poor), Grade 1 (Trace), and Grade 0 (Zero)

Position of Patient: Supine with head supported on table. Arms at sides.

Position of Therapist: Standing at head of table facing patient.

Test: Patient attempts to tuck chin (Fig. 2–20).

Instructions to Patient: "Try to tuck your chin into your neck."

Grading

Grade 2 (Poor): Patient completes partial range of motion.

Grade 1 (Trace): Contractile activity can be palpated in capital flexor muscles.

Grade 0 (Zero): No contractile activity.

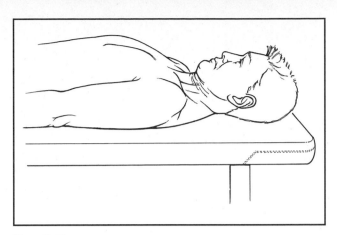

Figure 2–20

HELPFUL HINTS

1. Palpation of the small and deep muscles of capital flexion may be a difficult task unless the patient has severe atrophy. It is NOT recommended that much pressure be put on the neck in such attempts. Remember that the ascending arterial supply to the brain runs quite superficially in this region.

2. In patients with lower motor neuron lesions (including poliomyelitis) that do not affect the cranial nerves, capital flexion is seldom lost. This can possibly be attributed to the suprahyoid muscles, which are innervated by cranial nerves. Activity of the suprahyoid muscles can be identified by control of the floor of the mouth and the tongue as well as by the absence of impairment of swallowing or speech.[1]

3. When capital flexion is impaired or absent, there is usually serious impairment of the cranial nerves, and other CNS signs are present.

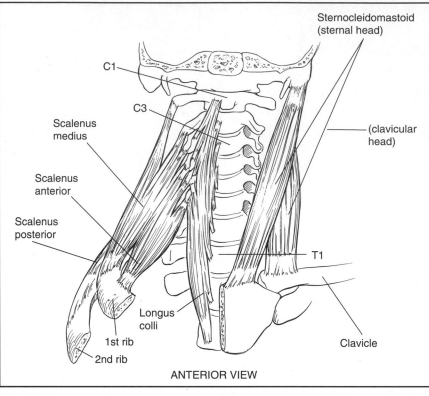

Figure 2–21

Table 2–4 CERVICAL FLEXION		
Muscle	**Origin**	**Insertion**
80. Scalenus anterior	C3–C6 vertebrae (ant. tubercle, transverse processes)	1st rib (scalene tubercle)
81. Scalenus medius	C2–C7 vertebrae (post. tubercle, transverse processes)	1st rib (upper surface)
82. Scalenus posterior	C4–C6 vertebrae (post. tubercle, transverse process)	2nd rib (outer surface)
83. Sternocleidomastoid		Both heads:
Sternal head	Manubrium (upper anterior)	Mastoid process (lateral) Occiput (lateral half of sup. nuchal line)
Clavicular head	Clavicle (superior, anterior surface)	

Range of Motion:

0° to 35°–45°
Note: Women have greater cervical lordosis than men, so it is likely that they will have a greater arc of motion.

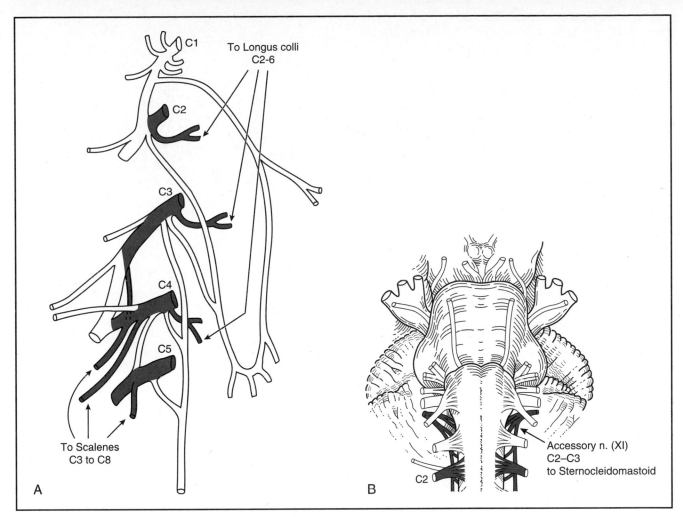

Figure 2–22

The muscles of cervical flexion act only on the cervical spine with the center of motion in the lower cervical spine.[2,3]

Grade 5 (Normal) and Grade 4 (Good)

Position of Patient: Refer to starting position for all flexion tests. Supine with arms at side. Head supported on table.

Position of Therapist: Standing next to patient's head. Hand for resistance is placed on patient's forehead. *Use two fingers only* (Fig. 2–23). Other hand may be placed on chest, but stabilization is needed only when the trunk is weak.

Test: Patient flexes neck by lifting head off table without tucking the chin. This is a weak muscle group.

Instructions to Patient: "Lift your head from the table; keep looking at the ceiling. Do not lift your shoulders off the table. Hold it. Don't let me push your head down."

Grading

Grade 5 (Normal): Patient completes available range of motion against moderate two-finger resistance.

Grade 4 (Good): Patient completes available range of motion against mild two-finger resistance.

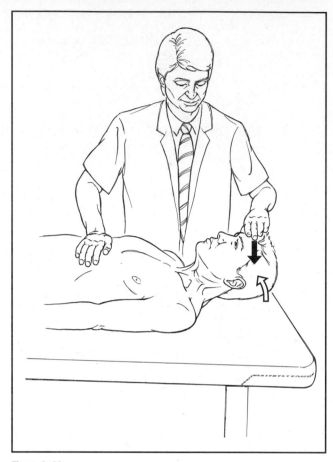

Figure 2–23

Grade 3 (Fair)

Positions of Patient and Therapist: Same as for Normal test. No resistance is used on the forehead.

Test: Patient flexes neck, keeping eyes on the ceiling (Fig. 2–24).

Instructions to Patient: "Bring your head off the table, keeping your eyes on the ceiling. Keep your shoulders completely on the table."

Grading

Grade 3 (Fair): Patient completes available range of motion correctly.

Grade 2 (Poor), Grade 1 (Trace), and Grade 0 (Zero)

Position of Patient: Supine with head supported on table. Arms at sides.

Position of Therapist: Standing at head of table facing patient. Fingers of both hands (or just the index finger) are placed over the Sternocleidomastoid muscles to palpate them during test (Fig. 2–25).

Test: Patient rolls head from side to side, keeping head supported on table.

Instructions to Patient: "Roll your head to the left and then to the right."

Grading

Grade 2 (Poor): Patient completes partial range of motion. The right Sternocleidomastoid produces the roll to the left side and vice versa.

Grade 1 (Trace): No motion occurs, but contractile activity in one or both muscles can be detected.

Grade 0 (Zero): No motion and no contractile activity detected.

HELPFUL HINTS

The Platysma may attempt to substitute for weak or absent Sternocleidomastoid muscles during cervical or combined flexion. When this occurs, the corners of the mouth pull down; a grimacing expression or "What do I do now?" expression is seen. Superficial muscle activity will be apparent over the anterior surface of the neck, with skin wrinkling.

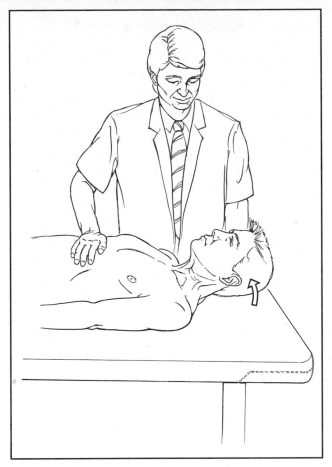

Figure 2–24

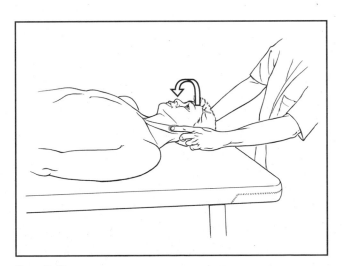

Figure 2–25

Grade 5 (Normal) and Grade 4 (Good)

Position of Patient: Supine with head supported on table. Arms at sides.

Position of Therapist: Standing at side of table at level of shoulder. Hand placed on forehead of patient to give resistance (Fig. 2–26). One arm may be used to provide stabilization of the thorax if there is trunk weakness. In such cases, the forearm is placed across the chest at the distal margin of the ribs. Although this arm does not offer resistance, considerable force may be required to maintain the trunk in a stable position. In a large patient, both arms may be required to provide such stabilization, the lower arm anchoring the pelvis. Examiner must use caution and not place too much weight or force over vulnerable nonbony areas like the abdomen.

Test: Patient flexes head and neck, bringing chin to chest.

Instructions to Patient: "Bring your head up until your chin is on your chest, and don't raise your shoulders. Hold it. Don't let me push it down."

Grading

Grade 5 (Normal): Patient completes available range of motion and tolerates strong resistance. (This combined flexion test is stronger than the capital or cervical component alone.)

Grade 4 (Good): Patient completes available range of motion and tolerates moderate resistance.

Grade 3 (Fair)

Position of Patient: Supine with head supported on table. Arms at sides.

Position of Therapist: Standing at side of table at about chest level. No resistance is given to the head motion. In the presence of trunk weakness, the thorax is stabilized.

Test: Patient flexes neck with chin tucked until the available range is completed (Fig. 2–27).

Instructions to Patient: "Bring your chin up on your chest. Don't raise your shoulders."

Grading

Grade 3 (Fair): Patient completes available range of motion without resistance.

Range of Motion:

0° to 45°–55°

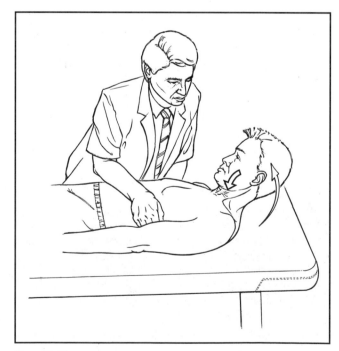

Figure 2–26

Figure 2–27

Grade 2 (Poor), Grade 1 (Trace), and Grade 0 (Zero)

Position of Patient: Supine with head fully supported on table. Arms at sides.

Position of Therapist: Standing at head of table facing the patient. Fingers of both hands, or preferably just the index finger, should be used to palpate the Sternocleidomastoid muscles bilaterally.

Test: Patient attempts to roll head from side to side. The Sternocleidomastoid on one side rotates the head to the opposite side. Most of the capital flexors rotate the head to the same side.

Instructions to Patient: "Try to roll your head to the right and then back and all the way to the left."

Grading

Grade 2 (Poor): Patient completes partial range of motion.

Grade 1 (Trace): Muscle contractile activity palpated, but no motion occurs.

Grade 0 (Zero): No palpable contractile activity.

⊙ HELPFUL HINTS

If the capital flexor muscles are weak and the Sternocleidomastoid is relatively strong, the latter muscle action will increase the extension of the cervical spine because its posterior insertion on the mastoid process makes it a weak extensor. This is true only if the capital flexors are not active enough to pre-fix the head in flexion. When the capital flexors are normal, they fix the spine in flexion, and the Sternocleidomastoid functions in its flexor mode. If the capital flexors are weak, the head can be raised off the table, but it will be in a position of capital extension with the chin leading.

This test should be performed when there is suspected or known asymmetry of strength in these neck flexor muscles.

Grade 5 (Normal), Grade 4 (Good), and Grade 3 (Fair)

Position of Patient: Supine with head supported on table and turned to the left (to test right Sternocleidomastoid).

Position of Therapist: Standing at head of table facing patient. One hand is placed on the temporal area above the ear for resistance (Fig. 2–28).

Test: Patient raises head from table.

Instructions to Patient: "Lift up your head, keeping your head turned and your ear up."

Grading

Grade 5 (Normal): Patient completes available range of motion and takes strong resistance. This is usually a very strong muscle group.

Grade 4 (Good): Patient completes available range of motion and takes moderate resistance.

Grade 3 (Fair): Patient completes available range of motion with no resistance (Fig. 2–29).

Grade 2 (Poor), Grade 1 (Trace), and Grade 0 (Zero)

Position of Patient: Supine with head supported on table.

Position of Therapist: Standing at head of table facing patient. Fingers are placed along the side of the head and neck so that they (or just the index finger) can palpate the Sternocleidomastoid (see Fig. 2–25).

Test: Patient attempts to roll head from side to side.

Instructions to Patient: "Roll your head to the right and then to the left."

Grading

Grade 2 (Poor): Patient completes partial range of motion.

Grade 1 (Trace): Palpable contractile activity in the Sternocleidomastoid, but no movement.

Grade 0 (Zero): No palpable contractile activity.

Range of Motion:
0° to 45°–55°

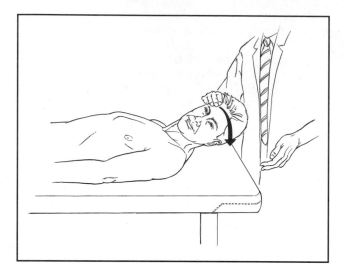

Figure 2–28

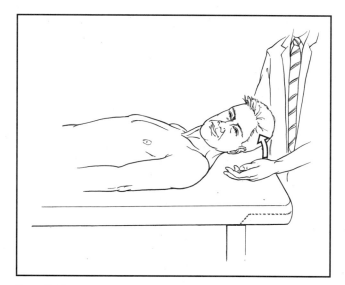

Figure 2–29

Grade 5 (Normal), Grade 4 (Good), and Grade 3 (Fair)

Position of Patient: Supine with cervical spine in neutral (flexion and extension). Head supported on table with face turned as far to one side as possible. Sitting is an alternative position for all tests.

Position of Therapist: Standing at head of table facing patient. Hand for resistance is placed over the side of head above ear. (Grades 5 and 4 only.)

Test: Patient rotates head to neutral against maximal resistance. This is a strong muscle group. Repeat for rotators on the opposite side. Alternatively, have patient rotate from left side of face on table to right side of face on table.

Instructions to Patient: "Turn your head and face the ceiling. Hold it. Do not let me turn your head back."

Grading

Grade 5 (Normal): Patient rotates head through full available range of motion to both right and left against maximal resistance.

Grade 4 (Good): Patient rotates head through full available range of motion to both right and left against moderate resistance.

Grade 3 (Fair): Patient rotates head through full available range to both right and left without resistance.

Grade 2 (Poor), Grade 1 (Trace), and Grade 0 (Zero)

Position of Patient: Sitting. Trunk and head may be supported against a high-back chair. Head posture neutral.

Position of Therapist: Standing directly in front of patient.

Test: Patient tries to rotate head from side to side, keeping the neck in neutral (chin neither down nor up).

Instructions to Patient: "Turn your head as far to the left as you can. Keep your chin level." Repeat for turn to right.

Grading

Grade 2 (Poor): Patient completes partial range of motion.

Grade 1 (Trace): Contractile activity in Sternocleidomastoid or posterior muscles, visible or by palpation evident. No movement.

Grade 0 (Zero): No palpable contractile activity.

Participating Muscles (with reference numbers)

56. Rectus capitis posterior major	67. Splenius cervicis
59. Obliquus capitis inferior	74. Longus capitis
60. Longissimus capitis	79. Longus colli (Inferior oblique)
61. Splenius capitis	
62. Semispinalis capitis	80. Scalenus anterior
65. Semispinalis cervicis	83. Sternocleidomastoid
66. Iliocostalis cervicis (accessory)	82. Scalenus posterior

REFERENCES

1. Perry J, Nickel VL. Total cervical-spine fusion for neck paralysis. J Bone Joint Surg 41A:37–60, 1959.
2. Fielding JW. Cineroentgenography of the normal cervical spine. J Bone Joint Surg 39A:1280–1288, 1957.
3. Ferlic D. The range of motion of the 'normal' cervical spine. Johns Hopkins Hosp Bull 110:59, 1962.

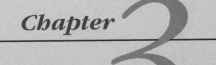

Testing the Muscles of the Trunk

Trunk Extension
Elevation of the Pelvis
Trunk Flexion
Trunk Rotation
Inspiration (Quiet)
Forced Expiration

Chapter 3

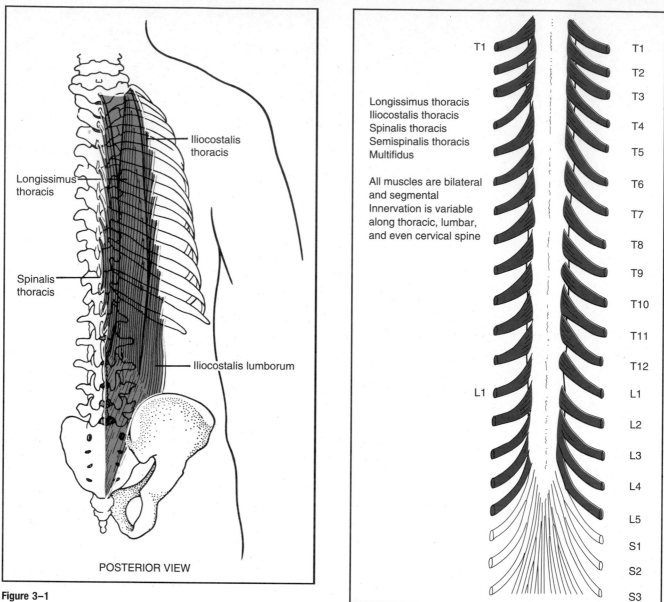

Longissimus
thoracis

Iliocostalis
thoracis

Spinalis
thoracis

Iliocostalis lumborum

POSTERIOR VIEW

Figure 3–1

Longissimus thoracis
Iliocostalis thoracis
Spinalis thoracis
Semispinalis thoracis
Multifidus

All muscles are bilateral
and segmental
Innervation is variable
along thoracic, lumbar,
and even cervical spine

T1 · T1
T2
T3
T4
T5
T6
T7
T8
T9
T10
T11
T12
L1 · L1
L2
L3
L4
L5
S1
S2
S3

Figure 3–2

Table 3–1	TRUNK EXTENSION		
Muscle	**Origin**		**Insertion**
89. Iliocostalis thoracis	Ribs 12 up to 7		Ribs 1 down to 6
90. Iliocostalis lumborum	Iliac crest		Ribs 5–12
	Sacrum		
91. Longissimus thoracis	Sacrum		L1–L3 vertebrae
	L1–L5 vertebrae		T1–T12 vertebrae
			Ribs 2–12
92. Spinalis thoracis	T11–T12 vertebrae		T1–T4 vertebrae (or up
	L1–L2 vertebrae		to T8)
93. Semispinalis thoracis	T6–T10 vertebrae (transverse		C6–T4 vertebrae
	processes)		(spinous processes)
94. Multifidus	Sacrum		Next highest vertebra
	Erector spinae (aponeurosis)		(may span 2–4
	Ilium (PSIS)		vertebrae before
	Sacroiliac ligaments		inserting)
	T1–T12 vertebrae		
95./96. Rotatores thoracis and lumborum	Thoracic and lumbar vertebrae (transverse processes)		Next highest vertebra (spinous processes)

Range of Motion:

Thoracic spine: 0° to 0°

Lumbar spine: 0° to 25°

LUMBAR SPINE

Grade 5 (Normal) and Grade 4 (Good)

Note: The Grade 5 and Grade 4 tests for spine extension are different for the lumbar and thoracic spines. Beginning at Grade 3, the tests for both levels are combined.

Position of Patient: Prone with hands clasped behind head.

Position of Therapist: Standing so as to stabilize the lower extremities just above the ankles if the patient has Normal hip strength (Fig. 3–3).

Alternate Position: Therapist stabilizes the lower extremities using body weight and both arms placed across the pelvis if the patient has hip extension weakness. It is very difficult to stabilize the pelvis adequately in the presence of significant hip weakness (Fig. 3–4).

Test: Patient extends the lumbar spine until the entire thorax is raised from the table (clears umbilicus).

Instructions to Patient: "Raise your head, shoulders, and chest off the table. Come up as high as you can."

Grading

Grade 5 (Normal) and Grade 4 (Good): The examiner distinguishes between Grade 5 and Grade 4 muscles by the nature of the response (see Figs. 3–3 and 3–4). The Grade 5 muscle holds like a lock; the Grade 4 muscle yields slightly because of an elastic quality at the end point. The patient with Normal back extensor muscles can quickly come to the end position and hold that position without evidence of significant effort. The patient with Grade 4 back extensors can come to the end position but may waver or display some signs of effort.

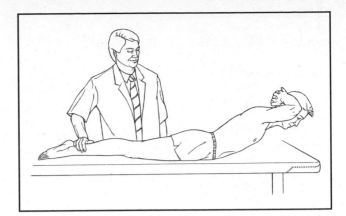

Figure 3–3

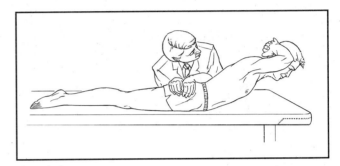

Figure 3–4

THORACIC SPINE

Grade 5 (Normal) and Grade 4 (Good)

Position of Patient: Prone with head and upper trunk extending off the table from about the nipple line (Fig. 3–5).

Position of Therapist: Standing so as to stabilize the lower limbs at the ankle.

Test: Patient extends thoracic spine to the horizontal.

Instructions to Patient: "Raise your head, shoulders, and chest to table level."

Grading

Grade 5 (Normal): Patient is able to raise the upper trunk quickly from its forward flexed position to the horizontal (or beyond) with ease and no sign of exertion (Fig. 3–6).

Grade 4 (Good): Patient is able to raise the trunk to the horizontal level but does it somewhat laboriously.

LUMBAR AND THORACIC SPINE

Grade 3 (Fair)

Position of Patient: Prone with arms at sides.

Position of Therapist: Standing at side of table. Lower extremities are stabilized just above the ankles.

Test: Patient extends spine, raising body from the table so that the umbilicus clears the table (Fig. 3–7).

Instructions to Patient: "Raise your head, arms, and chest from the table as high as you can."

Grading

Grade 3 (Fair): Patient completes the range of motion.

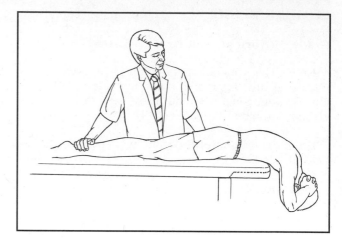

Figure 3–5

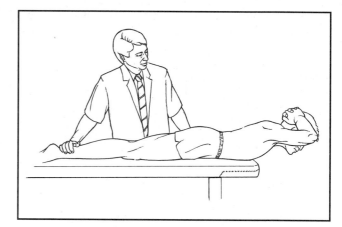

Figure 3–6

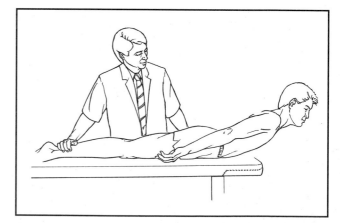

Figure 3–7

Grade 2 (Poor), Grade 1 (Trace), and Grade 0 (Zero)

These tests are identical to the Grade 3 test except that the examiner must palpate the lumbar (Fig. 3–8) and thoracic (Fig. 3–9) spine extensor muscle masses adjacent to both sides of the spine. The individual muscles cannot be isolated.

Grading

Grade 2 (Poor): Patient completes partial range of motion.

Grade 1 (Trace): Contractile activity is detectable but no movement.

Grade 0 (Zero): No contractile activity.

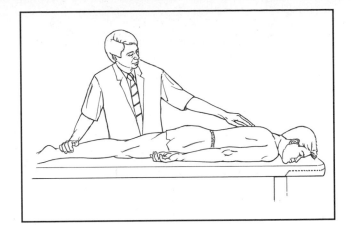

Figure 3–8

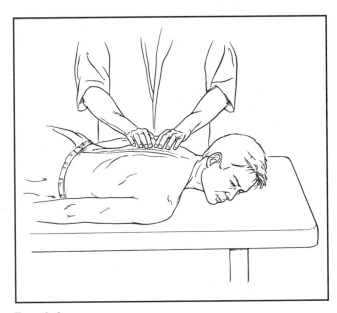

Figure 3–9

⊙ HELPFUL HINTS

1. Tests for hip extension and neck extension should precede tests for trunk extension.

   When the spine extensors are weak and the hip extensors are strong, the patient will be unable to raise the upper trunk from the table. Instead, the pelvis will tilt posteriorly while the lumbar spine moves into flexion (low back flattens).

   When the back extensors are strong and the hip extensors are weak, the patient can hyperextend the low back (increased lordosis) but will be unable to raise the trunk from the table without very strong stabilization of the pelvis by the examiner.

2. If the patient is a complete paraplegic, the test should be done on a mat table. Position the subject with both legs and pelvis off the mat. This allows the pelvis and limbs to contribute to stabilization, and the examiner holding the lower trunk has a chance to provide the necessary support. (If a mat table is not available, an assistant will be required, and the lower body may rest on a chair.)

3. If the neck extensors are weak, the examiner may need to support the head as the patient raises the trunk.

4. The position of the arms (clasped behind head) provides added resistance for Grades 5 and 4; the weight of the head and arms essentially substitutes for manual resistance by the examiner.

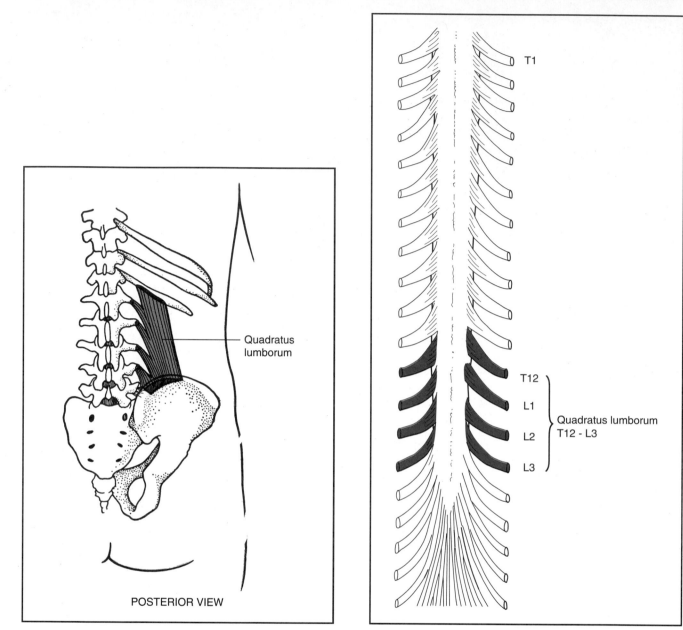

Quadratus
lumborum

POSTERIOR VIEW

Figure 3–10

T1

T12
L1
L2
L3

Quadratus lumborum
T12 - L3

Figure 3–11

Table 3–2	ELEVATION OF THE PELVIS		
Muscle		**Origin**	**Insertion**
100. Quadratus lumborum		Iliac crest	Rib 12
		Iliolumbar ligament	L1–L4 vertebrae

Others
130. Latissimus dorsi
110. Obliquus externus abdominus
111. Obliquus internus abdominus
 90. Iliocostalis lumborum

Range of Motion:

Approximates pelvis to lower ribs; range not precise

Grade 5 (Normal) and Grade 4 (Good)

Position of Patient: Supine or prone with hip and lumbar spine in extension. The patient grasps edges of the table to provide stabilization during resistance (not illustrated).

Position of Therapist: Standing at foot of table facing patient. Therapist grasps test limb with both hands just above the ankle and pulls caudally with a smooth, even pull (Fig. 3–12). Resistance is given as in traction.

Test: Patient hikes the pelvis on one side, thereby approximating the pelvic rim to the inferior margin of the rib cage.

Instructions to Patient: "Hike your pelvis to bring it up to your ribs. Hold it. Don't let me pull your leg down."

Grading

Grade 5 (Normal): This motion, certainly not attributable solely to the Quadratus lumborum, is one that tolerates a huge amount of resistance that is not readily broken when the muscles involved are Normal.

Grade 4 (Good): Patient tolerates very strong resistance. Testing this movement requires more than a bit of clinical judgment.

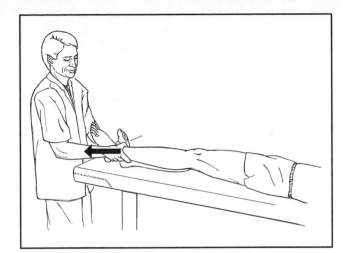

Figure 3–12

Grade 3 (Fair) and Grade 2 (Poor)

Position of Patient: Supine or prone. Hip in extension; lumbar spine neutral or extended.

Position of Therapist: Standing at foot of table facing patient. One hand supports the leg just above the ankle; the other is under the knee so the limb is slightly off the table to decrease friction (Fig. 3–13).

Test: Patient hikes the pelvis unilaterally to bring the rim of the pelvis closer to the inferior ribs.

Instructions to Patient: "Bring your pelvis up to your ribs."

Grading

Grade 3 (Fair): Patient completes available range of motion.

Grade 2 (Poor): Patient completes partial range of motion.

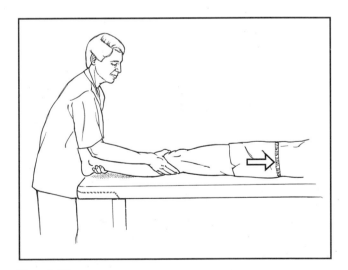

Figure 3–13

Grade 1 (Trace) and Grade 0 (Zero)

These grades should be avoided in the cause of clinical accuracy. The principal muscle to which pelvis elevation is attributable lies deep to the paraspinal muscle mass and can rarely be palpated. In persons who have extensive truncal atrophy or severe inanition, paraspinal muscle activity may be palpated, and possibly, but not necessarily convincingly, the Quadratus lumborum can be palpated.

Substitution: The patient may attempt to substitute with trunk lateral flexion, primarily using the abdominal muscles. The spinal extensors may be used without the Quadratus lumborum. In neither case can manual testing detect an inactive Quadratus lumborum.

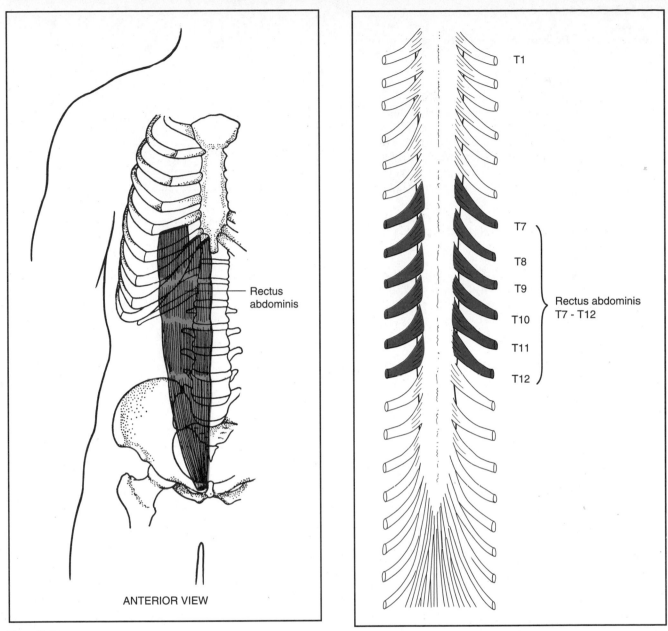

ANTERIOR VIEW

Rectus
abdominis

T1

T7
T8
T9
T10
T11
T12

Rectus abdominis
T7 - T12

Figure 3–14

Figure 3–15

Table 3–3 TRUNK FLEXION		
Muscle	**Origin**	**Insertion**
113. Rectus abdominis	Pubis (tubercle on crest and symphysis)	Ribs 5–7 Sternum

Others
111. Obliquus internus abdominis
110. Obliquus externus abdominus
174. Psoas major
175. Psoas minor

Range of Motion:
0° to 80°

Grade 5 (Normal)

Position of Patient: Supine with hands clasped behind head (Fig. 3–16).

Position of Therapist: Standing at side of table at level of patient's chest to be able to ascertain whether scapulae clear table during test (Fig. 3–16). For a patient with no other muscle weakness, the therapist does not need to touch the patient. If, however, the patient has weak hip flexors, the examiner should stabilize the pelvis by leaning across the patient on the forearms (Fig. 3–17).

Test: Patient flexes trunk through range of motion. A curl-up is emphasized, and trunk is curled until scapulae clear table (Fig. 3–17).

Instructions to Patient: "Tuck your chin and bring your head, shoulders, and arms off the table, as in a sit-up."

Grading

Grade 5 (Normal): Patient completes range of motion until inferior angles of scapulae are off the table. (Weight of the arms serves as resistance.)

Grade 4 (Good)

Position of Patient: Supine with arms crossed over chest (Fig. 3–18).

Test: Other than patient's position, all other aspects of the test are the same as for Grade 5.

Grading

Grade 4 (Good): Patient completes range of motion and raises trunk until scapulae are off the table. Resistance of arms is reduced in the cross-chest position.

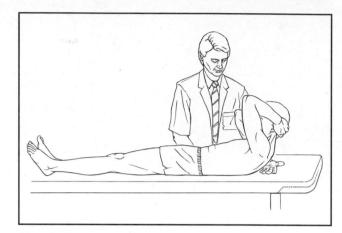

Figure 3–16

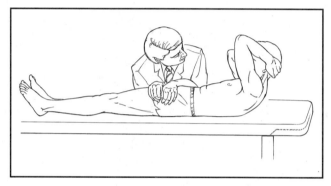

Figure 3–17

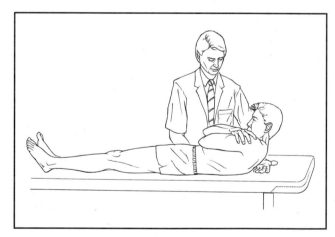

Figure 3–18

Grade 3 (Fair)

Position of Patient: Supine with arms outstretched in full extension above plane of body (Fig. 3–19).

Test: Other than patient's position, all other aspects of the test are the same as for Grade 5. Patient flexes trunk until inferior angles of scapulae are off the table. Position of the outstretched arms "neutralizes" resistance by bringing the weight of the arms closer to the center of gravity.

Instructions to Patient: "Raise your head, shoulders, and arms off the table."

Grading

Grade 3 (Fair): Patient completes range of motion and flexes trunk until inferior angles of scapulae are off the table.

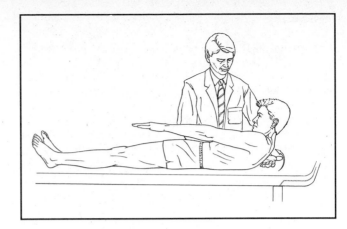

Figure 3–19

Grade 2 (Poor), Grade 1 (Trace), and Grade 0 (Zero)

Testing trunk flexion is rather clear-cut for Grades 5, 4, and 3. When testing Grade 2 and below, the results may be ambiguous, but observation and palpation are critical for defendable results. Sequenti-ally from 2 to 0, the patient will be asked to raise the head (Grade 2), do an assisted forward lean (Grade 1), or cough (Grade 1).

Position of Patient: Supine with arms at sides. Knees flexed.

Position of Therapist: Standing at side of table. The hand used for palpation is placed at the midline of the thorax over the linea alba, and the four fingers of both hands are used to palpate the Rectus abdominis (Fig. 3–20).

Test and Instructions to Patient: The examiner tests for Grades 2, 1, and 0 in a variety of ways to make certain that muscle contractile activity that may be present is not missed.

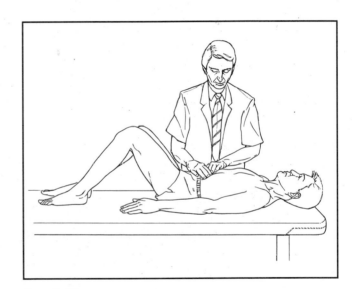

Figure 3–20

Grading

Sequence 1: Head raise (Fig. 3–21): Ask the patient to lift the head from the table. If the scapulae do not clear the table, the grade is 2 (Poor). If the patient cannot lift the head, proceed to Sequence 2.

Sequence 2: Assisted forward lean (Fig. 3–22): The examiner cradles the upper trunk and head off the table and asks the patient to lean forward. If there is depression of the rib cage, the grade is 2 (Poor). If there is no depression of the rib cage but visible or palpable contraction occurs, the grade assigned should be 1 (Trace). If there is no activity, the grade is 0; proceed to Sequence 3.

Sequence 3: Cough (Fig. 3–23): Ask the patient to cough. If he can cough to any degree and depression of the rib cage occurs, the grade is 2 (Poor). (If the patient coughs, regardless of its effectiveness, the abdominal muscles are automatically brought into play.) If the patient cannot cough but there is palpable Rectus abdominis activity, the grade is 1 (Trace). Lack of any demonstrable activity is grade 0 (zero).

HELPFUL HINTS

1. Tests for neck flexion should precede tests for trunk flexion. This will permit allowances to be made for neck weakness (should it exist), and support can be provided as required.
2. In all tests observe any deviations of the umbilicus. (This is not to be confused with the response to light stroking, which elicits superficial reflex activity.) In response to muscle testing, if there is a difference in the segments of the Rectus abdominis, the umbilicus will deviate toward the stronger part (i.e., cranially if the upper parts are stronger, caudally if the lower parts are stronger).
3. If the abdominal muscles are weak, reverse action of the hip flexors may cause lumbar lordosis. When this occurs, the patient should be positioned with the hips flexed with feet flat on the table to disallow the hip flexors to contribute to the test motion.
4. If the extensor muscles of the lumbar spine are weak, contraction of the abdominal muscles can cause posterior tilt of the pelvis. If this situation exists, tension of the hip flexor muscles would be useful to stabilize the pelvis; and, therefore, the examiner should position the patient in hip extension.

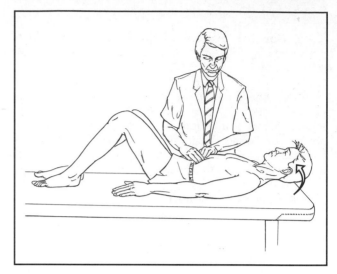

Figure 3–21

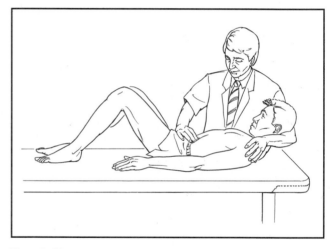

Figure 3–22

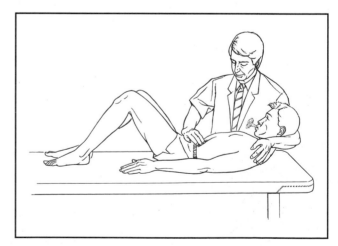

Figure 3–23

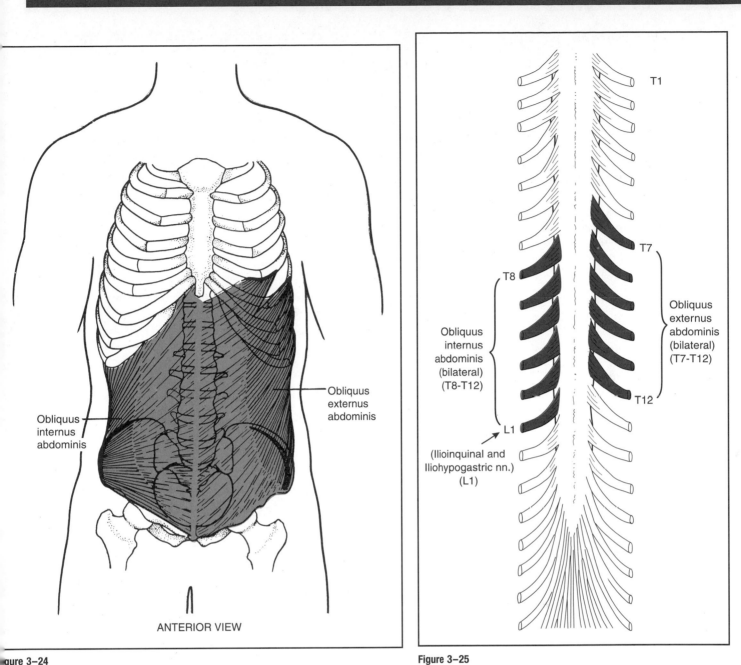

Obliquus externus abdominis

Obliquus internus abdominis

ANTERIOR VIEW

Figure 3-24

T1

T8

Obliquus internus abdominis (bilateral) (T8-T12)

L1

(Ilioinquinal and Iliohypogastric nn.) (L1)

T7

Obliquus externus abdominis (bilateral) (T7-T12)

T12

Figure 3-25

Table 3-4	TRUNK ROTATION		
Muscle		**Origin**	**Insertion**
110.	Obliquus externus abdominis	Ribs 4–12 (digitations)	Iliac crest
111.	Obliquus internus abdominis	Iliac crest Thoracolumbar fascia	Ribs 9–12 (inf. border) Ribs 7–9 (cartilages) Pubis (pectineal line)

Others
130. Latissimus dorsi
113. Rectus abdominis
Deep back muscles (one side)

Range of Motion:

0° to 45°

Grade 5 (Normal)

Position of Patient: Supine with hands clasped behind head.

Position of Therapist: Standing at patient's waist level.

Test: Patient flexes trunk and rotates to one side. This movement is then repeated on the opposite side so that the muscles on both sides can be examined.

Right elbow to left knee tests the right external obliques and the left internal obliques (Fig. 3–26). Left elbow to right knee tests the left external obliques and the right internal obliques (Fig. 3–27). When the patient rotates to one side, the internal oblique muscle is palpated on the side toward the turn; the external oblique muscle is palpated on the side away from the direction of turning.

Instructions to Patient: "Lift your head and shoulders from the table, taking your right elbow toward your left knee." Then, "Lift your head and shoulders from the table, taking your left elbow toward your right knee."

Grading

Grade 5 (Normal): The scapula corresponding to the side of the external oblique function must clear the table for a Normal grade.

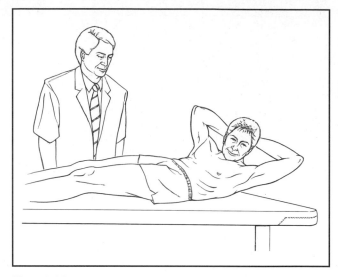

Figure 3–26

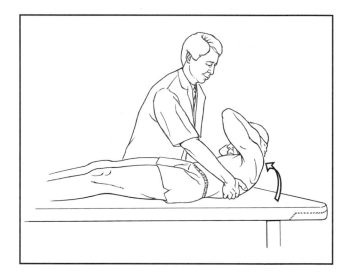

Figure 3–27

Grade 4 (Good)

Position of Patient: Supine with arms crossed over chest (Figs. 3–28 and 3–29).

Test: Other than patient's position, all other aspects of the test are the same as for Grade 5. The test is done first to one side and then to the other (Figs. 3–28 and 3–29).

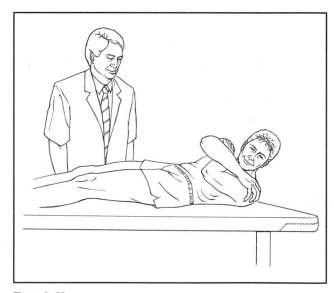

Figure 3–28

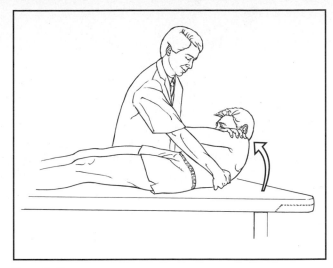

Figure 3–29

Grade 3 (Fair)

Position of Patient: Supine with arms outstretched above plane of body.

Test: All positions and instructions are the same as for Grade 5. The test is done first to the left (Fig. 3–30) and then to the right (Fig. 3–31).

Grading

Grade 3 (Fair): Patient is able to raise the scapula off the table. The therapist may use one hand to check for scapular clearance (Fig. 3–31).

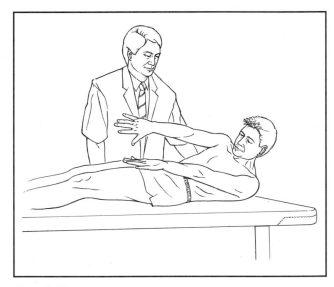

Figure 3–30

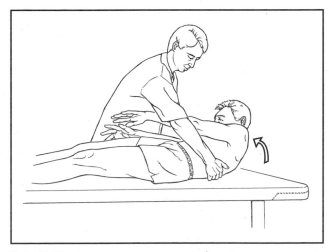

Figure 3–31

Grade 2 (Poor)

Position of Patient: Supine with arms outstretched above plane of body.

Position of Therapist: Standing at level of patient's waist. Therapist palpates the external oblique first on one side and then on the other, with one hand placed on the lateral part of the anterior abdominal wall distal to the rib cage (Fig. 3–32). Continue to palpate the muscle distally in the direction of its fibers until reaching the anterior superior iliac spine (ASIS).

At the same time, the internal oblique muscle on the opposite side of the trunk is palpated. The internal oblique muscle lies under the external oblique, and its fibers run in the opposite diagonal direction.

Examiners may remember this palpation procedure better if they think of positioning their two hands as if both hands were to be in the pants pockets or grasping the abdomen in pain. (The external obliques run from out to in; the internal obliques run from in to out.)

Instructions to Patient: "Lift your head and reach toward your right knee." (Repeat to left side for the opposite muscle.)

Test: Patient attempts to raise body and turn toward the right. Repeat toward left side.

Grading

Grade 2 (Poor): Patient is unable to clear the inferior angle of the scapula from the table on the side of the external oblique being tested. The examiner must, however, be able to observe depression of the rib cage during the test activity.

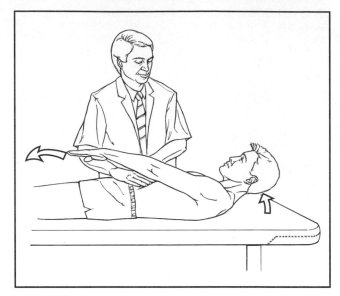

Figure 3–32

Grade 1 (Trace) and Grade 0 (Zero)

Position of Patient: Supine with arms at sides. Hips flexed with feet flat on table.

Position of Therapist: Head is supported as patient attempts to turn to one side (Fig. 3–33). (Turn to the other side in a subsequent test.) Under normal conditions, the abdominal muscles stabilize the trunk when the head is lifted. In patients with abdominal weakness, the supported head permits the patient to bring in abdominal muscle activity without having to overcome the entire weight of the head.

One hand palpates the internal obliques on the side toward which the patient turns (not illustrated) and the external obliques on the side away from the direction of turning (Fig. 3–33). The therapist assists the patient to raise the head and shoulders slightly and turn to one side. This procedure is used when abdominal muscle weakness is profound.

Instructions to Patient: "Try to lift up and turn to your right." (Repeat for turn to the left.)

Test: Patient attempts to flex trunk and turn to either side.

Grading

Grade 1 (Trace): The examiner can see or palpate muscular contraction.

Grade 0 (Zero): No response from the Obliquus internus or externus muscles.

Substitution by the Pectoralis Major: The shoulder will shrug or be raised from the table, and there is limited rotation of the trunk.

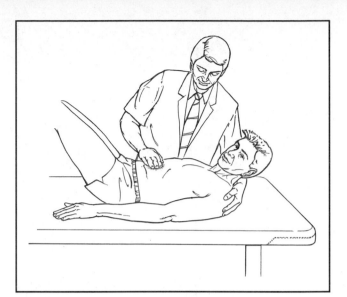

Figure 3–33

HELPFUL HINTS

1. In all tests observe any deviation of the umbilicus, which will move toward the strongest quadrant when there is unequal strength in the opposing oblique muscles.
2. Flaring of the rib cage denotes weakness of the external oblique muscles.
3. If the hip flexor muscles are weak, the examiner must stabilize the pelvis.
4. To cause the abdominals to come into action automatically, the examiner may resist a downward diagonal motion of the arm or a downward and outward movement of the lower limb.

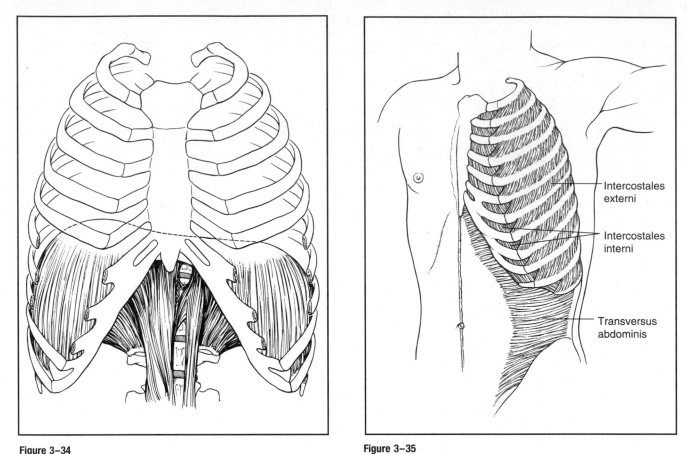

Figure 3–34

Figure 3–35

Intercostales externi

Intercostales interni

Transversus abdominis

Table 3–5 MUSCLES OF QUIET INSPIRATION		
Muscle	**Origin**	**Insertion**
101. Diaphragm		
Sternal	Xiphoid (posterior)	Central tendon of diaphragm
Costal	Ribs 7–12	
Lumbar	L1–L3	
102. Intercostales externi	Ribs 1–11 (lower border)	Ribs 2–12 (upper margins)

Others
102. Intercostales interni
103. Intercostales intimi
107. Levator costarum

Range of Motion:

Normal range of motion of the chest wall during quiet inspiration is about 0.75 inch (3/4″), with gender variations. Normal chest expansion in forced inspiration varies from *2.0 to 2.5* inches at the level of the xiphoid.[1]

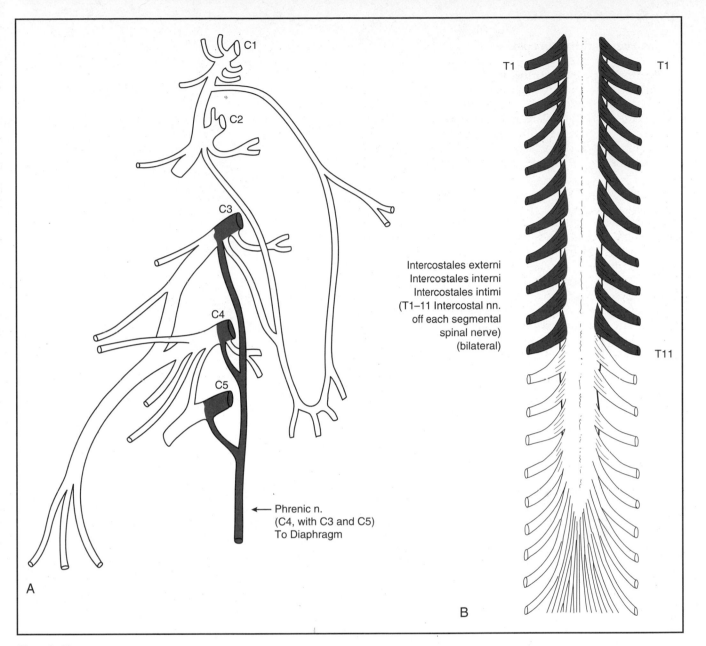

Figure 3–36

Intercostales externi
Intercostales interni
Intercostales intimi
(T1–11 Intercostal nn.
off each segmental
spinal nerve)
(bilateral)

Phrenic n.
(C4, with C3 and C5)
To Diaphragm

INSPIRATION (QUIET)

Diaphragm and Intercostals

Preliminary Examination

Uncover the patient's chest and abdominal areas so that the motions of the chest and abdominal walls can be observed. Watch the normal respiration pattern and observe differences in the motion of the chest wall and epigastric area and note any motion of the neck muscles and the abdominal muscles.

Epigastric rise and flaring of the lower margin of the rib cage during inspiration indicate that the Diaphragm is active. The rise on both sides of the linea alba should be symmetrical. During quiet inspiration, epigastric rise reflects the movement of the Diaphragm descending over one intercostal space.[2,3] In deeper inspiratory efforts the Diaphragm may move across three or more intercostal spaces.

An elevation and lateral expansion of the rib cage is indicative of intercostal activity during inspiration. Exertional chest expansion measured at the level of the xiphoid process is 2.0 to 2.5 inches (the expansion may exceed 3.0 inches in more active young people and athletes).[1]

THE DIAPHRAGM

All Grades (5 to 0)

Position of Patient: Supine.

Position of Therapist: Standing next to patient at chest level. One hand is placed lightly on the abdomen in the epigastric area just below the xiphoid process (Fig. 3–37). Resistance is given (by same hand) in a downward direction.

Test: Patient inhales with maximum effort and holds maximum inspiration.

Instructions to Patient: "Take in a deep breath . . . as much as you can . . . hold it. Push against my hand. Don't let me push you down."

Grading

Grade 5 (Normal): Patient completes full inspiratory (epigastric) excursion and holds against maximum resistance. A Grade 5 Diaphragm takes high resistance in the range of 100 pounds.[4]

Grade 4 (Good): Completes maximal inspiratory excursion but yields against heavy resistance.

Grade 3 (Fair): Completes maximal inspiratory expansion but cannot tolerate manual resistance.

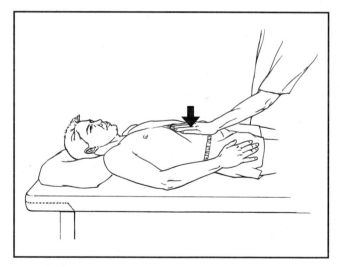

Figure 3–37

Grade 2 (Poor): Observable epigastric rise without completion of full inspiratory expansion.

Grade 1 (Trace): Palpable contraction is detected under the inner surface of the lower ribs, provided the abdominal muscles are relaxed (Fig. 3–38). Another way to detect minimal epigastric motion is by instructing the patient to "sniff" with the mouth closed.

Grade 0 (Zero): No epigastric rise and no palpable contraction of the diaphragm occur.

Substitution: Patient may attempt to substitute for an inadequate Diaphragm by hyperextending the lumbar spine in an effort to increase the response to the examiner's manual resistance.[4] The abdominal muscles also may contract, but both motions are improper attempts to follow the instruction to push up against the examiner's hand.

THE INTERCOSTALS

There is no method of direct assessment of the strength of the intercostal muscles. An indirect method measures the difference in magnitude of chest excursion between maximum inspiration and the girth of the chest at the end of full expiration.

Grading

There are no classic 5 to 0 grades given for the intercostal muscles. Instead, a flexible metal or new cloth tape is used to measure chest expansion.

Position of Patient: Supine on a firm surface. Arms at sides.

Position of Therapist: Standing at side of table. Tape measure placed lightly around thorax at level of xiphoid.

Test: Patient holds maximum inspiration for measurement and then holds maximum expiration for a second measurement. (A pneumograph may be used for the same purpose if one is available.) The difference between the two measurements is recorded as chest expansion.

Instructions to Patient: "Take a big breath in and hold it. Now blow it all out and hold it."

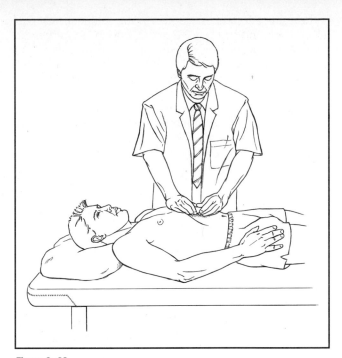

Figure 3–38

Coughing often is used as the clinical test for forced expiration. An effective cough requires the use of all muscles of active expiration in contrast to quiet expiration, which is the passive relaxation of the muscles of inspiration. It must be recognized, however, that a patient may not have an effective cough because of inadequate laryngeal control (refer to Chapter 7, Laryngeal muscles) or low vital capacity.

Grading

The usual muscle test grades do not apply here, and the following scale to assess cough is used.

Functional: Normal or only slight impairment

- Crisp or explosive expulsion of air
- Volume is sharp and clearly audible
- Able to clear airway of secretions

Weak Functional: Moderate impairment that affects the degree of active motion or endurance

- Decreased volume and diminished air movement
- Appears more labored
- May take several attempts to clear airway

Nonfunctional: Severe impairment

- No clearance of airway
- No expulsion of air
- Cough attempt may be nothing more than an effort to clear the throat

Zero: Cough is absent.

Table 3–6	MUSCLES OF FORCED EXPIRATION	
Muscle	**Origin**	**Insertion**
110. Obliquus externus abdominis	Ribs 4–12	Iliac crest
111. Obliquus internus abdominis	Iliac crest Inguinal ligament	Ribs 9–12 Pubis (pectineal line)
112. Transverse abdominis	Iliac crest Ribs 7–12	Linea alba Pubic crest
113. Rectus abdominis	Ribs 5–7 Sternum	Pubis (tubercle on crest and symphysis)
103. Intercostales interni	Ribs 1–11	Ribs 2–12

FORCED EXPIRATION

THE FUNCTIONAL ANATOMY OF COUGHING

Cough is an essential procedure to maintain airway patency and to clear the pharynx and bronchial tree when secretions accumulate. A cough may be a reflex or voluntary response to irritation anywhere along the airway downstream from the nose.

The cough reflex occurs as a result of stimulation of the mucous membranes of the pharynx, larynx, trachea, or bronchial tree. These tissues are so sensitive to light touch that any foreign matter or other irritation initiates the cough reflex. The sensory (afferent) limb of the reflex carries the impulses set up by the irritation via the glossopharyngeal and vagus cranial nerves to the fasciculus solitarius in the medulla, from which the motor impulses (efferent) then move out to the muscles of the pharynx, palate, tongue, and larynx and to the muscles of the abdominal wall and chest and the diaphragm. The reflex response is a deep inspiration (about 2.5 liters of air) followed quickly by a forced expiration, during which the glottis closes momentarily, trapping air in the lungs.[5] The diaphragm contracts spasmodically, as do the abdominal muscles and intercostal muscles. This raises the intrathoracic pressure (to above 200 mmHg) until the vocal cords are forced open, and the explosive outrush of air expels mucus and foreign matter. The expiratory airflow at this time may reach a velocity of 75 mph or higher.[5] Important to the reflex action is that the bronchial tree and laryngeal walls collapse because of the strong compression of the lungs, causing an invagination so that the high linear velocity of the airflow moving past and through these tissues dislodges mucus or foreign particles, thus producing an effective cough.

The three phases of cough—inspiration, compression, and forced expiration—are mediated by the muscles of the thorax and abdomen as well as those of the pharynx, larynx, and tongue. The deep inspiratory effort is supported by the Diaphragm, intercostals, and arytenoid abductor muscles (the posterior cricoarytenoids), permitting inhalation of upward of 1.5 liters of air.[6] The Palatoglossus and Styloglossus elevate the tongue and close off the oropharynx from the nasopharynx.

The compression phase requires the lateral cricoarytenoid muscles to adduct and close the glottis.

The strong expiratory movement is augmented by strong contractions of the thorax muscles, particularly the Latissimus dorsi and the oblique and transverse abdominal muscles. The abdominal muscles raise intra-abdominal pressure, forcing the relaxing Diaphragm up and drawing the lower ribs down and medially. Elevation of the Diaphragm raises the intrathoracic pressure to about 200 mmHg, and the explosive expulsion phase begins with forced abduction of the glottis.

REFERENCES

1. Carlson B. Normal chest excursion. Phys Ther 53:10–14, 1973.
2. Wade OL. Movements of the thoracic cage and diaphragm in respiration. J Physiol (Lond) 124:193–212, 1954.
3. Stone DJ, Keltz H. Effect of respiratory muscle dysfunction on pulmonary function. Am Rev Respir Dis 88:621–629, 1964.
4. Dail CW. Muscle breathing patterns. Med Art Sci 10:2–8, 1956.
5. Guyton AC. *Textbook of Medical Physiology.*, 8th. ed. Philadelphia: W.B. Saunders, 1991.
6. Starr JA. Manual techniques of chest physical therapy and airway clearance techniques. *In* Zadai CC. *Pulmonary Management in Physical Therapy.* New York: Churchill-Livingstone, 1992.

Testing the Muscles of the Upper Extremity

Scapular Abduction and Upward Rotation
Scapular Elevation
Scapular Adduction
Scapular Depression and Adduction
Scapular Adduction and Downward Rotation
Shoulder Flexion
Shoulder Extension
Shoulder Scaption
Shoulder Abduction
Shoulder Horizontal Abduction
Shoulder Horizontal Adduction
Shoulder External Rotation
Shoulder Internal Rotation
Elbow Flexion
Elbow Extension
Forearm Supination
Forearm Pronation
Wrist Flexion
Wrist Extension
Finger MP Flexion
Finger PIP and DIP Flexion
Finger MP Extension
Finger Abduction
Finger Adduction
Thumb MP and IP Flexion
Thumb MP and IP Extension
Thumb Abduction
Thumb Adduction
Thumb Opposition
Little Finger Opposition

Chapter 4

SCAPULAR ABDUCTION AND UPWARD ROTATION (Serratus anterior)

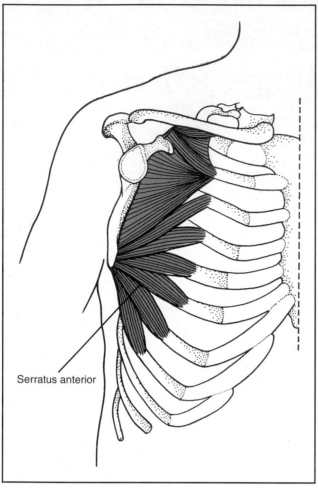

Serratus anterior

Figure 4–1

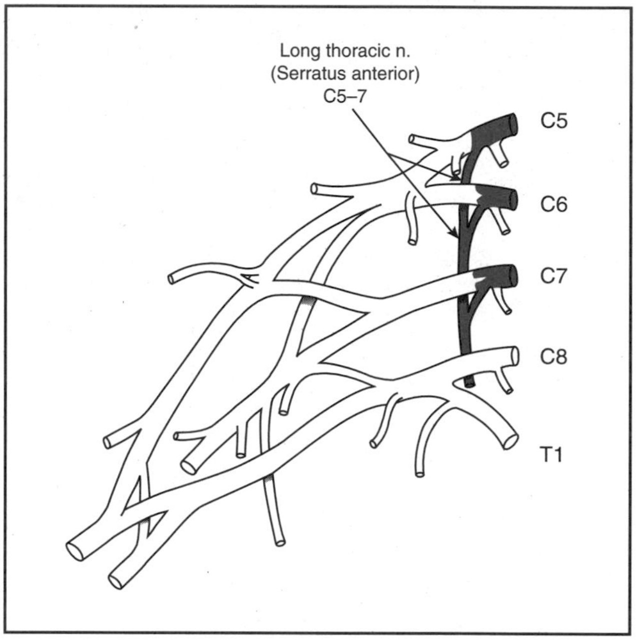

Long thoracic n.
(Serratus anterior)
C5–7
C5
C6
C7
C8
T1

Figure 4–2

Table 4–1 SCAPULAR ABDUCTION

Muscle	Origin	Insertion
128. Serratus anterior	Ribs 1–8 by digitations	Scapula (superior angle)
	Aponeurosis of intercostals	Scapula (vertebral border; inferior angle)
Others		
124. Trapezius		

Range of Motion

Reliable values not available

The Serratus often is graded incorrectly, perhaps because the muscle arrangement and the bony movement are unlike those of axial structures. The test procedure here is recommended as sound in that it is in keeping with known kinesiologic and pathokinesiologic principles. The scapular muscles, however, do need further dynamic testing with electromyography (EMG), magnetic resonance imaging (MRI), and other modern technology before completely reliable functional diagnoses can be made.

Preliminary Examination

Observation of the scapulae, both at rest and during active and passive shoulder flexion, is a routine part of the test. Examine the patient in short sitting position with hands in lap.

Palpate the vertebral borders of both scapulae with the thumbs; place the web of the thumb below the inferior angle; the fingers extend around the axillary borders (Fig. 4–3).

Check Before Testing:

1. *Position and symmetry of scapula.* Determine the position of the scapulae at rest and whether the two sides are symmetrical.

   The normal scapula lies close to the rib cage with the vertebral border nearly parallel to and from 1 to 3 inches lateral to the spinous processes. The inferior angle is tucked in. If the inferior angle of the scapula is tilted away from the rib cage, check for tightness of the Pectoralis minor, weakness of the Trapezius, and spinal deformity.

   The most prominent abnormal posture of the scapula is "winging," in which the vertebral border tilts away from the rib cage, a sign indicative of Serratus weakness. Other abnormal postures are adduction and downward rotation.

Figure 4–3

2. *Scapular range of motion.* Within the total arc of 180° of shoulder forward flexion, 120° is glenohumeral motion, and 60° is scapular motion. This is true, however, if the two motions are considered as isolated functions, but they do not work as such. It would be more correct to say that the glenohumeral and scapular motions are in synchrony after 60° and up to 150°.

Passively raise the test arm in forward flexion completely above the head to determine scapular mobility. The scapula should start to rotate at about 60°, although there is considerable individual variation. Scapular rotation continues until about −20° to −30° of full flexion.

Check that the scapula basically remains in its rest position at ranges of shoulder flexion less than 60° (the position is variable among subjects). If the scapula *moves* as the glenohumeral joint moves below 60°, that is, if in this range they move as a unit, there is limited glenohumeral motion, but the scapula may move through a complete or even excessive range. Above 60° and to about 150° or 160° in both active and passive motion, the scapula moves in concert with the humerus.

3. *The Serratus always should be tested in shoulder flexion to minimize the synergy with the Trapezius.*

If the scapular position at rest is normal, ask the patient to raise the test arm above the head in the sagittal plane. If the arm can be raised well above 90° (glenohumeral muscles must be at least Grade 3), observe the direction and amount of scapular motion that occurs. Normally, the scapula rotates forward in a motion that is controlled by the Serratus, and if erratic or "dis-coordinate" motion occurs, the Serratus is most likely weak. The normal amount of motion of the vertebral border from the start position is about the breadth of two fingers (Fig. 4–4). If the patient is able to raise the arm with simultaneous rhythmical scapular forward rotation, proceed with the test sequence for Grades 5 and 4.

4. If the scapula is positioned abnormally at rest (i.e., adducted or winging), the patient will not be able to flex the arm above 90°. Proceed to tests described for Grades 2, 1, and 0.

The Serratus anterior never can be graded higher than the grade given to shoulder flexion. If the patient has a weak Deltoid, the lever for testing is gone, and the arm cannot be used to apply resistance.

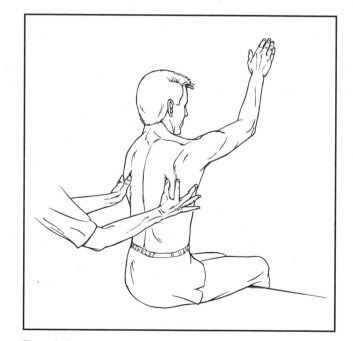

Figure 4–4

Grade 5 (Normal) and Grade 4 (Good)

Position of Patient (All Grades): Short sitting, over end or side of table. Hands on lap.

Position of Therapist: Standing at test side of patient. Hand giving resistance is on the arm just above the elbow (Fig. 4–5). The other hand uses the web space along with the thumb and index finger to palpate the edges of the scapula at the inferior angle and along the vertebral and axillary borders.

Test: Patient raises arm to approximately 130° of flexion with the elbow extended. (Examiner is reminded that the arm can be elevated up to 60° without using the Serratus.) The scapula should upwardly rotate (glenoid up) and abduct without winging.

Instructions to Patient: "Raise your arm forward over your head. Keep your elbow straight; hold it! Don't let me push your arm down."

Grading

Grade 5 (Normal): Scapula maintains its abducted and rotated position against maximal resistance given on the arm just above the elbow in a downward direction.

Grade 4 (Good): Scapular muscles "give" or "yield" against maximal resistance given on the arm. The glenohumeral joint is held rigidly in the presence of a strong Deltoid, but the Serratus yields, and the scapula moves in the direction of adduction and downward rotation.

Grade 3 (Fair)

Positions of Patient and Therapist: Same as for Grade 5 test.

Test: Patient raises the arm to approximately 130° of flexion with elbow extended (Fig. 4–6).

Instructions to Patient: "Raise your arm forward above your head."

Grading

Grade 3 (Fair): Scapula moves through full range of motion without winging but can tolerate no resistance other than the weight of the arm.

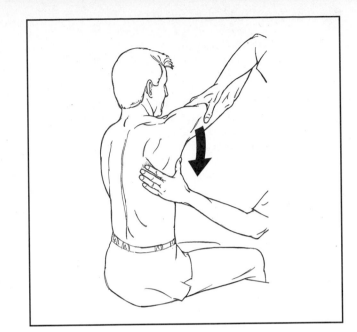

Figure 4–5

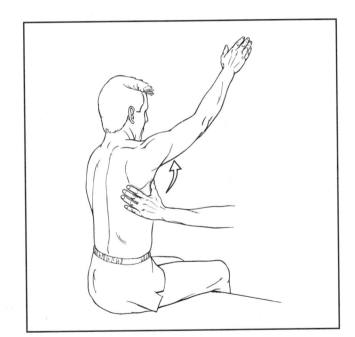

Figure 4–6

Alternate Test (Grades 5, 4, and 3)

Position of Patient: Short sitting with arm forward flexed to about 130° and then protracted in that plane as far as it can move.

Position of Therapist: Standing at test side of patient. Hand used for resistance grasps the forearm just above the wrist and gives resistance in a downward and backward direction. The other hand stabilizes the trunk just below the scapula on the same side; this prevents trunk rotation.

The examiner should select a spot on the wall or ceiling that can serve as a target for the patient to reach toward in line with about 130° of flexion.

Test: Patient abducts and upwardly rotates the scapula by protracting and elevating the arm to about 130° of flexion. The patient then holds against maximal resistance.

Instructions to Patient: "Bring your arm up, and reach for the target on the wall."

Grading: Same as for primary test.

Grade 2 (Poor)

Position of Patient: Short sitting with arm flexed above 90° and supported by examiner.

Position of Therapist: Standing at test side of patient. One hand supports the patient's arm at the elbow, maintaining it above the horizontal (Fig. 4–7). The other hand is placed at the inferior angle of the scapula with the thumb positioned along the axillary border and the fingers along the vertebral border (Fig. 4–7).

Test: Therapist monitors scapular motion by using a light grasp on the scapula at the inferior angle. Therapist must be sure not to restrict or resist motion. The scapula is observed to detect winging.

Instructions to Patient: "Hold your arm in this position" (i.e., above 90°). "Let it relax. Now hold your arm up again. Let it relax."

Grading

Grade 2 (Poor): If the scapula abducts and rotates upward as the patient attempts to hold the arm in the elevated position, the weakness is in the glenohumeral muscles. The Serratus is awarded a grade of 2. The Serratus is graded 2− (Poor−) if the scapula does not smoothly abduct and upwardly rotate without the weight of the arm or if the scapula moves toward the vertebral spine.

Figure 4–7

Grade 1 (Trace) and Grade 0 (Zero)

Position of Patient: Short sitting with arm forward flexed to above 90° (supported by therapist).

Position of Therapist: Standing in front of and slightly to one side of patient. Support the patient's arm at the elbow, maintaining it above 90° (Fig. 4–8). Use the other hand to palpate the Serratus with the tips of the fingers just in front of the inferior angle along the axillary border (Fig. 4–8).

Test: Patient attempts to hold the arm in the test position.

Instructions to Patient: "Try to hold your arm in this position."

Grading

Grade 1 (Trace): Muscle contraction is palpable.

Grade 0 (Zero): No contractile activity.

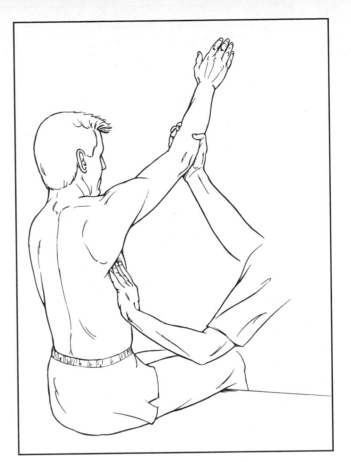

Figure 4–8

HELPFUL HINTS

1. The supine position, although best for isolating the Serratus, is not recommended at any grade level. The supine position allows too much substitution that may not be noticeable. The table gives added stabilization to the scapula so that it does not "wing," and protraction of the arm may be performed by the Pectoralis minor.

2. If the Triceps brachii is weak, supinate the forearm or manually assist the elbow to maintain the extended position. In either case, do not assist humeral flexion.

PLATE 2

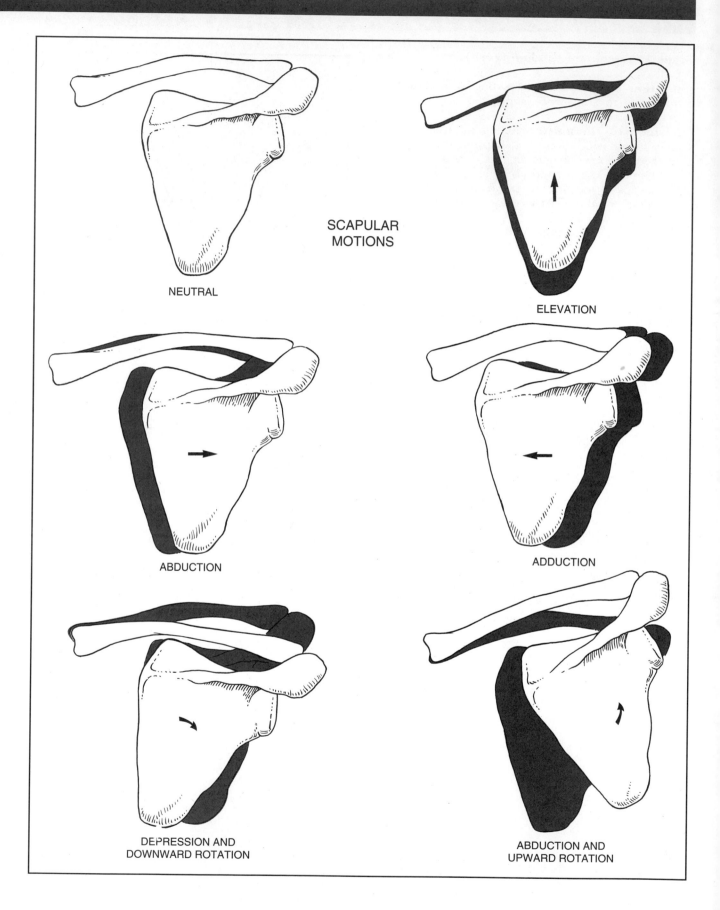

SCAPULAR
MOTIONS

NEUTRAL

ELEVATION

ABDUCTION

ADDUCTION

DEPRESSION AND
DOWNWARD ROTATION

ABDUCTION AND
UPWARD ROTATION

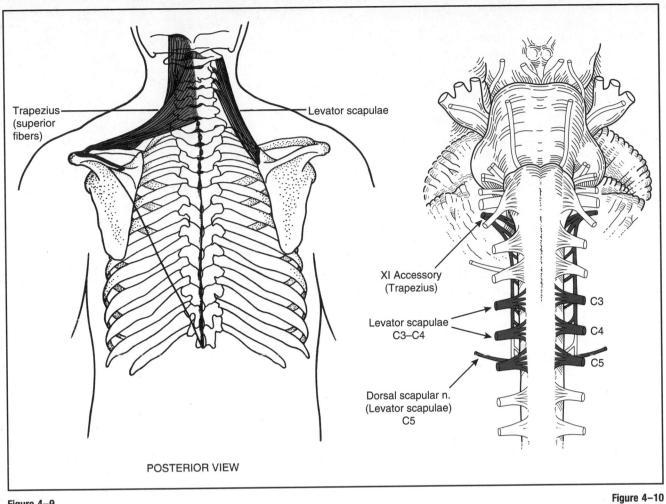

Trapezius
(superior
fibers)

Levator scapulae

XI Accessory
(Trapezius)

Levator scapulae
C3–C4

Dorsal scapular n.
(Levator scapulae)
C5

C3

C4

C5

POSTERIOR VIEW

Figure 4–9

Figure 4–10

Table 4–2 SCAPULAR ELEVATION		
Muscle	**Origin**	**Insertion**
124. Trapezius (upper fibers)	Occiput Ligamentum nuchae	Clavicle (lateral one-third of acromion)
127. Levator scapulae	C1–C4 vertebrae (transverse processes)	Scapula (vertebral border)

Others
125. Rhomboid major
126. Rhomboid minor

Range of Motion

Reliable data not available

Grade 5 (Normal) and Grade 4 (Good)

Position of Patient: Short sitting over end or side of table. Hands relaxed in lap.

Position of Therapist: Stand behind patient. Hands contoured over top of both shoulders to give resistance in a downward direction.

Test: Before testing, check patient for scapular asymmetry as well as for differences in bulk and height. This kind of asymmetry is common and can be caused by carrying purses or briefcases habitually on one side (Fig. 4–11).

Patient elevates ("shrugs") shoulders. Test is almost always performed on both sides simultaneously.

Instructions to Patient: "Shrug your shoulders" OR "Raise your shoulders toward your ears. Hold it. Don't let me push them down."

Grading

Grade 5 (Normal): Patient shrugs shoulders through available range of motion and holds against maximal resistance (Fig. 4–12).

Grade 4 (Good): Patient shrugs shoulders against strong to moderate resistance. The shoulder muscles may "give" at the end point.

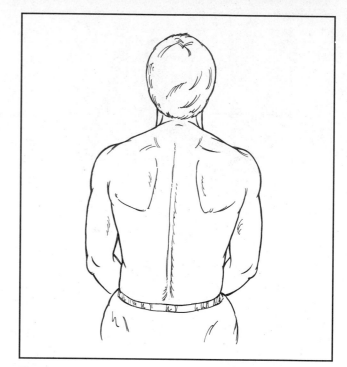

Figure 4–11

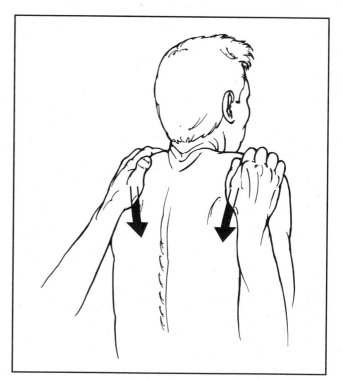

Figure 4–12

SCAPULAR ELEVATION
(Trapezius, upper fibers)

Grade 3 (Fair)

Position of Patient and Therapist: Same as those used for Grade 5 test except that no resistance is given (Fig. 4–13).

Test: Patient elevates shoulders through range of motion.

Instructions to Patient: "Raise your shoulders toward your ears" OR "Shrug your shoulders."

Grading

Grade 3 (Fair): Elevates shoulders through range but takes no resistance.

Grade 2 (Poor), Grade 1 (Trace), and Grade 0 (Zero)

Position of Patient: Prone or supine, fully supported on table. If prone, head is turned to one side (Fig. 4–14). If supine, head is in neutral position.

Position of Therapist: Stand at test side of patient. Support test shoulder in palm of one hand. The other hand palpates the upper Trapezius near its insertion above the clavicle. A second site for palpation is the upper Trapezius just adjacent to the cervical vertebrae.

Test: With the therapist supporting the shoulder, the patient elevates the shoulder (usually done unilaterally) toward the ear.

Instructions to Patient: "Raise your shoulder toward your ear."

Grading

Grade 2 (Poor): Patient completes full or partial range of motion in gravity-eliminated position.

Grade 1 (Trace): Upper Trapezius fibers can be palpated at clavicle or neck. The Levator muscle lies deep and is more difficult to palpate in the neck (between the Sternocleidomastoid and the Trapezius). It can be felt at its insertion on the vertebral border of the scapula superior to the scapular spine.

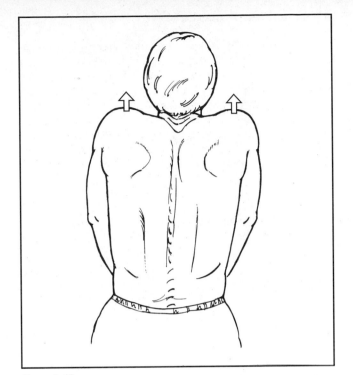

Figure 4–13

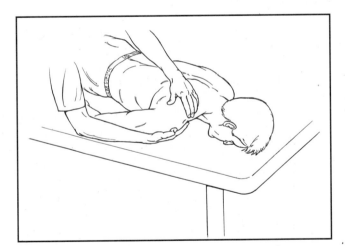

Figure 4–14

Alternate Test Procedure

In the sitting position, ask the patient to elevate one shoulder while the head, with the face turned away, is flexed laterally and down toward the shoulder (occiput leading). The occiput at full range will approximate the acromium. The examiner gives resistance at the shoulder in the direction of depression and simultaneously against the occiput in the anteromedial direction. If the upper Trapezius is weak, the acromium will not meet the occiput.[1]

SUBSTITUTION BY RHOMBOIDS

In patients with weak shoulder elevators, the Rhomboids may attempt to substitute (whereas normally they assist). In such cases, during unsuccessful attempts to shrug the shoulder the inferior angle of the scapula will move medially toward the vertebral spine (scapular adduction), and downward motion (rotation) also may occur.

HELPFUL HINTS

1. It is important to examine the patient's shoulders and scapula from a posterior view and to note any asymmetry in shoulder height, muscular bulk, or scapular winging.

2. If the sitting position for testing is contraindicated for any reason, the tests for Grade 5 and Grade 4 in the supine position will be quite inaccurate. If the Grade 3 test is done in the supine position at best it will require manual resistance because gravity is neutralized.

3. If the prone position is not comfortable, the tests for Grades 2, 1, and 0 may be performed with the patient supine, but palpation in such cases will be less than optimal.

4. In the prone position the turned head offers a disadvantage. When the face is turned to either side, there is more Trapezius activity and less Levator activity on that side.

5. Use the same lever in all scapular testing.

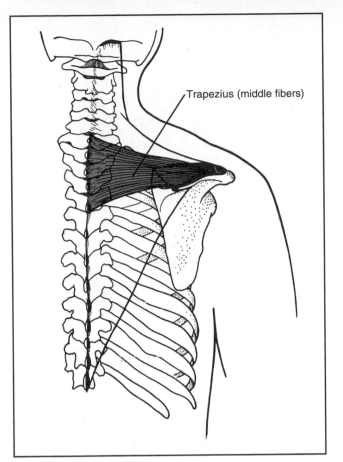

Trapezius (middle fibers)

Figure 4–15

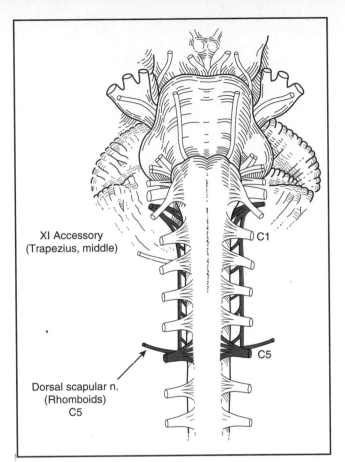

XI Accessory
(Trapezius, middle)

C1

C5

Dorsal scapular n.
(Rhomboids)
C5

Figure 4–16

Table 4–3 SCAPULAR ADDUCTION (RETRACTION)		
Muscle	**Origin**	**Insertion**
124. Trapezius (middle fibers)	Ligamentum nuchae T1–T6 vertebrae (spinous processes)	Scapula (acromium and spine)
125. Rhomboid major	T2–T5 vertebrae (spinous processes)	Scapula (medial border)
Others		
126. Rhomboid minor		
130. Latissimus dorsi		

Range of Motion
Reliable data not available

Grade 5 (Normal), Grade 4 (Good), and Grade 3 (Fair)

Position of Patient: Prone with shoulder at edge of table. Shoulder is abducted to 90° and externally rotated. Elbow is flexed to a right angle.

Alternatively, elbow may be fully extended provided the elbow extensor muscles are strong enough to stabilize the elbow on the humerus (Fig. 4–17). Head may be turned to either side for comfort.

Position of Therapist: Stand at test side close to patient's arm. Stabilize the contralateral scapular area to prevent trunk rotation. There are two ways to give resistance; one does not require as much strength as the other.

1. When the posterior Deltoid is Grade 3 or better: The hand for resistance is placed over the distal end of the humerus, and resistance is directed downward toward the floor. The wrist also may be used for a longer lever, but the lever selected should be maintained consistently throughout the test.

2. When the posterior Deltoid is Grade 2 or less: Resistance is given in a downward direction (toward floor) with the hand contoured over the shoulder joint (Fig. 4–18). This placement of resistance requires less adductor muscle strength by the patient than is needed in the test described earlier.

The fingers of the other hand can palpate the middle fibers of the Trapezius at the spine of the scapula from the acromion to the vertebral column if necessary.

Test: Patient horizontally abducts arm and adducts scapula.

Instructions to Patient: "Lift your elbow toward the ceiling. Hold it. Don't let me push it down."

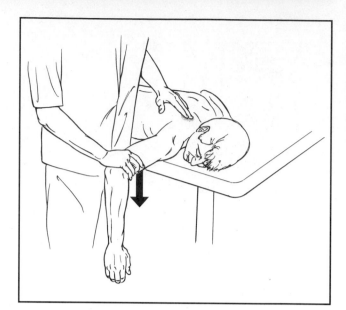

Figure 4–17

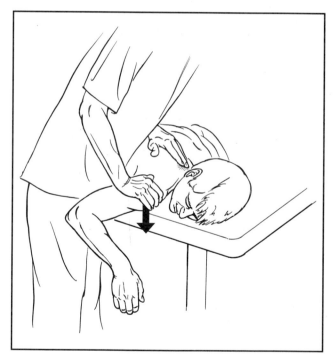

Figure 4–18

SCAPULAR ADDUCTION
(Trapezius, middle fibers)

Grading

Grade 5 (Normal): Completes available scapular adduction range and holds end position against maximal resistance.

Grade 4 (Good): Tolerates strong to moderate resistance.

Grade 3 (Fair): Completes available range but without manual resistance (Fig. 4–19).

Grade 2 (Poor), Grade 1 (Trace), and Grade 0 (Zero)

Position of Patient and Therapist: Same as for Normal test except that the therapist uses one hand to cradle the patient's shoulder and arm, thus supporting its weight (Fig. 4–20).

Test: Same as that for Grades 5 to 3.

Instruction to Patient: "Try to lift your elbow toward the ceiling."

Grading

Grade 2 (Poor): Completes partial range of motion without the weight of the arm.

Grade 1 (Trace) and Grade 0 (Zero): A Grade 1 (Trace) muscle exhibits contractile activity or slight movement. There will be neither motion nor contractile activity in the Grade 0 (Zero) muscle.

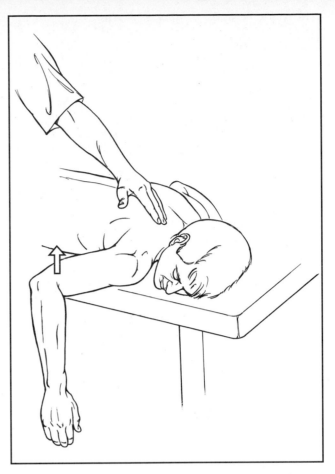

Figure 4–19

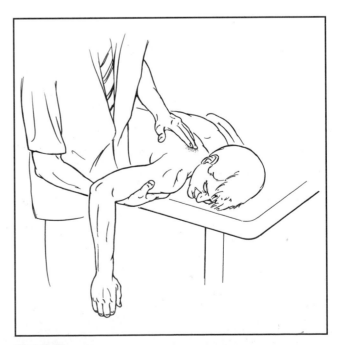

Figure 4–20

Alternate Test for Grades 5, 4, and 3

Position of Patient: Prone. Place scapula in full adduction. Arm is in horizontal abduction (90°) with shoulder externally rotated and elbow fully extended.

Position of Therapist: Stand near shoulder on test side. Stabilize the opposite scapular region to avoid trunk rotation. For Grades 5 and 4 give resistance toward the floor at the distal humerus or at the wrist, maintaining consistency of location of resistance.

Instructions to Patient: "Keep your shoulder blade close to the spine. Don't let me draw it away."

Test: Patient maintains scapular adduction.

SUBSTITUTIONS

1. *By the Rhomboids:* The Rhomboids can substitute for the Trapezius in adduction of the scapula. They cannot, however, substitute for the lateral rotation component. When substitution by the Rhomboids occurs, the scapula will adduct and downwardly rotate.

2. *By the posterior Deltoid:* If the scapular muscles are absent and the posterior Deltoid acts alone, horizontal abduction occurs at the shoulder joint but no scapular adduction.

HELPFUL HINTS

When the posterior Deltoid muscle is weak, support the patient's shoulder with the palm of one hand, and allow the patient's elbow to flex. Passively move the scapula into adduction via horizontal abduction of the arm. Have the patient hold the scapula in adduction as the examiner slowly releases the shoulder support. Observe whether the scapula maintains its adducted position. If it does, it is Grade 3.

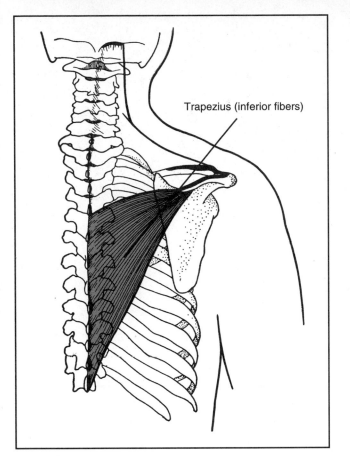

Trapezius (inferior fibers)

Figure 4–22

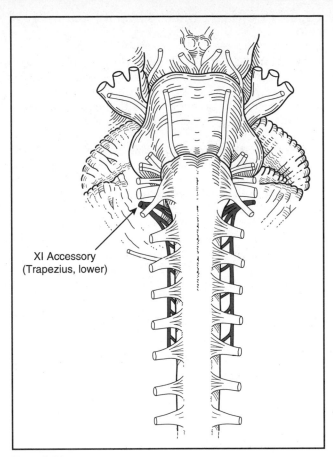

XI Accessory
(Trapezius, lower)

Figure 4–22

Table 4–4	SHOULDER DEPRESSION AND ADDUCTION	
Muscle	**Origin**	**Insertion**
124. Trapezius (lower fibers)	T7–T12 vertebrae (spinous processes)	Scapula (spine)
Others		
130. Latissimus dorsi		
131. Pectoralis major		
129. Pectoralis minor		

Range of Motion

Reliable data not available

Grade 5 (Normal), Grade 4 (Good), and Grade 3 (Fair)

Position of Patient: Prone with arms over head to about 145° of abduction (in line with the fibers of the lower Trapezius). Forearm is in midposition with the thumb pointing toward the ceiling. Head may be turned to either side for comfort.

Position of Therapist: Stand at test side. Hand giving resistance is contoured over the distal humerus just above the elbow (Fig. 4–23). Resistance will be given straight downward (toward the floor). For a less rigorous test, resistance may be given over the axillary border of the scapula.

Fingertips of the opposite hand palpate (for Grade 3) below the spine of the scapula and across to the thoracic vertebrae, following the muscle as it curves down to the lower thoracic vertebrae.

Test: Patient raises arm from the table to at least ear level and holds it strongly against resistance. Alternatively, preposition the arm diagonally over the head and ask the patient to hold it strongly against resistance.

Instructions to Patient: "Raise your arm from the table as high as possible. Hold it. Don't let me push it down."

Grading

Grade 5 (Normal): Completes available range and holds it against maximal resistance. This is a strong muscle.

Grade 4 (Good): Takes strong to moderate resistance.

Grade 3 (Fair): Same procedure is used, but patient tolerates no manual resistance (Fig. 4–24).

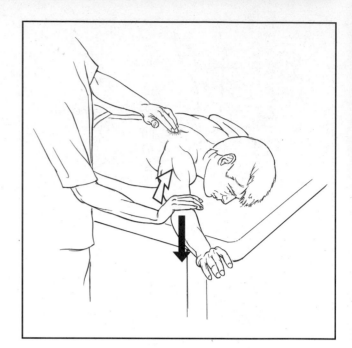

Figure 4–23

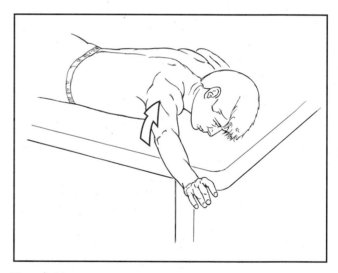

Figure 4–24

Grade 2 (Poor), Grade 1 (Trace), and Grade 0 (Zero)

Position of Patient: Same as for Grade 5.

Position of Therapist: Stand at test side. Support patient's arm under the elbow (Fig. 4–25).

Test: Patient attempts to lift the arm from the table. If the patient is unable to lift the arm because of a weak posterior and middle Deltoid, the examiner should lift and support the weight of the arm.

Instructions to Patient: "Try to lift your arm from the table past your ear."

Grading

Grade 2 (Poor): Completes full scapular range of motion without the weight of the arm.

Grade 1 (Trace): Contractile activity can be palpated in the triangular area between the root of the spine of the scapula and the lower thoracic vertebra (T7–T12), that is, the course of the fibers of the lower Trapezius.

Grade 0 (Zero): No palpable contractile activity.

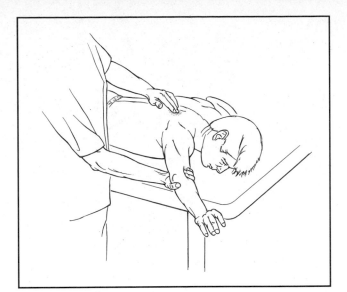

Figure 4–25

HELPFUL HINTS

1. If shoulder range of motion is limited in flexion and abduction, the patient's arm should be positioned over the side of the table and supported by the examiner at its maximal range of elevation as the start position.

2. Examiners are reminded of the test principle that the same lever arm must be used in sequential testing (over time) for valid comparison of results.

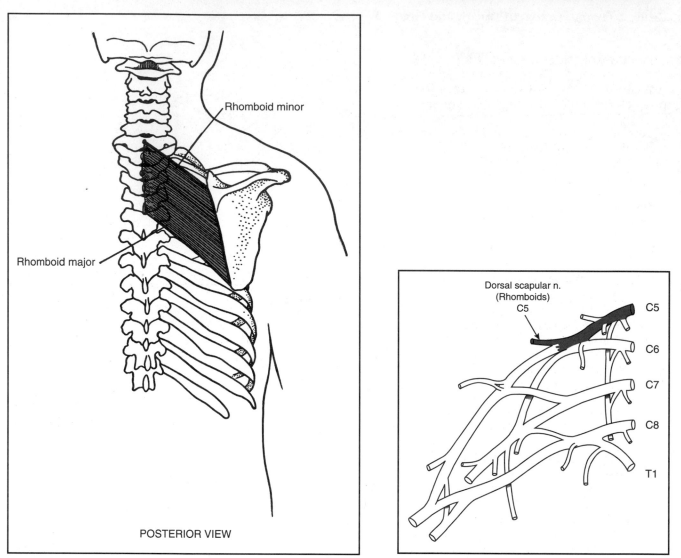

Figure 4–26

POSTERIOR VIEW

Figure 4–27

Table 4–5	SCAPULAR ADDUCTION AND DOWNWARD ROTATION	
Muscle	**Origin**	**Insertion**
125. Rhomboid major	T2–T5 vertebrae (spinous processes)	Scapula (spine)
126. Rhomboid minor	C7–T1 vertebrae (spinous processes) Ligamentum nuchae	Scapula (root of spine)

Others
130. Latissimus dorsi
127. Levator scapulae
131. Pectoralis major
129. Pectoralis minor

Range of Motion
Reliable data not available

The test for the Rhomboid muscles has become the focus of some clinical debate. Kendall and coworkers claim, with good evidence, that these muscles frequently are underrated, that is, they are too often graded at a level less than their performance.[1] At issue also is the confusion that can occur in separating the function of the Rhomboids from that of other scapular or shoulder muscles, particularly the Trapezius and the Pectoralis minor. Innervated only by C5, a test for the Rhomboids, correctly done, can confirm or rule out a cord lesion at this level. With these issues in mind, the authors present first their method and then, with the generous permission of Mrs. Kendall, her Rhomboid test as another method of assessment.

Grade 5 (Normal), Grade 4 (Good), and Grade 3 (Fair)

Position of Patient: Prone. Head may be turned to either side for comfort. Shoulder is internally rotated and the arm is adducted across the back with the elbow flexed and hand resting on the back (Fig. 4–28).

Position of Therapist: Stand at test side. When the shoulder extensor muscles are Grade 3 or above, the hand used for resistance is placed on the humerus just above the elbow, and resistance is given in a downward and outward direction (Fig. 4–29).

When the shoulder extensors are weak, place the hand for resistance along the axillary border of the scapula (Fig. 4–30). Resistance is applied in a downward and outward direction.

The fingers of the hand used for palpation are placed deep under the vertebral border of the scapula.

Test: Patient lifts the hand off the back and adducts the scapula, maintaining the arm position across the back. With strong muscle activity the therapist's fingers will "pop" out from under the edge of the scapular vertebral border. (See Fig. 4–28.)

Instructions to Patient: "Lift your hand. Hold it. Don't let me push it down."

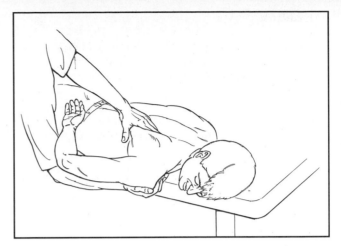

Figure 4–28

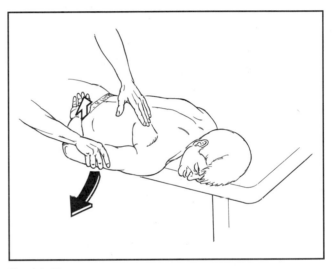

Figure 4–29

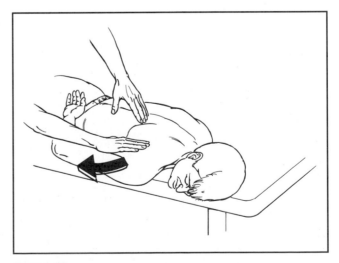

Figure 4–30

Grading

Grade 5 (Normal): Completes available range and holds against maximal resistance (Fig. 4–31). The fingers will "pop out" from under the scapula when strong Rhomboids contract.

Grade 4 (Good): Completes range and holds against strong to moderate resistance. Fingers usually will "pop out."

Grade 3 (Fair): Completes range but tolerates no manual resistance (Fig. 4–32).

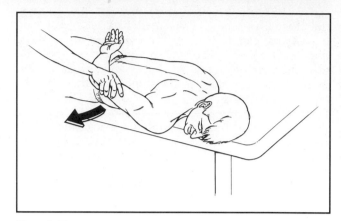

Figure 4–31

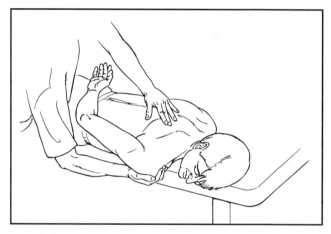

Figure 4–32

Grade 2 (Poor), Grade 1 (Trace), and Grade 0 (Zero)

Position of Patient: Short sitting with shoulder internally rotated and arm extended and adducted behind back (Fig. 4–33).

Position of Therapist: Stand at test side, support arm by grasping the wrist. The fingertips of one hand palpate the muscle under the vertebral border of the scapula.

Test: Patient attempts to move hand away from back.

Instructions to Patient: "Try to move your hand off your back."

Grading

Grade 2 (Poor): Completes partial range of scapular motion.

Grades 1 (Trace) and 0 (Zero): A Grade 1 muscle has palpable contractile activity. A Grade 0 muscle shows no response.

Alternate Test for Grades 2, 1, and 0

Position of Patient: Prone with shoulder in about 45° of abduction and elbow at about 90° of flexion with the hand on the back.

Position of Therapist: Stand at test side and support test arm by cradling it under the shoulder (Fig. 4–34). Fingers used for palpation are placed firmly under the vertebral border of the scapula.

Test: Patient attempts to lift hand from back.

Instructions to Patient: "Try to lift your hand away from your back" OR "Lift your hand toward the ceiling."

Grading

Grade 2 (Poor): Completes partial range of scapular motion.

Grades 1 (Trace) and 0 (Zero): A Grade 1 (Trace) muscle has some palpable contractile activity. A Grade 0 muscle shows no contractile response.

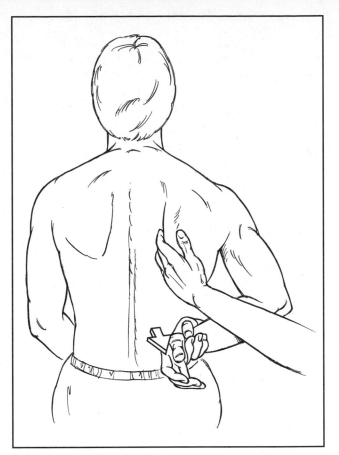

Figure 4–33

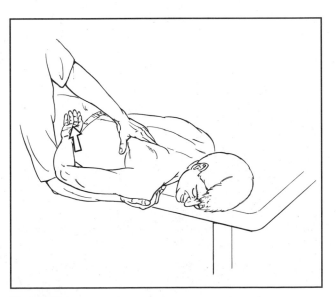

Figure 4–34

Alternate Rhomboid Test After Kendall[1]

As a preliminary to the Rhomboid test, the shoulder adductors should be tested and found sufficiently strong to allow the arm to be used as a lever.

Position of Patient: Prone with head turned to side of test. Nontest arm is abducted with elbow flexed.

Test arm is near the edge of the table. Arm (humerus) is fully adducted and held firm to the side of the trunk in external rotation and some extension with elbow fully flexed. In this position the scapula is in adduction, elevation, and downward rotation (glenoid down).

Position of Therapist: Stand at test side. One hand used for resistance is cupped around the flexed elbow. The resistance applied by this hand will be in the direction of scapular abduction and upward rotation (out and up; Fig. 4–35). The other hand is used to give resistance simultaneously. It is contoured over the shoulder joint and gives resistance caudally in the direction of shoulder depression.

Test: Examiner tests the ability of the patient to hold the scapula in its position of adduction, elevation, and downward rotation (glenoid down).

Instructions to Patient: "Hold your arm as I have placed it. Do not let me pull your arm forward" OR "Hold the position you are in; keep your shoulder blade against your spine as I try to pull it away."

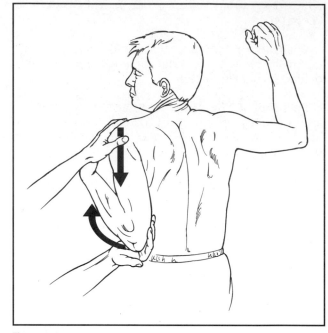

Figure 4–35

SUBSTITUTION BY MIDDLE TRAPEZIUS

The middle fibers of the Trapezius can substitute for the adduction component of the Rhomboids. The Trapezius cannot, however, substitute for the downward rotation component. When substitution occurs, the patient's scapula will adduct with no downward rotation (no glenoid down occurs). Only palpation can detect this substitution for sure.

◔ HELPFUL HINTS

When the Rhomboid test is performed with the hand behind the back, never allow the patient to lead the lifting motion with the elbow because this will activate the humeral extensors.

SHOULDER FLEXION
(Anterior Deltoid, Supraspinatus, and Coracobrachialis)

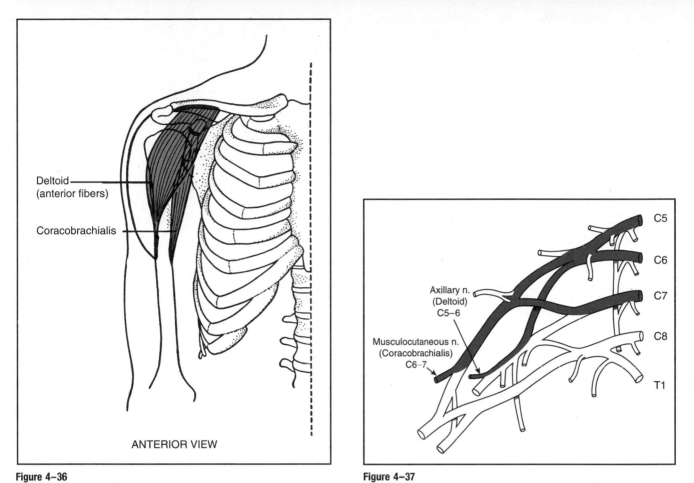

Deltoid (anterior fibers)

Coracobrachialis

ANTERIOR VIEW

Figure 4–36

Axillary n. (Deltoid) C5–6

Musculocutaneous n. (Coracobrachialis) C6–7

C5
C6
C7
C8
T1

Figure 4–37

Table 4–6 SHOULDER FLEXION		
Muscle	**Origin**	**Insertion**
133. Deltoid (anterior and middle)	Clavicle (front) Scapula (acromium)	Humerus (deltoid tuberosity)
135. Supraspinatus	Scapula (supraspinous fossa)	Humerus (greater tubercle)

Others
131. Pectoralis major (upper)
139. Coracobrachialis
140. Biceps brachii

Range of Motion
0° to 180°

The Coracobrachialis muscle cannot be isolated, nor is it readily palpable. It has no unique function. It is included here because classically it is considered a shoulder flexor and adductor.

Grade 5 (Normal) and Grade 4 (Good)

Position of Patient: Short sitting with arms at sides, elbow slightly flexed, forearm pronated (to avoid substitution by long head of Biceps).

Position of Therapist: Stand at test side. Hand giving resistance is contoured over the distal humerus just above the elbow. The other hand may stabilize the shoulder (Fig. 4–38).

Test: Patient flexes shoulder to 90° without rotation or horizontal movement (Fig. 4–38). The scapula should be allowed to abduct and upwardly rotate. The normal ratio of scapular to humeral movement after the first 20° is 2:1—i.e., there are 2° of glenohumeral motion to each 1° of scapulothoracic motion.

Instructions to Patient: "Raise your arm forward to shoulder height. Hold it. Don't let me push it down."

Grading

Grade 5 (Normal): Holds end position (90°) against maximal resistance.

Grade 4 (Good): Holds end position against strong to moderate resistance.

Grade 3 (Fair)

Position of Patient: Short sitting with arm at side and elbow slightly flexed.

Position of Therapist: Stand at test side.

Test: Patient flexes shoulder to 90° (Fig. 4–39).

Instructions to Patient: "Raise your arm forward to shoulder height."

Grading

Grade 3 (Fair): Completes test range (90°) but tolerates no resistance.

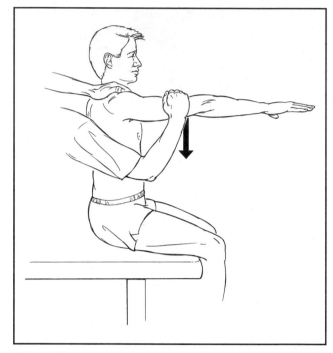

Figure 4–38

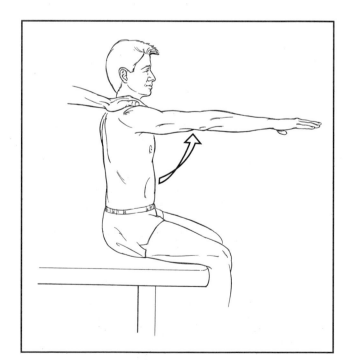

Figure 4–39

Grade 2 (Poor), Grade 1 (Trace), and Grade 0 (Zero)

Position of Patient: Short sitting with arm at side and elbow slightly flexed.

Position of Therapist: Stand at test side. Fingers used for palpation are placed on the anterior surface of the Deltoid over the shoulder joint (Fig. 4–40).

Test: Patient attempts to flex shoulder to 90°.

Instructions to Patient: "Try to raise your arm."

Grading

Grade 2 (Poor): Completes partial range of motion.

Grade 1 (Trace): Examiner feels or sees contractile activity in the anterior Deltoid, but no motion occurs.

Grade 0 (Zero): No contractile activity.

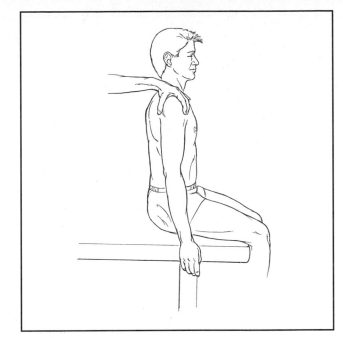

Figure 4–40

Alternate Test for Grades 2, 1, and 0

If for any reason the patient is unable to sit, the test can be conducted in the side-lying position. In this posture, the examiner cradles the test arm at the elbow before asking the patient to flex the shoulder.

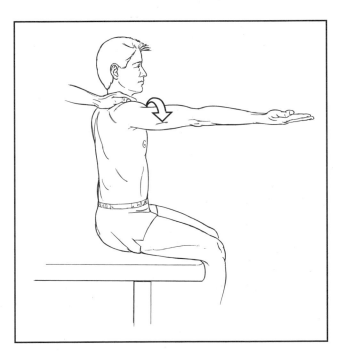

Figure 4–41

SUBSTITUTIONS

1. The patient may attempt to flex the shoulder with the Biceps brachii by first externally rotating the shoulder (Fig. 4–41). To avoid this, the arm should be kept in the midposition between internal and external rotation.

2. Attempted substitution by the upper Trapezius results in shoulder elevation.

3. Attempted substitution by the Pectoralis major results in horizontal adduction.

4. The patient may lean backward or try to elevate the shoulder girdle to assist in flexion.

HELPFUL HINTS

Although the Coracobrachialis is a contributor to shoulder flexion, it is deep-lying and may be difficult or impossible to palpate within a reasonable range of comfort for the patient.

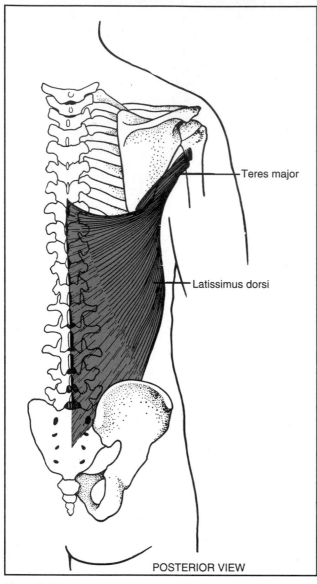

Figure 4–42

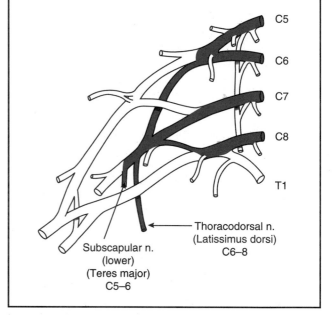

Figure 4–43

Table 4–7 SHOULDER EXTENSION		
Muscle	**Origin**	**Insertion**
130. Latissimus dorsi	T6–T12 vertebrae L1–L5 vertebrae Sacral vertebrae Ribs 9–12 Scapula (inferior angle) Iliac crest	Humerus (intertibercular groove)
133. Deltoid (posterior)	Scapula (spine)	Humerus (deltoid tuberosity)
138. Teres major	Scapula (inferior angle)	Humerus (intertubercular sulcus)
Others **142.** Triceps brachii (long head)		

Range of Motion
0° to 45°

SHOULDER EXTENSION
(Latissimus dorsi, Teres major, Posterior Deltoid)

Grade 5 (Normal) and Grade 4 (Good)

There are three tests for Grades 5 and 4 that should be used routinely. The first is the traditional way of testing shoulder extension in the prone position. The other two tests are used to isolate the Latissimus dorsi to the extent possible and to simulate function.

Test 1: Generic Shoulder Extension

Position of Patient: Prone with arms at sides and shoulder internally rotated (palm up) (Fig. 4–44).

Position of Therapist: Stand at test side. Hand used for resistance is contoured over the posterior arm just above the elbow.

Test: Patient raises arm off the table, keeping the elbow straight (Fig. 4–45).

Instructions to Patient: "Lift your arm as high as you can. Hold it. Don't let me push it down."

Grading

Grade 5 (Normal): Completes available range and holds against maximal resistance.

Grade 4 (Good): Completes available range but yields against strong resistance.

Test 2: To Isolate Latissimus dorsi

Position of Patient: Prone with head turned to test side; arms are at sides and shoulder is internally rotated (palm up). Test shoulder is "hiked" to the level of the chin.

Position of Therapist: Stand at test side. Grasp forearm above patient's wrist with both hands (Fig. 4–46).

Test: Patient depresses arm caudally and in so doing approximates the rib cage to the pelvis.

Instructions to Patient: "Reach toward your feet. Hold it. Don't let me push your arm upward toward your head."

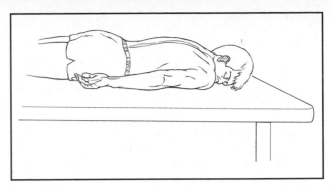

Figure 4–44

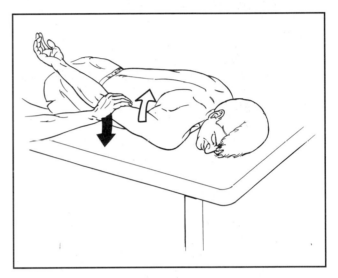

Figure 4–45

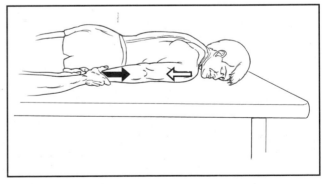

Figure 4–46

Grading

Grade 5 (Normal): Patient completes available range against maximal resistance. If the therapist is unable to push the arm upward using both hands for resistance, test the patient in the sitting position as described in Test 3.

Grade 4 (Good): Patient completes available range of motion, but the shoulder yields at end point against strong resistance.

Test 3: To Isolate Latissimus dorsi

Position of Patient: Short sitting, with hands flat on table adjacent to hips (Fig. 4–47).

If the patient's arms are too short to assume this position, provide a push-up block for each hand.

Position of Therapist: Stand behind patient. Fingers are used to palpate fibers of the Latissimus dorsi on the lateral aspects of the thoracic wall (bilaterally) just above the waist (Fig. 4–47). (In this test the sternal head of the Pectoralis major is equally active.)

Test: Patient pushes down on hands (or blocks) and lifts buttocks from table (Fig. 4–47).

Instructions to Patient: "Lift your bottom off the table."

Grading

Grade 5 (Normal): Patient is able to lift buttocks clear of table.

Grade 4 (Good): There is no Grade 4 in this sequence because the prone test (Test 2) determines a grade of less than 5.

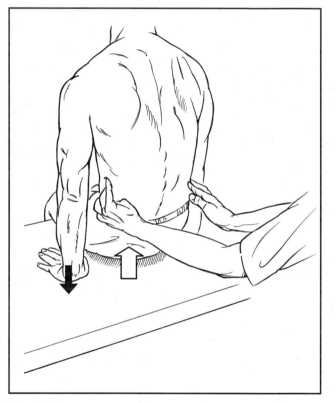

Figure 4–47

Grade 3 (Fair) and Grade 2 (Poor)

Position of Patient: Prone with head turned to one side. Arms at sides; test arm is internally rotated (palm up) (Fig. 4–48).

Position of Therapist: Stand at test side.

Test: *Test 1* (generic extension): Patient raises arm off table (Fig. 4–48). *Test 2* (isolation of Latissimus): Patient pushes arm toward feet.

Instructions to Patient: *Test 1:* "Lift your arm as high as you can." *Test 2* (Latissimus): "Reach down toward your feet."

Grading

Grade 3 (Fair): Completes available range of motion with no manual resistance.

Grade 2 (Poor): Completes partial range of motion.

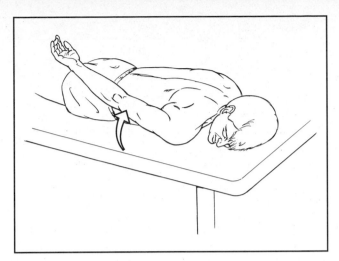

Figure 4–48

Grade 1 (Trace) and Grade 0 (Zero)

Position of Patient: Prone with arms at sides and shoulder internally rotated (palm up).

Position of Therapist: Stand at test side. Fingers for palpation (Latissimus) are placed on the side of the thoracic wall (Fig. 4–49) below and lateral to the inferior angle of the scapula.
Palpate over the posterior shoulder just superior to the axilla for posterior Deltoid fibers. Palpate the Teres major on the lateral border of the scapula just below the axilla. The Teres major is the lower of the two muscles that enter the axilla at this point; it forms the lower posterior rim of the axilla.

Test and Instructions to Patient: Patient attempts to lift arm from table on request.

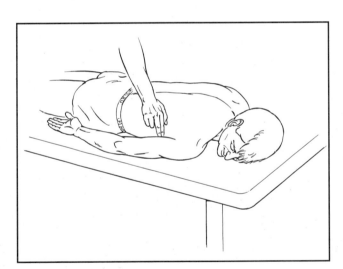

Figure 4–49

Grading

Grade 1 (Trace): Palpable contractile activity in any of the participating muscles but no movement of the shoulder.

Grade 0 (Zero): No contractile response in participating muscles.

Table 4–8 SCAPTION			Range of Motion
Muscle	**Origin**	**Insertion**	0° to 170°
133. Deltoid			
Anterior fibers	Clavicle (lateral third)	Humerus (deltoid tuberosity)	
Middle fibers	Scapula (acromium)		
135. Supraspinatus	Scapula (supraspinous fossa)	Humerus (greater tubercle)	

Others
128. Serratus anterior
140. Biceps brachii (long head)
124. Trapezius (lower)
139. Coracobrachialis

SHOULDER SCAPTION
(Deltoid and Supraspinatus)

This newly minted motion is arm elevation in the plane of the scapula, i.e., 30° to 45° anterior to the coronal plane about halfway between shoulder flexion and shoulder abduction.[2] This position, called scaption, is a position more frequently used for function than either forward flexion or abduction.

Grade 5 (Normal) to Grade 0 (Zero)

Position of Patient (All Grades): Short sitting.

Position of Therapist: Stand in front of and slightly to the test side of patient. Hand used for resistance is contoured over the arm above the elbow (Grades 5 and 4 only).

Test: Patient elevates arm halfway between flexion and abduction (30° to 45° anterior to coronal plane) (Fig. 4–50).

Instructions to Patient: "Raise your arm to shoulder height halfway between straight ahead and out to the side. Hold it. Don't let me push your arm down." (Demonstrate this motion to the patient.)

Grading

Grade 5 (Normal): Completes available range of motion and holds against maximal resistance.

Grade 4 (Good): Completes available range and holds against strong resistance, but there will be some yielding at the end of the range.

Grade 3 (Fair): Completes available range but tolerates no resistance other than the weight of the arm.

Grade 2 (Poor): Moves only through partial range of motion. The therapist's fingers for palpation are positioned on the anterior and medial aspect of the shoulder (for Grades 2 and below).

Grade 1 (Trace) and Grade 0 (Zero): Palpable or visible contractile activity for Grade 1; no activity detected for Grade 0.

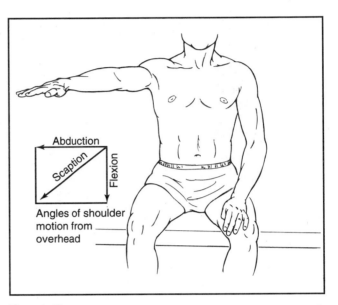

Figure 4–50

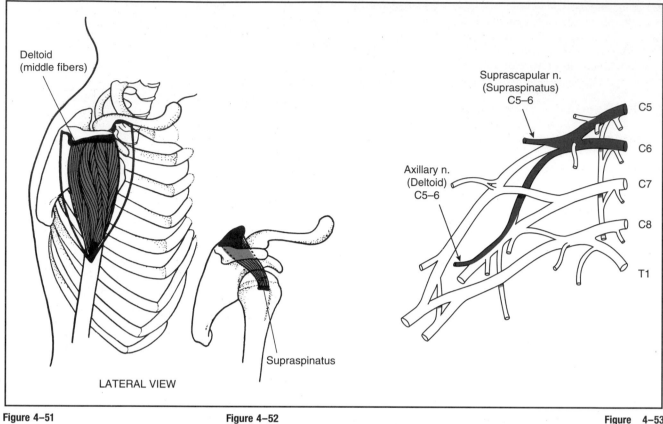

Deltoid
(middle fibers)

Suprascapular n.
(Supraspinatus)
C5–6

Axillary n.
(Deltoid)
C5–6

C5

C6

C7

C8

T1

Suprapinatus

LATERAL VIEW

Figure 4–51

Figure 4–52

Figure 4–53

Table 4–9	SHOULDER ABDUCTION	
Muscle	**Origin**	**Insertion**
133. Deltoid (middle fibers)	Scapula (acromium)	Humerus (deltoid tubercle)
135. Supraspinatus	Scapula (supraspinous fossa)	Humerus (greater tubercle)

Others
140. Biceps brachii (long head)

Range of Motion
0° to 180°

SHOULDER ABDUCTION
(Middle Deltoid and Supraspinatus)

Grade 5 (Normal), Grade 4 (Good), and Grade 3 (Fair)

Preliminary Evaluation: Examiner should check for full range of shoulder motion in all planes and should observe scapula for stability and smoothness of movement. (Refer to test for Scapular abduction and upward rotation.)

Position of Patient: Short sitting with arm at side and elbow slightly flexed.

Position of Therapist: Stand behind patient. Hand giving resistance is contoured over arm just above elbow (Fig. 4–54).

Test: Patient abducts arm to 90°.

Instructions to Patient: "Lift your arm out to the side to shoulder level. Hold it. Don't let me push it down."

Grading

Grade 5 (Normal): Holds end test position against maximal downward resistance.

Grade 4 (Good): Holds end test position against strong to moderate downward resistance.

Grade 3 (Fair): Completes range of motion to 90° with no manual resistance (Fig. 4–55).

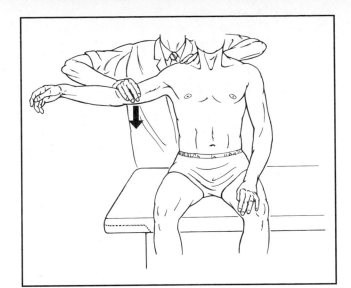

Figure 4–54

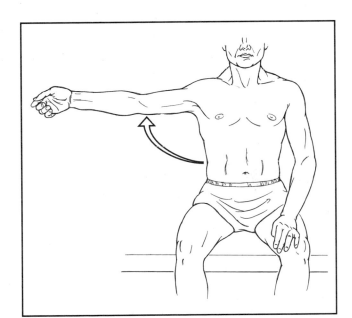

Figure 4–55

Grade 2 (Poor)

Position of Patient: Short sitting with arm at side and slight elbow flexion.

Position of Therapist: Stand behind patient to palpate muscles on test side. Palpate the Deltoid (Fig. 4–56) lateral to the acromium process on the superior aspect of the shoulder. The Supraspinatus can be palpated by placing the fingers deep under the Trapezius in the supraspinous fossa of the scapula.

Test: Patient attempts to abduct arm.

Instructions to Patient: "Try to lift your arm out to the side."

Alternate Test for Grade 2

Position of Patient: Supine. Arm is abducted to 90° but is supported on table with elbow slightly flexed (Fig. 4–57).

Position of Therapist: Stand at test side of patient. Hand used for palpation is positioned as described for Grade 2 test.

Test: Patient attempts to abduct shoulder by sliding arm on table without rotating it.

Instructions to Patient: "Take your arm out to the side."

Grading

Grade 2 (Poor) (For Both Sitting and Supine Tests): Completes partial range of motion.

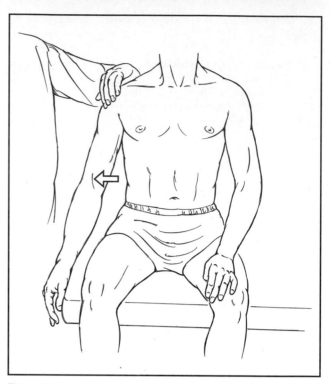

Figure 4–56

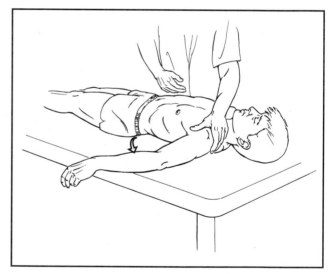

Figure 4–57

SHOULDER ABDUCTION
(Middle Deltoid and Supraspinatus)

Grade 1 (Trace) and Grade 0 (Zero)

Position of Patient: Short sitting.

Position of Therapist: Stand behind and to the side of patient. Therapist cradles test arm with the shoulder in about 90° of abduction providing limb support at the elbow (Fig. 4–58).

Test: Patient tries to maintain the arm in abduction.

Instructions to Patient: "Try to hold your arm in this position."

Alternate Test for Grade 1 and Grade 0 (Supine)

Position of Patient: Supine with arm at side and elbow slightly flexed.

Position of Therapist: Stand at side of table at a place where the Deltoid can be reached. Palpate the Deltoid on the lateral surface of the upper one-third of the arm (Fig. 4–59).

Grading

Grade 1 (Trace): Palpable or visible contraction of Deltoid with no movement.

Grade 0 (Zero): No contractile activity.

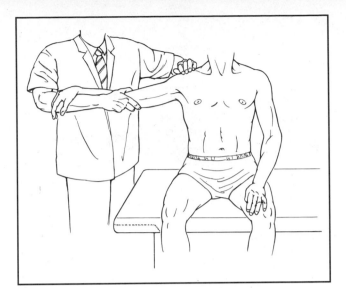

Figure 4–58

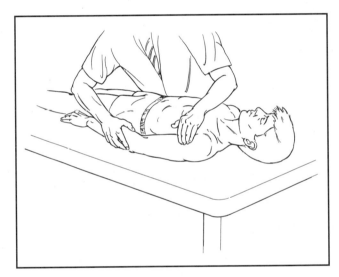

Figure 4–59

SUBSTITUTION BY BICEPS BRACHII

When a patient uses the Biceps to substitute, the shoulder will externally rotate and the elbow will flex. The arm will be raised but not by the action of the abductor muscles. To avoid this substitution begin the test with the arm in a few degrees of elbow flexion, but do not allow active contraction of the Biceps during the test.

⬤ HELPFUL HINTS

1. Turning the face to the opposite side and extending the neck will put the Trapezius on slack and make the Supraspinatus more accessible for palpation.

2. The Deltoid and Supraspinatus work in tandem, and when one is active in abduction the other will also be active. Only when Supraspinatus weakness is suspected is it necessary to palpate.

3. Do not allow shoulder elevation or lateral flexion of the trunk to the opposite side because these movements can create an illusion of abduction.

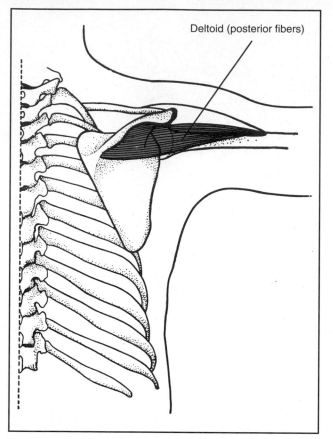

Figure 4–60

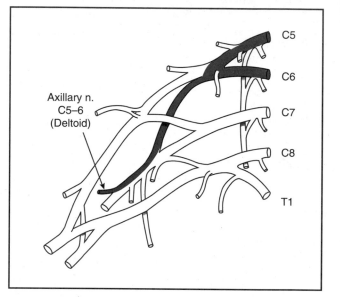

Figure 4–61

Table 4–10	SHOULDER HORIZONTAL ABDUCTION	
Muscle	**Origin**	**Insertion**
133. Deltoid (posterior fibers)	Scapula (spine on posterior border)	Humerus (deltoid tuberosity)
Others		
136. Infraspinatus		
137. Teres minor		

Range of Motion

When starting from a position of 90° of forward flexion: 0° to 90° (range, 90°)

When starting with the arm in full horizontal adduction: −40° to 90° (range, 130°)

SHOULDER HORIZONTAL ABDUCTION
(Posterior Deltoid)

Grade 5 (Normal), Grade 4 (Good), and Grade 3 (Fair)

Position of Patient: Prone. Shoulder abducted to 90° and forearm off edge of table with elbow flexed.

Position of Therapist: Stand at test side. Hand giving resistance is contoured over posterior arm just above the elbow (Fig. 4–62).

Test: Patient horizontally abducts shoulder against maximal resistance.

Instructions to Patient: "Lift your elbow up toward the ceiling. Hold it. Don't let me push it down."

Grading

Grade 5 (Normal): Completes range and holds end position against maximum resistance.

Grade 4 (Good): Completes range and holds end position against strong to moderate resistance.

Grade 3 (Fair): Completes range of motion with no manual resistance (Fig. 4–63).

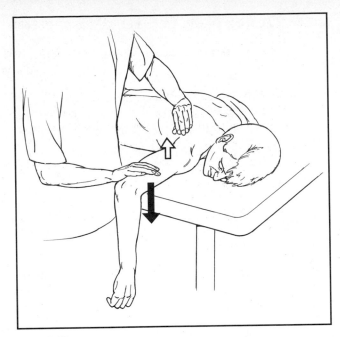

Figure 4–62

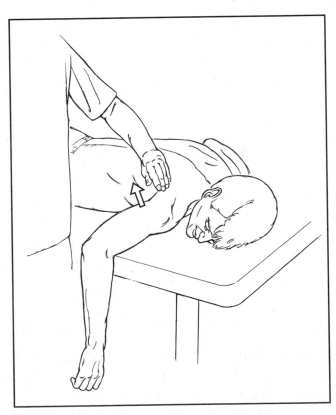

Figure 4–63

Grade 2 (Poor), Grade 1 (Trace), Grade 0 (Zero)

Position of Patient: Short sitting over end or side of table.

Position of Therapist: Stand at test side. Support forearm under distal surface (Fig. 4–64) and palpate over the posterior surface of the shoulder just superior to the axilla.

Alternate Test for Grades 2, 1, and 0

Position of Patient: Short sitting with arm supported on table (smooth surface) in 90° of abduction; elbow partially flexed.

Position of Therapist: Stand behind patient. Stabilize by contouring hand over the superior aspect of the shoulder and over the scapula (Fig. 4–65). Palpate the fibers of the posterior Deltoid below and lateral to the spine of the scapula and on the posterior aspect of the proximal arm adjacent to the axilla.

Test: Patient slides (or tries to move) the arm across the table in horizontal abduction.

Instructions to Patient: "Slide your arm backward."

Grading

Grade 2 (Poor): Moves through partial range of motion.

Grade 1 (Trace): Palpable contraction; no motion.

Grade 0 (Zero): No contractile activity.

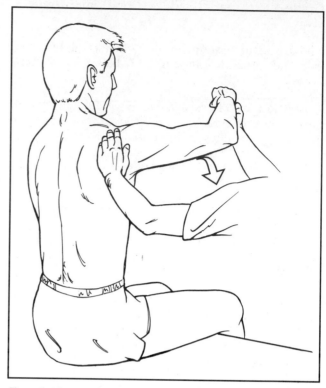

Figure 4–64

SUBSTITUTION BY TRICEPS BRACHII (LONG HEAD)

Maintain the elbow in flexion to avoid substitution by the long head of the Triceps.

⊙ **HELPFUL HINTS**

If the scapular muscles are weak, the examiner must manually stabilize the scapula to avoid scapular abduction.

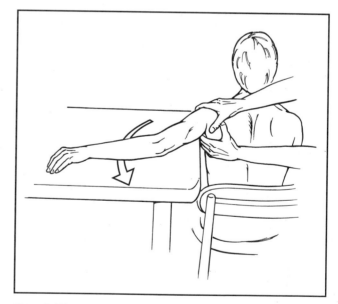

Figure 4–65

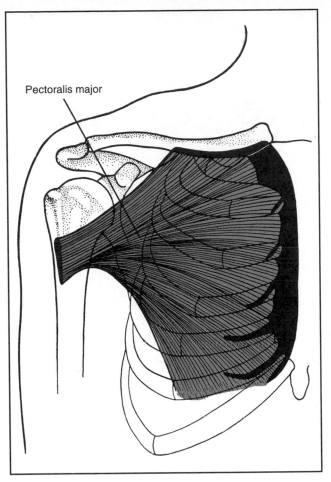

Figure 4–66

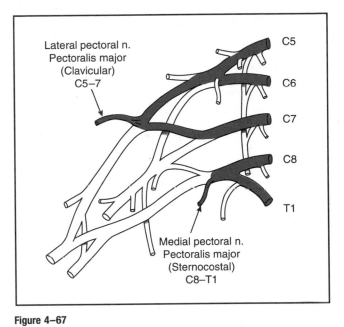

Figure 4–67

Table 4–11 SHOULDER HORIZONTAL ADDUCTION		
Muscle	**Origin**	**Insertion**
131. Pectoralis major	Clavicle (sternal half)	Humerus (greater tubercle)
	Sternum (anterior surface down to rib 6)	
	Ribs 1–7 (cartilages)	
Others		
133. Deltoid (anterior fibers)		

Range of Motion
0° to 130°
When starting from a position of 90° of forward flexion: 0° to −40° (range, 40°)
When starting with the arm in full horizontal abduction: 0° passing across the midline to −40° (range, 130°)

Preliminary Examination

The examiner begins with the patient supine and checks the range of motion and then tests both heads of the Pectoralis major simultaneously. The patient is asked to move the arm in horizontal adduction, keeping it parallel to the floor without rotation.

If the arm moves across the body in a diagonal motion, test the sternal and clavicular heads of the muscle separately. Testing both heads of the Pectoralis major separately should be routine in any patient with cervical spinal cord injury.

Grade 5 (Normal) and Grade 4 (Good)

Position of Patient

Whole Muscle: Supine. Shoulder abducted to 90°; elbow flexed to 90° (Fig. 4–68).

Clavicular Head: Patient begins test with shoulder in 60° of abduction, elbow flexed. Patient then is asked to horizontally adduct the shoulder.

Sternal Head: Patient begins test with shoulder in about 120° of abduction with elbow flexed.

Position of Therapist: Stand at side of shoulder to be tested. Hand used for resistance is contoured around the forearm just proximal to the wrist. The other hand is used to check the activity of the Pectoralis major on the upper aspect of the chest just medial to the shoulder joint (Fig. 4–69). (Palpation is not needed in a Grade 5 test, but it is prudent to assess activity in the muscle being tested.)

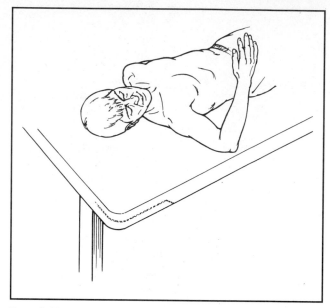

Figure 4–68

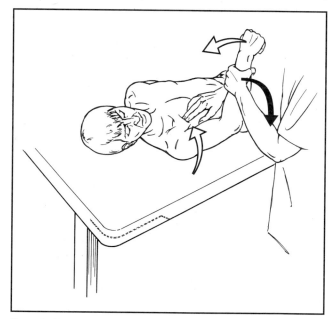

Figure 4–69

Palpate the clavicular fibers of the Pectoralis major up under the medial half of the clavicle (Fig. 4–70). Palpate the sternal fibers on the chest wall at the lower anterior border of the axilla (Fig. 4–71).

Test: When the *whole muscle* is tested, the patient horizontally adducts the shoulder through the available range of motion.

To test the *clavicular head,* the patient's motion begins at 60° of abduction and moves up and in. The examiner applies resistance above the wrist in a downward direction (toward floor) and outward (i.e., opposite to the direction of the fibers of the clavicular head, which moves the arm diagonally up and inward; see Fig. 4–70).

To test the *sternal head,* the motion begins at 120° of shoulder abduction and moves down and in. Resistance is given above the wrist in an up and outward direction (i.e., opposite to the motion of the sternal head, which is diagonally down and inward; see Fig. 4–71).

Instructions to Patient

Both Heads: "Move your arm across your chest. Hold it. Don't let me pull it back."

Clavicular Head: "Move your arm up and in."

Sternal Head: "Move your arm down and in."

Grading

Grade 5 (Normal): Completes range of motion and takes maximal resistance.

Grade 4 (Good): Completes range of motion and takes strong to moderate resistance, but muscle exhibits some "give" at end of range.

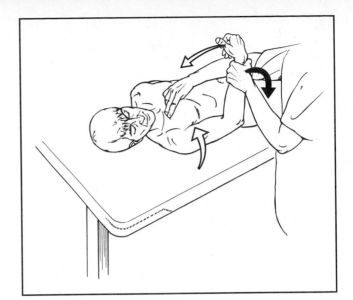

Figure 4–70

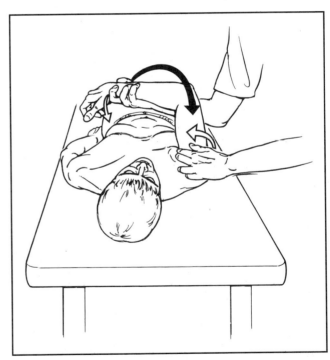

Figure 4–71

Grade 3 (Fair)

Position of Patient: Supine. Shoulder at 90° of abduction and elbow at 90° of flexion.

Position of Therapist: Same as for Grade 5.

Test

Both Heads: Patient horizontally adducts extremity across chest in a straight pattern with no diagonal motion (Fig. 4–72).

Clavicular Head: Direction of motion by the patient is diagonally up and inward.

Sternal Head: Direction of motion is diagonally down and inward.

Instructions to Patient: Same as for the Grade 5 (Normal) test, but no resistance is offered.

Grading

Grade 3 (Fair): Patient completes available range of motion in all three tests with no resistance other than the weight of the extremity.

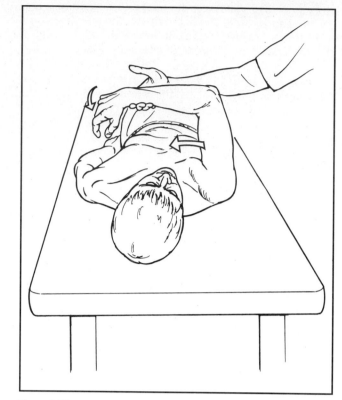

Figure 4–72

Grade 2 (Poor), Grade 1 (Trace), and Grade 0 (Zero)

Position of Patient: Supine. Arm is supported in 90° of abduction with elbow flexed to 90°.

Alternate Position: Patient is seated with test arm supported on table (at level of axilla) with arm in 90° of abduction (or in scaption) and elbow slightly flexed (Fig. 4–73). Friction of the table surface should be minimized.

Position of Therapist: Stand at side of shoulder to be tested or behind the sitting patient. When the patient is supine, support the full length of the forearm and hold the limb at the wrist (Fig. 4–71).
For both tests palpate the Pectoralis major muscle on the anterior aspect of the chest medial to the shoulder joint (see Fig. 4–69).

Test: Patient attempts to horizontally adduct the shoulder. The use of the alternate test position, in which the arm moves across the table, precludes individual testing for the two heads.

Instructions to Patient: "Try to move your arm across your chest."
In seated position: "Move your arm forward."

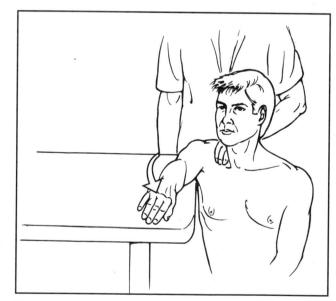

Figure 4–73

Grading

Grade 2 (Poor): Patient horizontally adducts shoulder through available range of motion with the weight of the arm supported by the examiner or the table.

Grade 1 (Trace): Palpable contractile activity.

Grade 0 (Zero): No contractile activity.

HELPFUL HINTS

This test requires resistance on the forearm, which in turn requires that the elbow flexors be strong. If they are weak, provide resistance on the arm just proximal to the elbow.

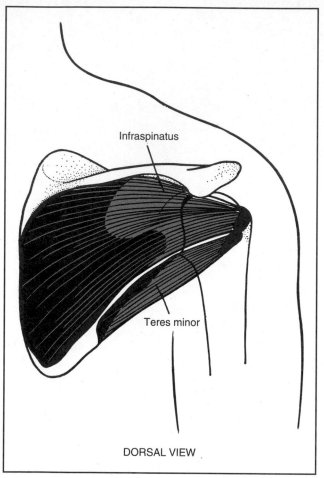

Infraspinatus

Teres minor

DORSAL VIEW

Figure 4–74

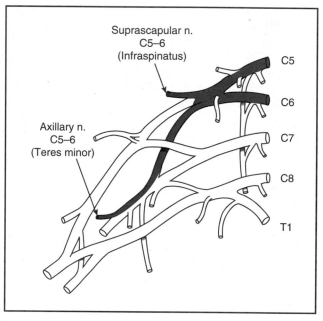

Suprascapular n.
C5–6
(Infraspinatus)

Axillary n.
C5–6
(Teres minor)

C5

C6

C7

C8

T1

Figure 4–75

Table 4–12 SHOULDER EXTERNAL ROTATION		
Muscle	**Origin**	**Insertion**
136. Infraspinatus	Scapula (infraspinous fossa)	Humerus (greater tubercle)
137. Teres minor	Scapula (axillary border)	Humerus (greater tubercle)

Others
133. Deltoid (posterior)

Range of Motion

0° to 60°
(In literature, range varies between 0° and 90°. Range also varies with elevation of arm.)

Grade 5 (Normal), Grade 4 (Good), and Grade 3 (Fair)

Position of Patient: Prone with head turned toward test side. Shoulder abducted to 90° with arm fully supported on table; forearm hanging vertically over edge of table. Place a folded towel under the arm at the edge of the table if it has a sharp edge.

Alternate Position: Short sitting with shoulder abducted to 90°. The amount of resistance tolerated in this position may be much greater for Grades 5 and 4.

Position of Therapist: Stand at test side at level of patient's waist (Fig. 4–76). Two fingers of one hand are used to give resistance at the wrist for Grades 5 and 4. The other hand supports the elbow to provide some counterpressure at the end of the range.

Test: Patient moves forearm upward through the range of external rotation.

Instructions to Patient: "Raise your arm to the level of the table. Hold it. Don't let me push it down." Therapist may need to demonstrate the desired motion.

Grading

Grade 5 (Normal): Completes available range of motion and holds firmly against two-finger resistance.

Grade 4 (Good): Completes available range, but the muscle at end range yields or gives way.

Grade 3 (Fair): Completes available range of motion but is unable to take any manual resistance (Fig. 4–77).

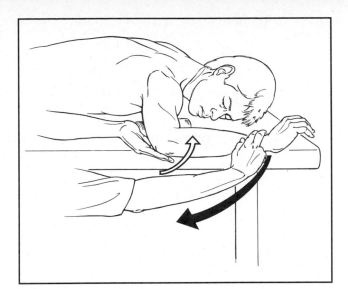

Figure 4–76

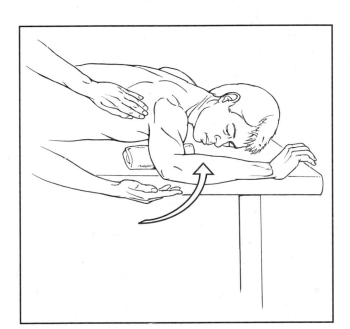

Figure 4–77

Grade 2 (Poor), Grade 1 (Trace), and Grade 0 (Zero)

Position of Patient: Prone with head turned to test side, trunk at edge of table. The entire limb hangs down loosely from the shoulder in neutral rotation, palm facing table (Fig. 4–78).

Position of Therapist: Stand or sit on a low stool at test side of patient at shoulder level. Palpate the Infraspinatus over the body of the scapula below the spine in the infraspinous fossa (see Fig. 4–77). Palpate the Teres minor on the inferior margin of the axilla and along the axillary border of the scapula (see Fig. 4–78).

Test: Patient attempts to externally rotate the shoulder. Alternatively, place the patient's arm in lateral rotation and ask the patient to hold the end position (Fig. 4–79).

Instructions to Patient: "Turn your palm outward."

Grading

Grade 2 (Poor): Completes available range (i.e., palm faces forward) in this gravity-eliminated position.

Grade 1 (Trace): Palpation of either or both muscles reveals contractile activity but no motion.

Grade 0 (Zero): No palpable or visible activity.

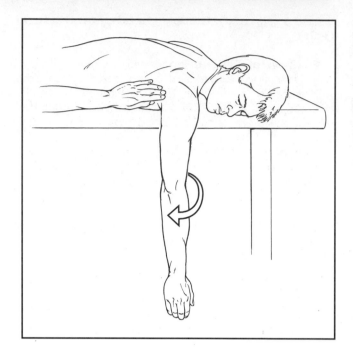

Figure 4–78

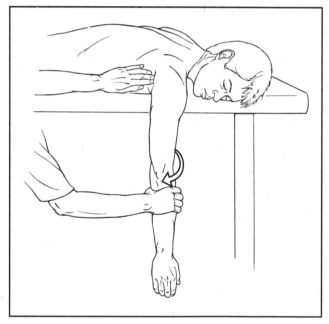

Figure 4–79

⊙ HELPFUL HINTS

1. Resistance in tests of shoulder rotation should be administered gradually and slowly, with great care taken to prevent injury, which can occur readily because the shoulder lacks inherent stability. This is particularly important for the elderly patient.

2. The therapist must be careful to discern whether supination occurs instead of the requested external rotation during the testing of Grades 2 and 1 muscles because this motion can be mistaken for lateral rotation.

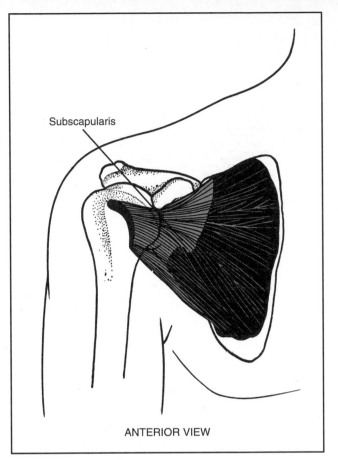

Subscapularis

ANTERIOR VIEW

Figure 4–80

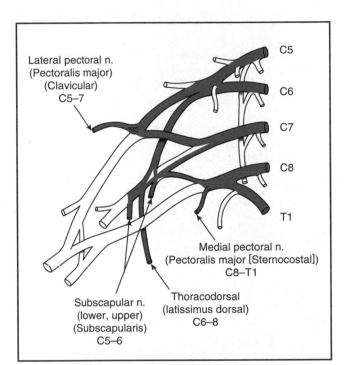

Lateral pectoral n.
(Pectoralis major)
(Clavicular)
C5–7

C5

C6

C7

C8

T1

Medial pectoral n.
(Pectoralis major [Sternocostal])
C8–T1

Subscapular n.
(lower, upper)
(Subscapularis)
C5–6

Thoracodorsal
(latissimus dorsal)
C6–8

Figure 4–81

Table 4–13	SHOULDER INTERNAL ROTATION	
Muscle	**Origin**	**Insertion**
134. Subscapularis	Scapula (subscapular margin)	Humerus (lesser tubercle)
131. Pectoralis major	Clavicle (sternal half) Sternum (anterior surface down to rib 6) Ribs 1–7 (cartilages)	Humerus (greater tubercle)
130. Latissimus dorsi	T6–T12 vertebrae L1–L5 vertebrae Sacral vertebrae Ribs 9–12 Scapula (inferior angle) Iliac crest	Humerus (intertubercular groove)
138. Teres major	Scapula (inferior angle)	Humerus (intertubercular sulcus)

Others
133. Deltoid (anterior)

Range of Motion

0° to 80°
(In literature, range varies from
0° to 45° and to as high as 90°.
Range also varies with elevation
of arm.

Grade 5 (Normal), Grade 4 (Good), Grade 3 (Fair)

Position of Patient: Prone with head turned toward test side. Shoulder is abducted to 90° with folded towel placed under distal arm and forearm hanging vertically over edge of table. Short sitting is a common alternate position.

Position of Therapist: Stand at test side. Hand giving resistance is placed on the volar side of the forearm just above the wrist. The other hand provides counterforce at the elbow (Fig. 4–82). The resistance hand applies resistance in a downward and forward direction; the counterforce is applied backward and slightly upward. Stabilize the scapular region if muscles are weak.

Test: Patient moves forearm through available range of internal rotation (backward and upward).

Instructions to Patient: "Move your forearm up and back. Hold it. Don't let me push it down." Demonstrate the desired motion to the patient.

Grading

Grade 5 (Normal): Completes available range and holds firmly against strong resistance.

Grade 4 (Good): Completes available range, but there is a "spongy" feeling against strong resistance.

Grade 3 (Fair): Completes available range with no manual resistance (Fig. 4–83).

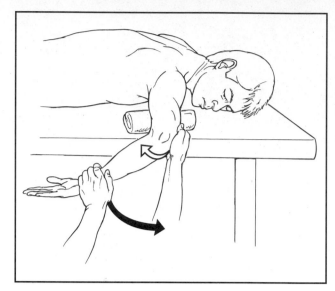

Figure 4–82

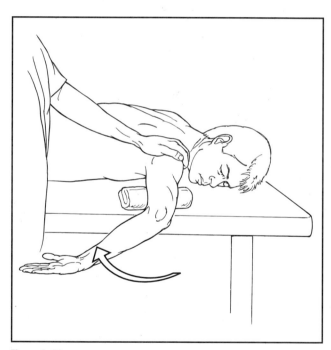

Figure 4–83

Grade 2 (Poor), Grade 1 (Trace), and Grade 0 (Zero)

Position of Patient: Prone with head turned toward test side. Patient must be near the edge of the table on test side so that entire arm can hang down freely over the edge (Fig. 4–84). Arm is in neutral with palm facing the table.

Position of Therapist: Stand at test side or sit on low stool. Hand used for palpation must find the tendon of the Subscapularis deep in the central area of the axilla (Fig. 4–85). Therapist may have to stabilize test arm at the shoulder.

Test: Patient internally rotates arm with thumb leading so that the palm faces out or away from the table.

Instructions to Patient: "Turn your arm so that the palm faces away from the table." (Not Shown.)

Grading

Grade 2 (Poor): Completes available range.

Grade 1 (Trace): Palpable contraction occurs.

Grade 0 (Zero): No palpable contraction.

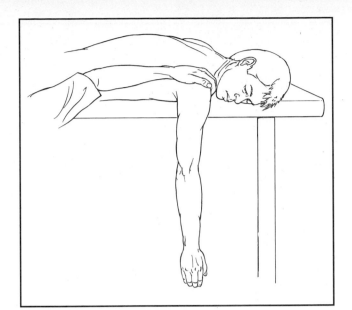

Figure 4–84

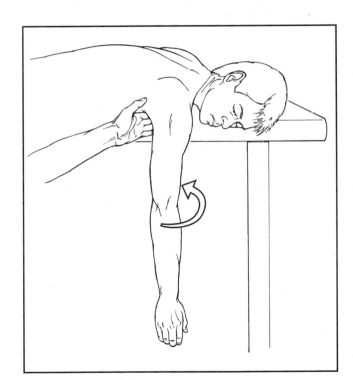

Figure 4–85

◉ HELPFUL HINTS

1. The therapist should be wary of pronation in this test. Forearm pronation is rather easily mistaken for internal rotation.

2. Internal rotation is a much stronger motion than external rotation. This is largely a factor of differing muscle mass.

3. If you cannot palpate the Subscapularis, try the Pectoralis major, which, as a surface muscle, is more readily felt.

4. The hand of the examiner may substitute for a towel roll under the distal arm, the purpose being to protect the patient from the discomfort of moving against a hard table and to keep the arm horizontal to the floor.

5. The prone position is preferred to the supine or sitting position in tests for Grades 2, 1, and 0 because a weak patient has a tendency to use trunk rotation as a substitute.

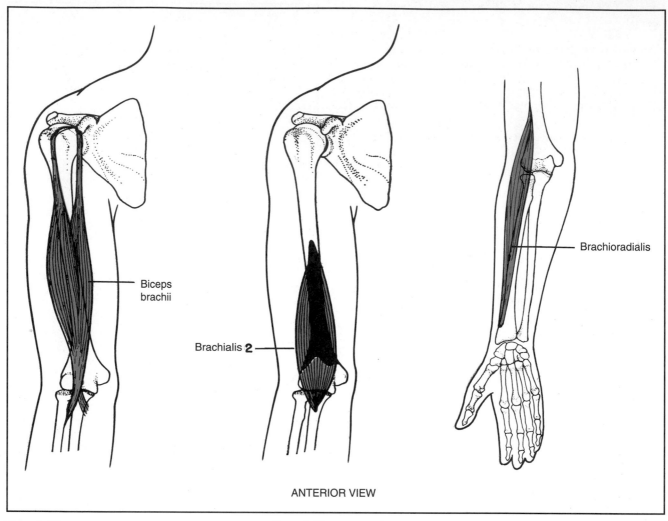

Biceps brachii

Brachialis 2

Brachioradialis

ANTERIOR VIEW

Figure 4–86　　　　　　　　　**Figure 4–87**　　　　　　　　　**Figure 4–88**

Table 4–14　ELBOW FLEXION		
Muscle	**Origin**	**Insertion**
140. Biceps brachii		
Short head	Scapula (coracoid apex)	Radius (radial tuberosity)
Long head	Scapula (supraglenoid tubercle)	
141. Brachialis	Humerus (anterior shaft of distal 2/3)	Ulna (ulnar tuberosity)
143. Brachioradialis	Humerus (supracondylar ridge)	Radius (proximal to styloid process)

Range of Motion
0° to 150°

ELBOW FLEXION
(Biceps, Brachialis, and Brachioradialis)

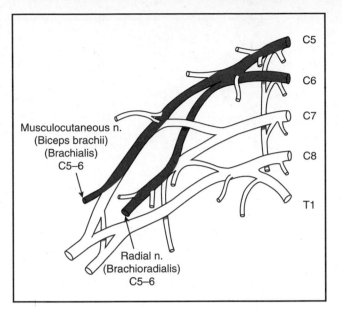

Musculocutaneous n.
(Biceps brachii)
(Brachialis)
C5–6

C5
C6
C7
C8
T1

Radial n.
(Brachioradialis)
C5–6

Figure 4–89

ELBOW FLEXION
(Biceps, Brachialis, and Brachioradialis)

Grade 5 (Normal), Grade 4 (Good), and Grade 3 (Fair)

Position of Patient: Short sitting with arms at sides. The following are the positions of choice, but it is doubtful whether the individual muscles can be separated when strong effort is used. The Brachialis in particular is independent of forearm position.

Biceps brachii: forearm in supination (Fig. 4–90)
Brachialis: forearm in pronation (Fig. 4–91)
Brachioradialis: forearm in midposition between pronation and supination (Fig. 4–92)

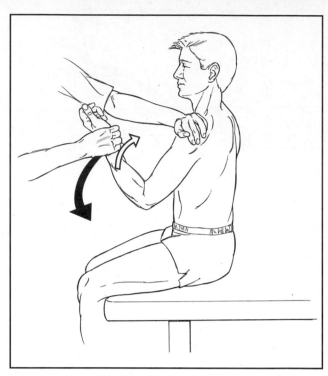

Figure 4–90

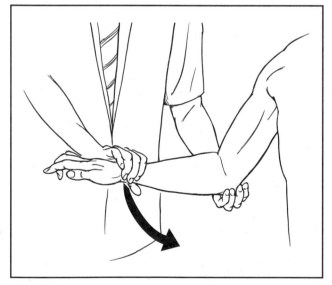

Figure 4–91

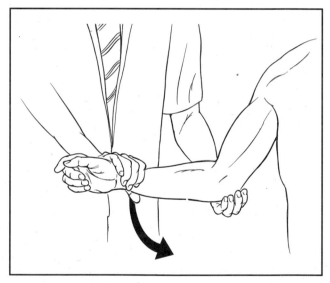

Figure 4–92

ELBOW FLEXION
(Biceps, Brachialis, and Brachioradialis)

Position of Therapist: Stand in front of patient toward the test side. Hand giving resistance is contoured over the flexor surface of the forearm proximal to the wrist (see Fig. 4–90). The other hand applies counterforce by cupping the palm over the anterior superior surface of the shoulder.

No resistance is given in a Grade 3 test, but the test elbow is cupped by the examiner's hand (Fig. 4–93, Biceps illustrated at end range).

Test (All Three Forearm Positions): Patient flexes elbow through range of motion.

Instructions to Patient (All Three Tests)

Grades 5 and 4: "Bend your elbow. Hold it. Don't let me pull it down."

Grade 3: "Bend your elbow."

Grading

Grade 5 (Normal): Completes available range and holds firmly against maximal resistance.

Grade 4 (Good): Completes available range against strong to moderate resistance, but the end point may not be firm.

Grade 3 (Fair): Completes available range with each forearm position with no manual resistance.

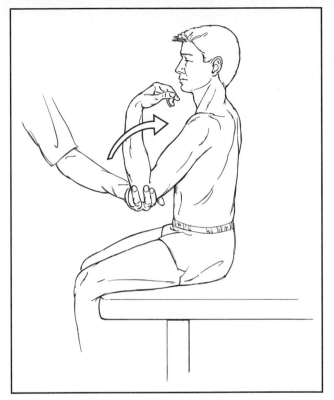

Figure 4–93

Grade 2 (Poor)

Position of Patient

All Elbow Flexors: Short sitting with arm abducted to 90° and supported by examiner (Fig. 4–94). Forearm is supinated (Biceps), pronated (Brachialis), and in midposition (Brachioradialis).

Alternate Position: Supine. Elbow is flexed to about 45° with forearm in position described for each of the muscles (Fig. 4–95, Biceps illustrated).

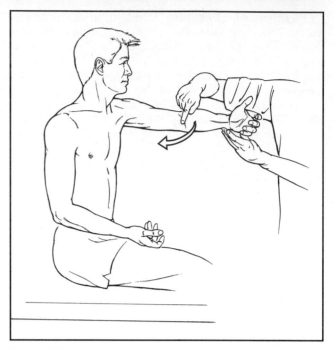

Figure 4–94

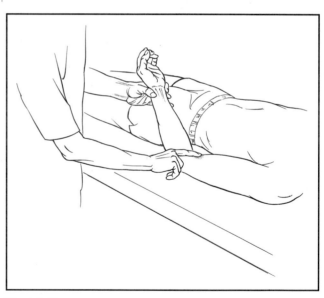

Figure 4–95

Position of Therapist

All Three Flexors: Stand in front of patient and support abducted arm under the elbow and wrist if necessary (see Fig. 4–94). Palpate the tendon of the Biceps in the antecubital space (see Fig. 4–95). On the arm the muscle fibers may be felt on the anterior surface of the middle two-thirds with the short head lying medial to the long head.

Palpate the Brachialis in the distal arm medial to the tendon of the Biceps. Palpate the Brachioradialis on the proximal volar surface of the forearm, where it forms the lateral border of the cubital fossa (Fig. 4–96).

Test: Patient attempts to flex the elbow.

Instructions to Patient: "Try to bend your elbow."

Grading

Grade 2 (Poor): Completes partial range of motion (in each of the muscles tested).

Grade 1 (Trace) and Grade 0 (Zero)

Positions of Patient and Therapist: Supine for all three muscles with therapist standing at test side (see Fig. 4–96). All other aspects are the same as Grade 2 test.

Test: Patient attempts to bend elbow with hand supinated, pronated, and in midposition.

Grading

Grade 1 (Trace): Examiner can palpate a contractile response in each of the three muscles for which a Trace grade is given.

Grade 0 (Zero): No palpable contractile activity.

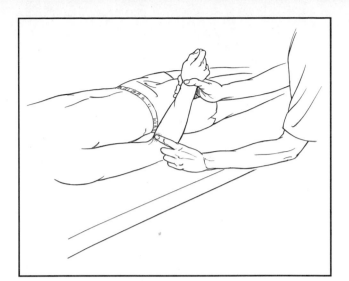

Figure 4–96

◯ HELPFUL HINTS

1. The patient's wrist flexor muscles should remain relaxed throughout the test because strongly contracting wrist flexors may assist in elbow flexion.

2. If the sitting position is contraindicated for any reason, all tests for these muscles may be performed in the supine position, but in that case manual resistance should be part of the Grade 3 test (gravity compensation).

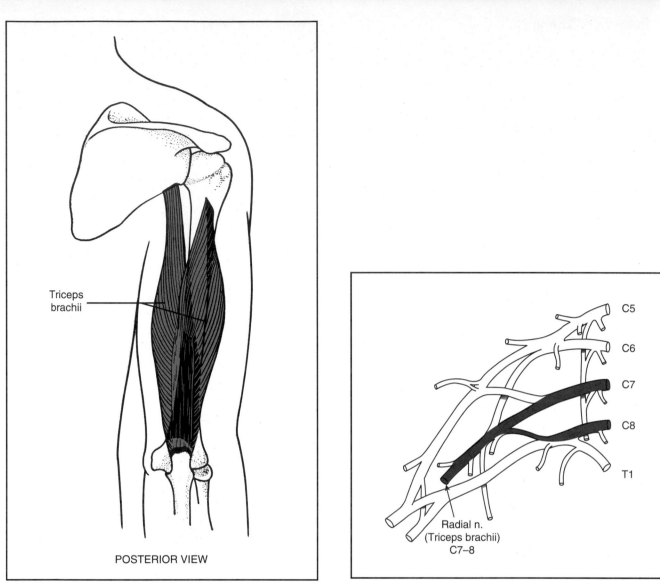

Triceps
brachii

POSTERIOR VIEW

Figure 4–97

Radial n.
(Triceps brachii)
C7–8

C5
C6
C7
C8
T1

Figure 4–98

Table 4–15 ELBOW EXTENSION		
Muscle	**Origin**	**Insertion**
142. Triceps brachii		
Long head	Scapula (infraglenoid tuberosity)	Ulna (olecranon)
Lateral head	Humerus (posterior)	
Medial head	Humerus (posterior, below lateral head)	

Range of Motion
150° to 0°

ELBOW EXTENSION
(Triceps brachii)

Grade 5 (Normal), Grade 4 (Good), and Grade 3 (Fair)

Position of Patient: Prone on table. The patient starts the test with the arm in 90° of abduction and the forearm flexed and hanging vertically over the side of the table (Fig. 4–99).

Position of Therapist: For the prone patient the therapist provides support just above the elbow. The other hand is used to apply downward resistance on the dorsal surface of the wrist (Fig. 4–100, illustrates end position).

Test: Patient extends elbow to end of available range or until the forearm is horizontal to the floor.

Instructions to Patient: "Straighten your elbow. Hold it. Don't let me bend it."

Grading

Grade 5 (Normal): Completes available range and holds firmly against maximal resistance.

Grade 4 (Good): Completes available range against strong resistance, but there is a "give" to the resistance at the end range.

Grade 3 (Fair): Completes available range with no manual resistance (Fig. 4–101).

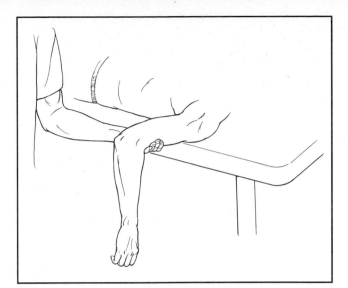

Figure 4–99

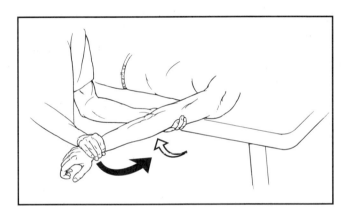

Figure 4–100

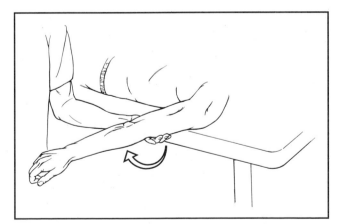

Figure 4–101

Grade 2 (Poor), Grade 1 (Trace), and Grade 0 (Zero)

Position of Patient: Short sitting. The arm is abducted to 90° with the shoulder in neutral rotation and the elbow flexed to about 45°. The entire limb is horizontal to the floor (Fig. 4–102).

Position of Therapist: Stand at test side of patient. For the Grade 2 test, support the limb at the elbow. For a Grade 1 or 0 test, support the limb under the forearm and palpate the Triceps on the posterior surface of the arm just proximal to the olecranon process (Fig. 4–103).

Test: Patient attempts to extend the elbow.

Instructions to Patient: "Try to straighten your elbow."

Grading

Grade 2 (Poor): Completes available range in the absence of gravity.

Grade 1 (Trace): Examiner can feel tension in the Triceps tendon just proximal to the olecranon (Fig. 4–103) or contractile activity in the muscle fibers on the posterior surface of the arm.

Grade 0 (Zero): No evidence of any muscle activity.

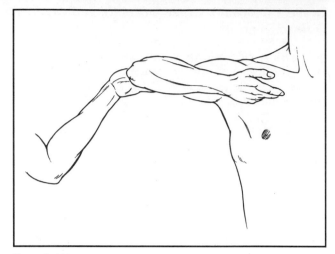

Figure 4–102

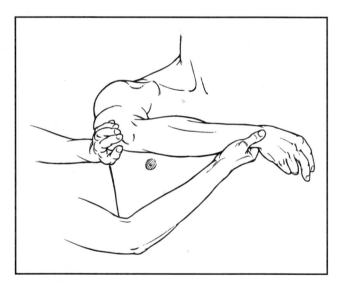

Figure 4–103

ELBOW EXTENSION
(Triceps brachii)

SUBSTITUTIONS

1. *Via external rotation.* When the patient is sitting with the arm abducted, elbow extension can be accomplished with a Grade 0 Triceps (Fig. 4–104). This can occur when the patient externally rotates the shoulder, thus dropping the arm below the forearm. As a result, the elbow literally falls into extension.

2. *Via horizontal adduction.* This substitution can accomplish elbow extension and is done purposefully by patients with a cervical cord injury and a Grade 0 Triceps. With the distal segment fixed (as when the examiner stabilizes the hand or wrist), the patient horizontally adducts the arm, and the thrust pulls the elbow into extension (Fig. 4–105). The therapist, therefore, should provide support at the elbow for testing purposes rather than at the wrist.

ⓒ HELPFUL HINTS

1. The therapist should confirm that muscle activity is seen and felt (i.e., Triceps activity is actually present!) because patients can become very adept at substituting. In fact, substitutions frequently are taught and encouraged as a functional movement but are not allowed for the purpose of testing.

2. Give resistance in Grades 5 and 4 tests with the elbow slightly flexed to avoid enabling the patient to "lock" the elbow joint by hyperextending it.

3. Ideally, elbow extension should not be tested in the prone position because when the shoulder is horizontally abducted to conduct the test, the two-joint muscle is less effective, and the test grade may well be lower than it should be.

4. An alternate position for Grades 5, 4, and 3 is with the patient short sitting. The examiner stands behind the patient, supporting the arm in 90° of abduction just above the flexed elbow (Fig. 4–106). The patient straightens the elbow against resistance given at the wrist.

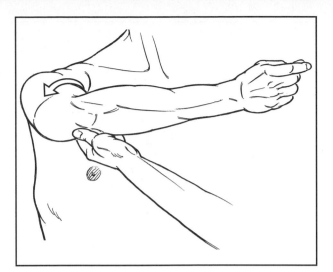

Figure 4–104

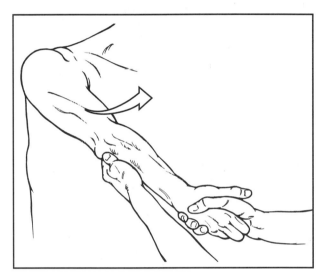

Figure 4–105

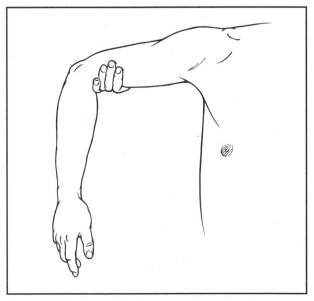

Figure 4–106

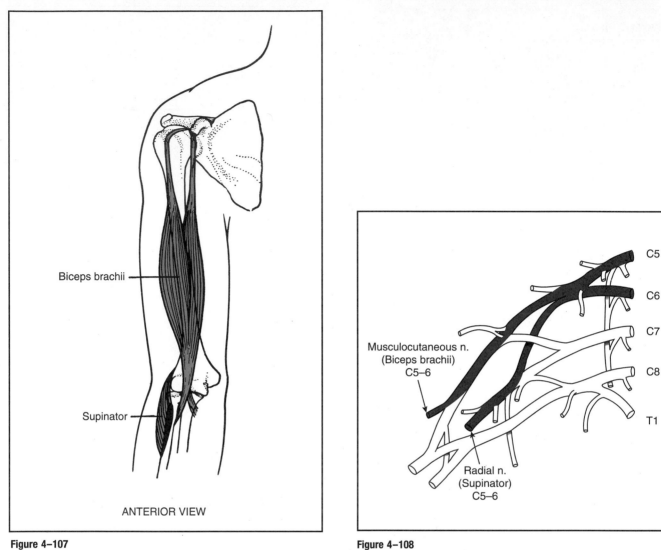

Biceps brachii

Supinator

ANTERIOR VIEW

Figure 4–107

Musculocutaneous n.
(Biceps brachii)
C5–6

Radial n.
(Supinator)
C5–6

C5
C6
C7
C8
T1

Figure 4–108

Table 4–16 FOREARM SUPINATION		
Muscle	**Origin**	**Insertion**
145. Supinator	Humerus (lateral condyle) Ulna (dorsal shaft)	Radius (dorsal and lateral body)
140. Biceps brachii Short head	Scapula (coracoid apex)	Radius (radial tuberosity)
Long head	Scapula (supraglenoid tubercle)	

Range of Motion

0° to 80°

Grade 5 (Normal), Grade 4 (Good), and Grade 3 (Fair)

Position of Patient: Short sitting; arm at side and elbow flexed to 90°; forearm in neutral or midposition (Fig. 4–109, showing end range). Alternatively, patient may sit at a table.

Position of Therapist: Stand at side or in front of patient. One hand supports the elbow (Fig. 4–109). For resistance, grasp the forearm on the volar surface at the wrist.

Test: Patient begins at neutral wrist position and supinates the forearm until the palm faces the ceiling. Therapist resists motion in the direction of pronation. (No resistance is given for Grade 3.)

Alternate Test: Grasp patient's hand as if shaking hands; cradle the elbow and resist via the hand grip (Fig. 4–110). This test is used if the patient has Grade 5 or 4 wrist and hand strength. If wrist flexion is painful, give resistance at the wrist (a more difficult level).

Instructions to Patient: "Turn your palm up. Hold it. Don't let me turn it down. Keep your wrist and fingers relaxed."
For Grade 3: "Turn your palm up."

Grading

Grade 5 (Normal): Completes full available range of motion and holds against maximal resistance.

Grade 4 (Good): Completes full range of motion against strong to moderate resistance.

Grade 3 (Fair): Completes available range of motion without resistance (Fig. 4–111, showing end range).

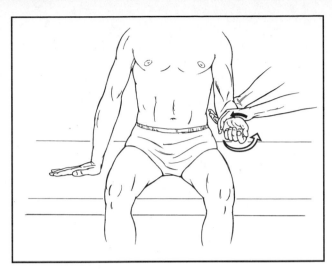

Figure 4–109

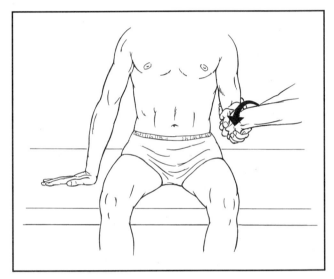

Figure 4–110

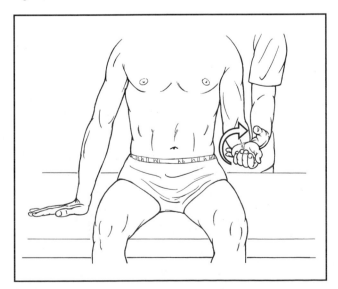

Figure 4–111

Grade 2 (Poor)

Position of Patient: Short sitting with shoulder flexed between 45° and 90° and elbow flexed to 90°. Forearm in neutral.

Position of Therapist: Support the test arm by cupping the hand under the elbow.

Test: Patient supinates forearm (Fig. 4–112) through partial range of motion.

Instructions to Patient: "Turn your palm toward your face."

Grading

Grade 2 (Poor): Completes a partial range of motion.

Grade 1 (Trace) and Grade 0 (Zero)

Position of Patient: Short sitting. Arm and elbow are flexed as for the Grade 3 test.

Position of Therapist: Support the forearm just distal to the elbow. Palpate the Supinator distal to the head of the radius on the dorsal aspect of the forearm (Fig. 4–113).

Test: Patient attempts to supinate the forearm.

Instructions to Patient: "Try to turn your palm so it faces the ceiling."

Grading

Grade 1 (Trace): Slight contractile activity but no limb movement.

Grade 0 (Zero): No contractile activity.

SUBSTITUTION

1. Patient may externally rotate and adduct the arm across the body (Fig. 4–114) as forearm supination is attempted. When this occurs, the forearm rolls into supination with no activity of the Supinator muscle.

2. Patient should be instructed to keep the wrist and fingers as relaxed as possible to avoid substitutions, particularly by the wrist extensors.

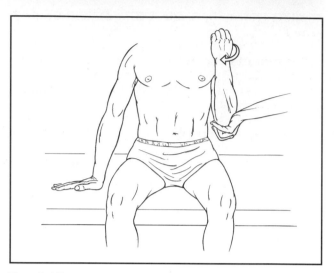

Figure 4–112

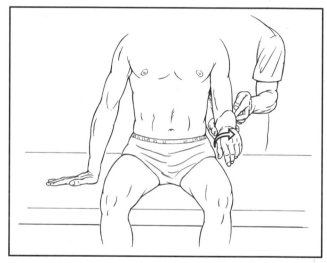

Figure 4–113

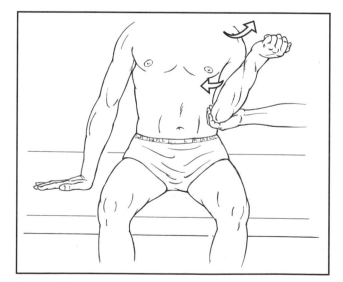

Figure 4–114

FOREARM PRONATION
(Pronator teres and Pronator quadratus)

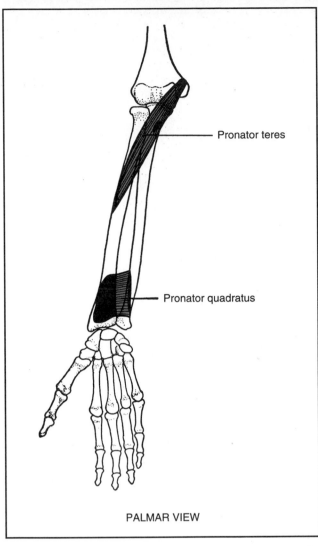

PALMAR VIEW

Figure 4–115

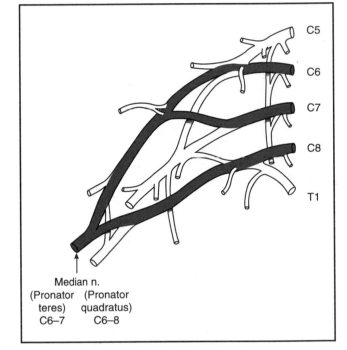

Median n.
(Pronator   (Pronator
teres)   quadratus)
C6–7   C6–8

Figure 4–116

Table 4-17 FOREARM PRONATION		
Muscle	**Origin**	**Insertion**
146. Pronator teres		
Humeral head	Humerus (supracondylar ridge; medial epicondyle)	Radius (shaft, lateral surface of middle)
Ulnar head	Ulna (coronoid proccess)	
147. Pronator quadratus	Ulna (distal 1/4 of anterior surface)	Radius (anterior surface distally)
Others		
151. Flexor carpi radialis		

Range of Motion
0° to 80°

Grade 5 (Normal), Grade 4 (Good), and Grade 3 (Fair)

Position of Patient: Short sitting or may sit at a table. Arm at side with elbow flexed to 90° and forearm in neutral.

Position of Therapist: Standing at side or in front of patient. Support the elbow (Fig. 4–117, showing end range). Hand used for resistance grasps the forearm over the dorsal surface at the wrist.

Test: Patient begins with a neutral forearm position and pronates the forearm until the palm faces downward. Therapist resists motion at the wrist in the direction of supination for Grades 4 and 5. (No resistance is given for Grade 3.)

Alternate Test: Grasp patient's hand as if to shake hands, cradling the elbow with the other hand and resisting pronation via the hand grip. This alternate test may be used if the patient has Normal or Good wrist and hand strength.

Instructions to Patient: "Turn your palm down. Hold it. Don't let me turn it up. Keep your wrist and fingers relaxed."

Grading

Grade 5 (Normal): Completes available range of motion and holds against maximal resistance.

Grade 4 (Good): Completes all available range against strong to moderate resistance.

Grade 3 (Fair): Completes available range without resistance (Fig. 4–118, showing end range).

Grade 2 (Poor)

Position of Patient: Short sitting with shoulder flexed between 45° and 90° and elbow flexed to 90°. Forearm in neutral (not illustrated).

Position of Therapist: Support the test arm by cupping the hand under the elbow.

Test: Patient pronates forearm.

Instructions to Patient: "Turn your palm facing outward away from your face."

Grading

Grade 2 (Poor): Completes partial range of motion (Fig. 4–119, showing end range).

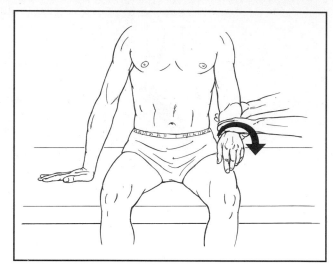

Figure 4–117

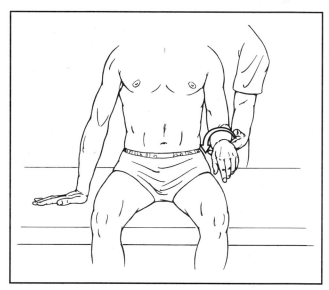

Figure 4–118

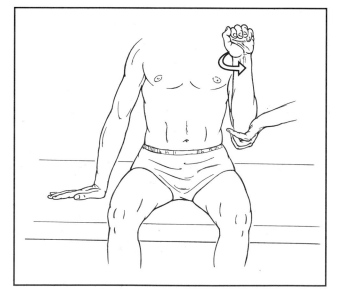

Figure 4–119

Grade 1 (Trace) and Grade 0 (Zero)

Position of Patient: Short sitting. Arm is positioned as for the Grade 3 test.

Position of Therapist: Support the forearm just distal to the elbow. The fingers of the other hand are used to palpate the Pronator teres over the upper third of the volar surface of the forearm on a diagonal line from the medial condyle of the humerus to the lateral border of the radius (Fig. 4–120).

Test: Patient attempts to pronate the forearm.

Instructions to Patient: "Try to turn your palm down."

Grading

Grade 1 (Trace): Visible or palpable contractile activity with no motion of the part.

Grade 0 (Zero): No contractile activity.

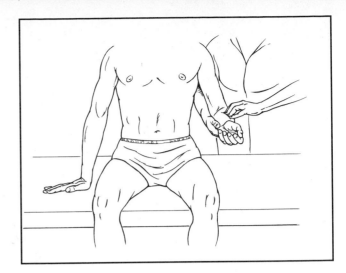

Figure 4–120

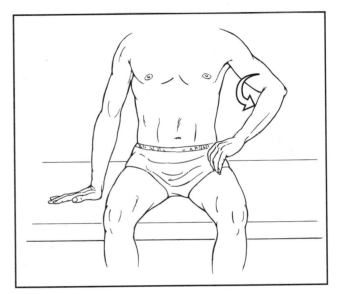

Figure 4–121

SUBSTITUTION

Patient may internally rotate the shoulder or abduct it during attempts at pronation (Fig. 4–121). When this occurs, the forearm rolls into pronation without the benefit of activity by the pronator muscles.

⊙ **HELPFUL HINTS**

Patient should be instructed to keep the wrist and fingers relaxed to avoid substitution by the Flexor carpi radialis and the finger flexors.

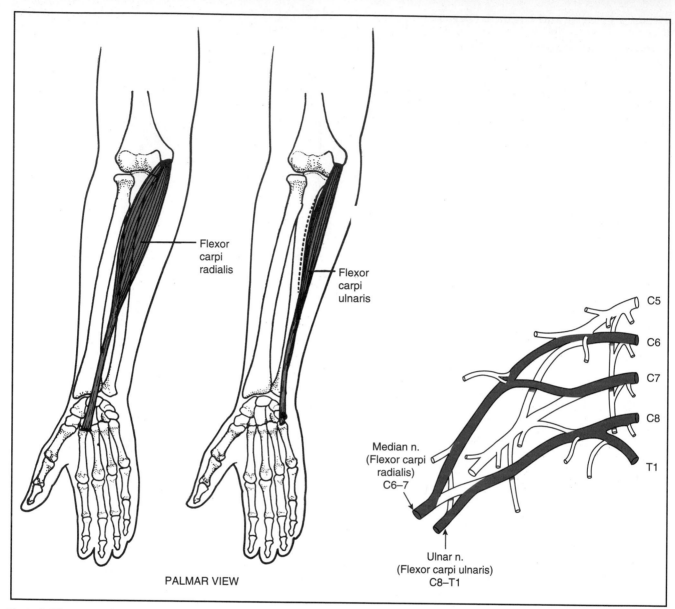

Flexor carpi radialis

Flexor carpi ulnaris

PALMAR VIEW

C5
C6
C7
C8
T1

Median n.
(Flexor carpi radialis)
C6–7

Ulnar n.
(Flexor carpi ulnaris)
C8–T1

Figure 4–122

Figure 4–123

Figure 4–124

Table 4–18	WRIST FLEXION	
Muscle	**Origin**	**Insertion**
151. Flexor carpi radialis	Humerus (medial epicondyle)	2nd and 3rd metacarpals (base)
153. Flexor carpi ulnaris	Humerus (medial epicondyle) Ulna (olecranon)	Pisiform Hamulus of hamate 5th metacarpal base

Range of Motion
0° to 80°

Others
152. Palmaris longus
166. Abductor pollicis longus
156. Flexor digitorum superficialis
169. Flexor pollicis longus
157. Flexor digitorum profundus

Grade 5 (Normal) and Grade 4 (Good)

Position of Patient (All Tests): Short sitting. Forearm is supported on its dorsal surface on a table. To start, forearm is supinated (Fig. 4–125). Wrist is in neutral or slightly extended.

Position of Therapist: One hand supports the patient's forearm under the wrist (Fig. 4–125).

To Test Both Wrist Flexors: The examiner grasps the palm of the test hand with the thumb circling around to the dorsal surface (Fig. 4–126). Resistance is given evenly across the hand in a straight-down direction into wrist extension.

To Test the Flexor carpi radialis: Resistance is focused over the 2nd metacarpal (radial side of the hand) in the direction of extension and ulnar deviation.

To Test the Flexor carpi ulnaris: Resistance is focused over the 5th metacarpal (ulnar side of the hand) in the direction of extension and radial deviation.

Test: Patient flexes the wrist, keeping the digits and thumb relaxed.

Instructions to Patient (All Tests): "Bend your wrist. Hold it. Don't let me pull it down. Keep your fingers relaxed."

Grading

Grade 5 (Normal): Completes available range of wrist flexion and holds against maximal resistance.

Grade 4 (Good): Completes available range and holds against strong to moderate resistance.

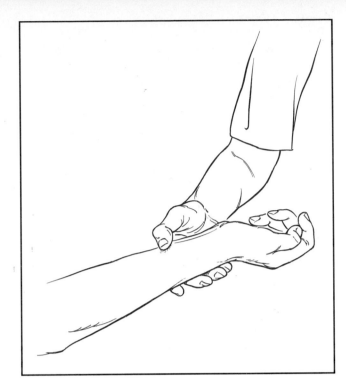

Figure 4–125

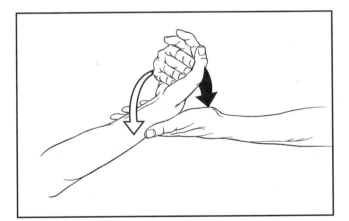

Figure 4–126

WRIST FLEXION
(Flexor carpi radialis and Flexor carpi ulnaris)

Grade 3 (Fair)

Position of Patient: Starting position with forearm supinated and wrist neutral as in Grade 5 and 4 tests.

Position of Therapist: Support the patient's forearm under the wrist.

Test

For Both Wrist Flexors: Patient flexes the wrist straight up without resistance and without radial or ulnar deviation.

For Flexor carpi radialis: Patient flexes the wrist in radial deviation (Fig. 4–127).

For Flexor carpi ulnaris: Patient flexes the wrist in ulnar deviation (Fig. 4–128).

Instructions to Patient

For Both Wrist Flexors: "Bend your wrist. Keep it straight with your fingers relaxed."

For Flexor carpi radialis: "Bend your wrist leading with the thumb side."

For Flexor carpi ulnaris: "Bend your wrist leading with the little finger."

Grading

Grade 3 (Fair) (All Tests): Completes available range without resistance.

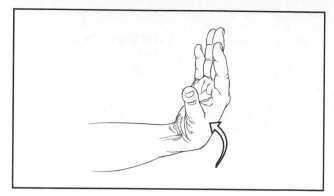

Figure 4–127

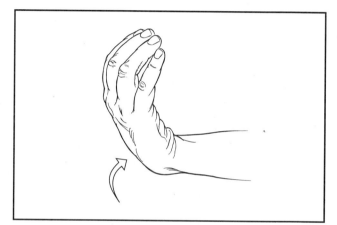

Figure 4–128

Grade 2 (Poor)

Position of Patient: Short sitting with elbow supported on table. Forearm in midposition with hand resting on ulnar side (Fig. 4–129).

Position of Therapist: Support patient's forearm proximal to the wrist.

Test: Patient flexes wrist with the ulnar surface gliding across or not touching the table (Fig. 4–129). To test the two wrist flexors separately, hold the forearm so that the wrist does not lie on the table and ask the patient to perform the flexion motion while the wrist is in ulnar and then radial deviation.

Instructions to Patient: "Bend your wrist, keeping your fingers relaxed."

Grading

Grade 2 (Poor): Completes available range of wrist flexion without assistance of gravity.

Grade 1 (Trace) and Grade 0 (Zero)

Position of Patient: Supinated forearm supported on table.

Position of Therapist: Support the wrist in flexion; the index finger of the other hand is used to palpate the appropriate tendons.
Palpate the tendons of the Flexor carpi radialis (Fig. 4–130) and the Flexor carpi ulnaris (Fig. 4–131) in separate tests.
The Flexor carpi radialis lies on the lateral palmar aspect of the wrist (Fig. 4–130) lateral to the Palmaris longus, if patient has one!
The tendon of the Flexor carpi ulnaris (Fig. 4–131) lies on the medial palmar aspect of the wrist.

Test: Patient attempts to flex the wrist.

Instructions to Patient: "Try to bend your wrist. Relax. Bend it again."
Patient should be asked to repeat the test so the examiner can feel the tendons during both relaxation and contraction.

Grading

Grade 1 (Trace): One or both tendons may exhibit visible or palpable contractile activity, but the part does not move.

Grade 0 (Zero): No contractile activity.

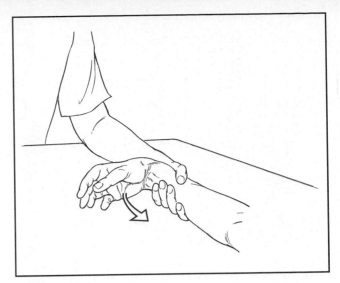

Figure 4–129

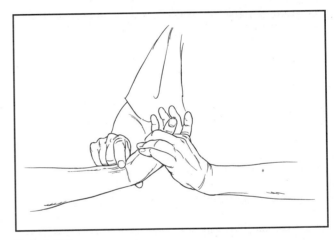

Figure 4–130

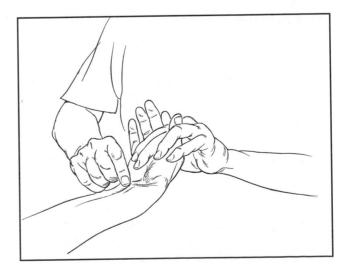

Figure 4–131

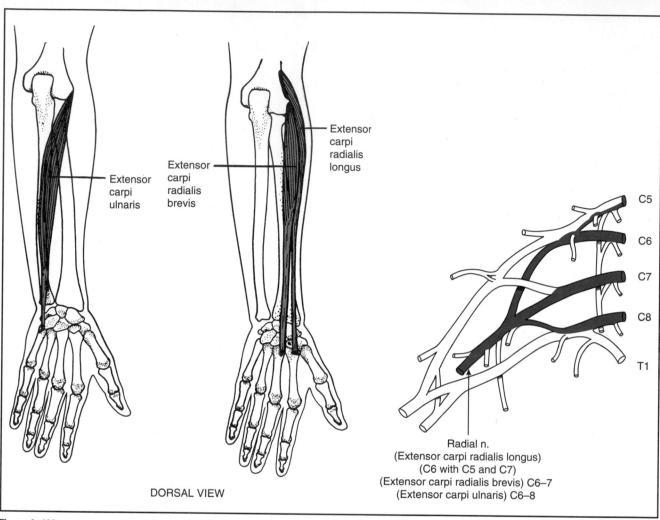

Extensor carpi ulnaris

Extensor carpi radialis brevis

Extensor carpi radialis longus

DORSAL VIEW

C5
C6
C7
C8
T1

Radial n.
(Extensor carpi radialis longus)
(C6 with C5 and C7)
(Extensor carpi radialis brevis) C6–7
(Extensor carpi ulnaris) C6–8

Figure 4–132   Figure 4–133   Figure 4–134

Table 4–19	WRIST EXTENSION	
Muscle	**Origin**	**Insertion**
148. Extensor carpi radialis longus	Humerus (lateral supracondylar ridge)	2nd metacarpal (base)
149. Extensor carpi radialis brevis	Humerus (lateral epicondyle)	3rd metacarpal (base)
150. Extensor carpi ulnaris	Humerus (lateral epicondyle) Ulna (dorsal border)	5th metacarpal

Others
154. Extensor digitorum
147. Extensor digiti minimi
155. Extensor indicis
167. Extensor pollicis longus

Range of Motion
0° to 70°

WRIST EXTENSION
(Extensor carpi radialis longus, Extensor carpi radialis brevis, and Extensor carpi ulnaris)

Grade 5 (Normal), Grade 4 (Good), and Grade 3 (Fair)

Position of Patient: Short sitting. Elbow and forearm are supported on table and forearm is in full pronation.

Position of Therapist: Sit or stand at a diagonal in front of patient. Support the patient's forearm under the wrist. The hand used for resistance is placed over the dorsal surface of the metacarpals.

To test all three muscles, the patient extends the wrist without deviation. Resistance for Grades 4 and 5 is given in a forward and downward direction over the 2nd to 5th metacarpals (Fig. 4–135).

To test the Extensor carpi radialis longus and brevis (for extension with radial deviation), resistance is given on the dorsal surface of the 2nd and 3rd metacarpals (radial side of hand) in the direction of flexion and ulnar deviation.

To test the Extensor carpi ulnaris (for extension and ulnar deviation), resistance is given on the dorsal surface of the 5th metacarpal (ulnar side of hand) in the direction of flexion and radial deviation.

Test: For the combined test of the three wrist extensor muscles, the patient extends the wrist straight up through the full available range. Do not permit extension of the fingers.

To test the two radial extensors, the patient extends the wrist leading with the thumb side of the hand. The wrist may be prepositioned in some extension and radial deviation to direct the patient's motion.

To test the Extensor carpi ulnaris, the patient extends the wrist leading with the ulnar side of the hand. The therapist may preposition the wrist in this attitude to direct the movement ulnarward.

Instructions to Patient: "Bring your wrist up. Hold it. Don't let me push it down." For Grade 3: "Bring your wrist up."

Grading

Grade 5 (Normal): Completes full wrist extension (when testing all three muscles) against maximal resistance. Full extension is not required for the tests of radial and ulnar deviation.

Grade 4 (Good): Completes full wrist extension against strong to moderate resistance when all muscles are being tested. When testing the individual muscles, full wrist extension will not be achieved.

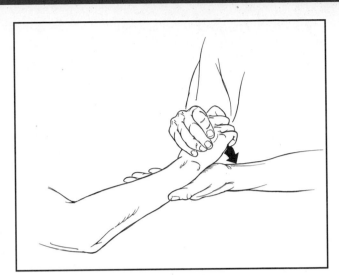

Figure 4–135

Grade 3 (Fair): Completes full range of motion with no resistance in the test for all three muscles. In the separate tests for the radial and ulnar extensors, the deviation required precludes a large range of motion.

Grade 2 (Poor)

Position of Patient: Forearm supported on table in neutral position.

Position of Therapist: Support the patient's wrist. This elevates the hand from the table and removes friction (Fig. 4–136).

Test: Patient extends the wrist.

Instructions to Patient: "Bend your wrist back."

Grading

Grade 2 (Poor): Completes full range with gravity eliminated. If the patient completes a partial range he may be given a grade of 2− (one of the few instances when a minus grade is acceptable).

Grade 1 (Trace) and Grade 0 (Zero)

Position of Patient: Hand and forearm supported on table with hand fully pronated.

Position of Therapist: Support the patient's wrist in extension. The other hand is used for palpation. Use one finger to palpate one muscle in a given test.

Extensor carpi radialis longus: Palpate this tendon on the dorsum of the wrist in line with the 2nd metacarpal (Fig. 4–137).

Extensor carpi radialis brevis: Palpate this tendon on the dorsal surface of the wrist in line with the 3rd metacarpal bone (Fig. 4–138).

Extensor carpi ulnaris: Palpate this tendon on the dorsal wrist surface proximal to the 5th metacarpal and just distal to the ulnar styloid process (Fig. 4–139).

Test: Patient attempts to extend the wrist.

Instructions to Patient: "Try to bring your wrist back."

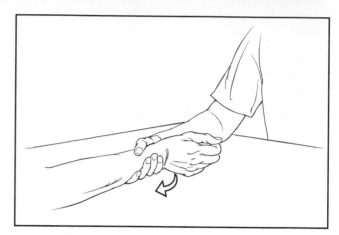

Figure 4–136

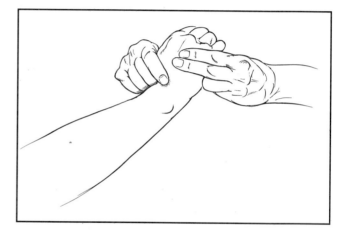

Figure 4–137

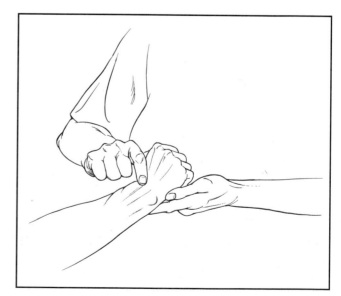

Figure 4–138

WRIST EXTENSION
(Extensor carpi radialis longus, Extensor carpi radialis brevis, and Extensor carpi ulnaris)

Grading

Grade 1 (Trace): For any given muscle there is visible or palpable contractile activity, but no wrist motion ensues.

Grade 0 (Zero): No contractile activity.

SUBSTITUTION

The most common substitution occurs when the finger extensors are allowed to participate. This can be avoided to a large extent by ensuring that the fingers are relaxed and are not permitted to extend.

C HELPFUL HINTS

---

1. The radial wrist extensors are considerably stronger than the Extensor carpi ulnaris.
2. A patient with complete quadriplegia at C5–C6 will have only the radial wrist extensors remaining. Radial deviation during extension is therefore the prevailing extensor motion at the wrist.

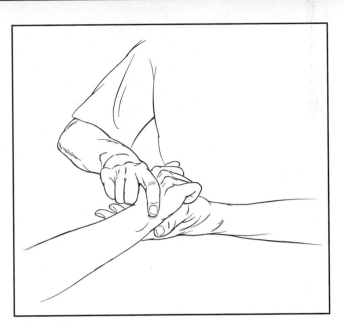

Figure 4–139

HAND TESTING REQUIRES JUDGMENT AND EXPERIENCE

When evaluating the muscles of the hand, care must be taken to use graduated resistance that takes into consideration the relatively small mass of the muscles. In general, the examiner should not use the full thrust of the fist, wrist, or arm but rather one or two fingers to resist hand motions.

The degree of resistance offered to hand muscles is an issue, particularly when testing a postoperative hand. Similarly, the amount of motion allowed or encouraged should be monitored so that sudden or excessive excursions that could "tear out" a surgical reconstruction are avoided.

Applying resistance in a safe fashion requires experience in assessing hand injuries or repair and a large amount of clinical judgment to avoid dislodging a tendon transfer or other surgical reconstruction. The neophyte examiner would be wise to err in the direction of caution.

Considerable practice in testing normal hands and comparing injured hands with their normal contralateral sides should give some of the necessary judgment with which to approach the fragile hand.

This text remains true to the principles of testing in the ranges of 5, 4, and 3 with respect to gravity. It is admitted, however, that the influence of gravity on the fingers is inconsequential, so the gravity and antigravity positions need not be followed carefully.

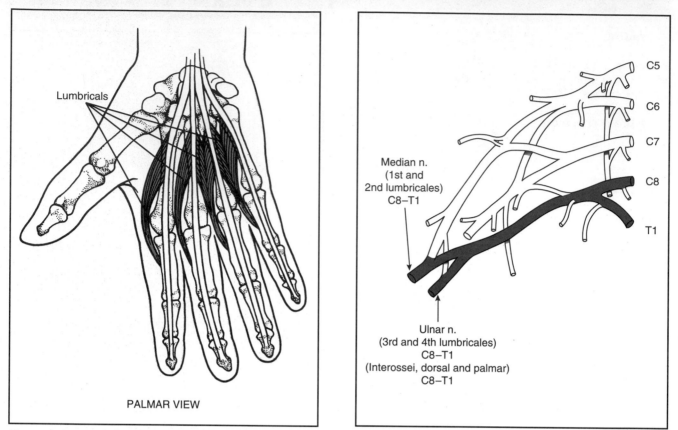

Lumbricals

PALMAR VIEW

Figure 4–140

Median n.
(1st and
2nd lumbricales)
C8–T1

Ulnar n.
(3rd and 4th lumbricales)
C8–T1
(Interossei, dorsal and palmar)
C8–T1

C5
C6
C7
C8
T1

Figure 4–141

Table 4–20 MP FLEXION OF FINGERS		
Muscle	**Origin**	**Insertion**
163. Lumbricales	Tendons of flexor digitorum profundus	Radial side of corresponding digit into extensor expansion
1st and 2nd	Index and middle fingers (radial and palmar sides)	
3rd and 4th	Ring and little fingers (double heads from adjacent sides of tendon)	
164. Dorsal interossei	Metacarpals:	Extensor expansion and proximal phalanges:
1st	1st and 2nd	Index finger (radial side)
2nd	2nd and 3rd	Long finger (radial side)
3rd	3rd and 4th	Long finger (ulnar side)
4th	4th and 5th	Ring finger (ulnar side)
165. Palmar interossei	Metacarpals: phalanges:	Dorsal expansion and proximal
1st	2nd	Index finger (ulnar side)
2nd	4th	Ring finger (radial side)
3rd	5th	Little finger (radial side)

Others
156. Flexor digitorum superficialis
157. Flexor digitorum profundus

Range of Motion

MP joints: 0° to 90°

Grade 5 (Normal), Grade 4 (Good), and Grade 3 (Fair)

Position of Patient: Short sitting or supine with forearm in supination. Wrist is maintained in neutral. The metacarpophalangeal (MP) joints should be fully extended; all interphalangeal (IP) joints are flexed (Fig. 4–142).

Position of Therapist: Stabilize the metacarpals proximal to the MP joint. Resistance is given on the palmar surface of the proximal row of phalanges in the direction of MP extension (Fig. 4–143).

Test: Patient simultaneously flexes the MP joints and extends the IP joints. Fingers may be tested separately. Do not allow fingers to curl; they must remain extended.

Instructions to Patient: "Uncurl your fingers while flexing your knuckles. Hold it. Don't let me straighten your knuckles." The final position is a right angle at the MP joints. Demonstrate motion to patient and insist on practice to get the motions performed correctly simultaneously.

Grading

Grade 5 (Normal): Patient completes simultaneous MP flexion and finger extension and holds against maximal resistance. Resistance is given to fingers individually because of the variant strength of the different Lumbricales. The Lumbricales also have different innervations.

Grade 4 (Good): Patient completes range of motion against moderate to strong resistance.

Grade 3 (Fair): Patient completes both motions correctly and simultaneously without resistance.

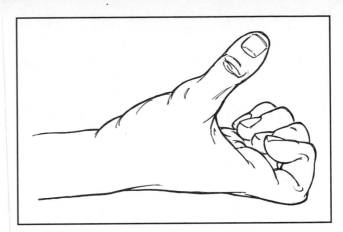

Figure 4–142

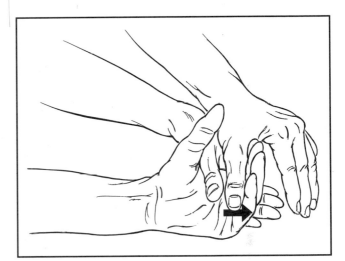

Figure 4–143

Grade 2 (Poor), Grade 1 (Trace), and Grade 0 (Zero)

Position of Patient: Forearm and wrist in midposition to remove influence of gravity. MP joints are fully extended; all IP joints are flexed.

Position of Therapist: Stabilize metacarpals.

Test: Patient attempts to flex MP joints through full available range while extending IP joints (Fig. 4–144).

Instructions to Patient: "Try to uncurl your fingers while bending your knuckles." Demonstrate motion to patient and allow practice.

Grading

Grade 2 (Poor): Completes full range of motion in gravity-eliminated position.

Grade 1 (Trace): Except in the hand that is markedly atrophied, the Lumbricales cannot be palpated. A grade of 1 may be awarded for minimal motion.

Grade 0 (Zero): A Grade 0 must be given in the absence of any movement.

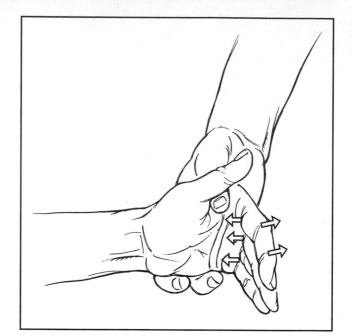

Figure 4–144

SUBSTITUTION

The long finger flexors may substitute for the Lumbricales. To avoid this pattern, make sure that the IP joints fully extend.

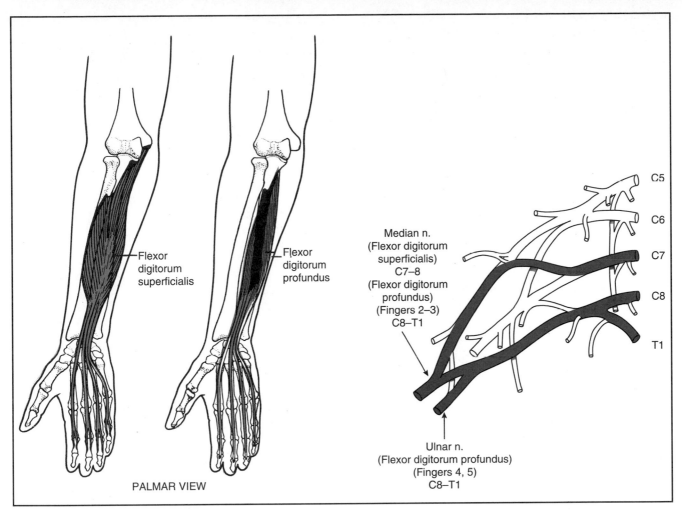

Flexor digitorum superficialis

Flexor digitorum profundus

Median n.
(Flexor digitorum superficialis)
C7–8
(Flexor digitorum profundus)
(Fingers 2–3)
C8–T1

Ulnar n.
(Flexor digitorum profundus)
(Fingers 4, 5)
C8–T1

PALMAR VIEW

C5
C6
C7
C8
T1

Figure 4–145   **Figure 4–146**   **Figure 4–147**

Table 4–21 PIP AND DIP FINGER FLEXION		
Muscle	**Origin**	**Insertion**
156. Flexor digitorum superficialis	Humerus (medial epicondyle)	Digits 2–5 (middle phalanx)
157. Flexor digitorum profundus	Ulna (proximal 3/4 of shaft plus coronoid)	Digits 2–5 (terminal phalanges at base)

Range of Motion
PIP joints: 0° to 100°
DIP joints: 0° to 90°

PIP TESTS
(Flexor digitorum superficialis)

Grade 5 (Normal), Grade 4 (Good), and Grade 3 (Fair)

Position of Patient: Forearm supinated, wrist in neutral. Finger to be tested is in slight flexion at the MP joint (Fig 4–148).

Position of Therapist: Hold all fingers (except the one being tested) in extension at all joints (Fig. 4–148). Isolation of the index finger may not be complete. The other hand is used to resist the head of the middle phalanx of the test finger in the direction of extension (not illustrated).

Test: Each of the four fingers is tested separately. Patient flexes the PIP joint without flexing the DIP joint. Do not allow motion of any joints of the other fingers.

Flick the terminal end of the finger being tested with the thumb to make certain that the Flexor digitorum profundus is not active; that is, the DIP joint goes into extension. The distal phalanx should be floppy.

Instructions to Patient: "Bend your index (long, ring, or little) finger; hold it. Don't let me straighten it. Keep your other fingers relaxed."

Grading

Grade 5 (Normal): Completes range of motion and holds against maximal finger resistance.

Grade 4 (Good): Completes range against moderate resistance.

Grade 3 (Fair): Completes range of motion with no resistance (Fig. 4–149).

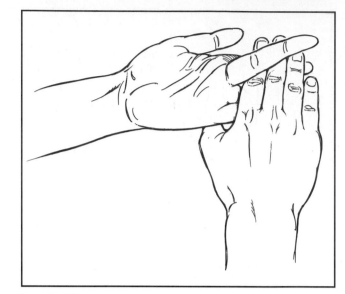

Figure 4–148

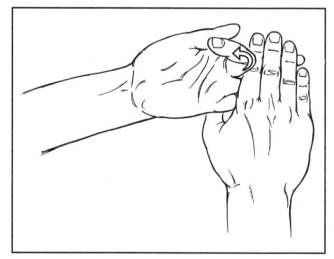

Figure 4–149

FINGER PIP AND DIP FLEXION
(Flexor digitorum superficialis and Flexor digitorum profundus)

Grade 2 (Poor), Grade 1 (Trace), and Grade 0 (Zero)

Position of Patient: Forearm is in midposition to eliminate the influence of gravity on finger flexion.

Position of Therapist: Same as for Grades 5, 4, and 3.
Palpate the Flexor digitorum superficialis on the palmar surface of the wrist between the Palmaris longus and the Flexor carpi ulnaris (Fig. 4–150).

Test: Patient flexes the PIP joint.

Instructions to Patient: "Bend your middle finger." (Select other fingers individually.)

Grading

Grade 2 (Poor): Completes range of motion.

Grade 1 (Trace): Palpable or visible contractile activity, which may or may not be accompanied by a flicker of motion.

Grade 0 (Zero): No contractile activity.

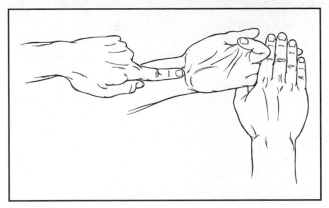

Figure 4–150

SUBSTITUTIONS

1. The major substitution for this motion is offered by the Flexor digitorum profundus, and this will occur if the DIP joint is allowed to flex.

2. If the wrist is allowed to extend, tension increases in the long finger flexors, and may result in passive flexion of the IP joints. This is referred to as a "tenodesis" action.

3. Relaxation of IP extension will result in passive IP flexion.

HELPFUL HINTS

Many persons cannot isolate the little finger. When this is the case, test the little and ring fingers at the same time.

DIP TESTS
(Flexor digitorum profundus)

Grade 5 (Normal), Grade 4 (Good), and Grade 3 (Fair)

Position of Patient: Forearm in supination, wrist in neutral, and proximal PIP joint in extension.

Position of Therapist: Stabilize the middle phalanx in extension by grasping it on either side (Fig. 4–151). Resistance is provided on the distal phalanx in the direction of extension (not illustrated).

Test: Test each finger individually. Patient flexes distal phalanx of each finger.

Instructions to Patient: "Bend the tip of your finger. Hold it. Don't let me straighten it."

Grading

Grade 5 (Normal): Completes maximal available range against a carefully assessed maximal level of resistance (See Sidebar, p. 131).

Grade 4 (Good): Completes maximal available range against some resistance.

Grade 3 (Fair): Completes maximum available range with no resistance (See Fig. 4–151).

Grade 2 (Poor), Grade 1 (Trace), and Grade 0 (Zero)

All aspects of testing these grades are the same as those used for the higher grades except that the position of the forearm is in neutral to eliminate the influence of gravity.

Grades are assigned as for the PIP tests.

The tendon of the Flexor digitorum profundus can be palpated on the palmar surface of the middle phalanx of each finger.

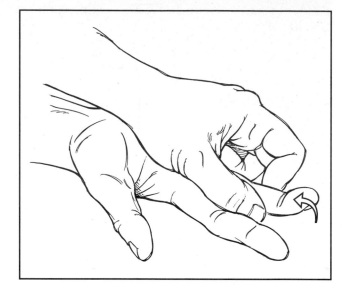

Figure 4–151

SUBSTITUTIONS

1. The wrist must be kept in a neutral position and must not be allowed to extend to rule out the tenodesis effect of the wrist extensors.

2. Don't be fooled if the patient extends the DIP joint and then relaxes, which can give the impression of active finger flexion.

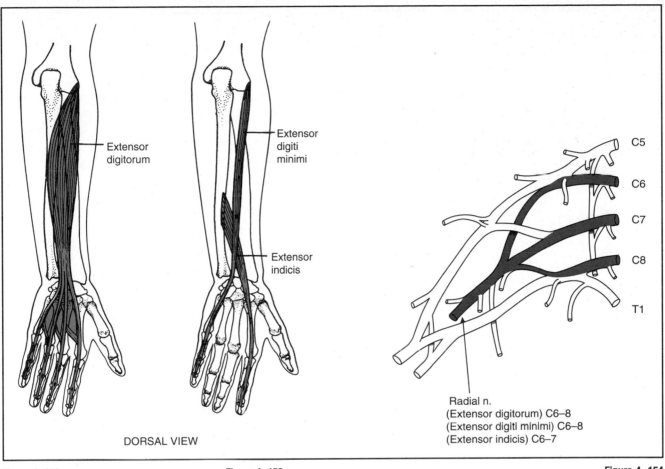

Extensor digitorum

Extensor digiti minimi

Extensor indicis

DORSAL VIEW

C5

C6

C7

C8

T1

Radial n.
(Extensor digitorum) C6–8
(Extensor digitl minimi) C6–8
(Extensor indicis) C6–7

Figure 4–152   Figure 4–153   Figure 4–154

Table 4–22	MP FINGER EXTENSION	
Muscle	**Origin**	**Insertion**
154. Extensor digitorum	Humerus (lateral epicondyle)	Four tendons to digits 2–5 (dorsum of middle phalanges and distal phalanges)
155. Extensor indicis	Ulna (posterior surface of shaft)	2nd digit (via extensor hood)
158. Extensor digit minimi	Common extensor tendon	5th digit (proximal phalanx)

Range of Motion
0° to 15°

Grade 5 (Normal), Grade 4 (Good), and Grade 3 (Fair)

Position of Patient: Forearm in pronation, wrist in neutral. MP joints and IP joints are in relaxed flexion posture.

Position of Therapist: Stabilize the wrist in neutral. Place the index finger of the resistance hand across the dorsum of all proximal phalanges just distal to the MP joints. Give resistance in the direction of flexion.

Test

Extensor digitorum: Patient extends MP joints (all fingers simultaneously), allowing the IP joints to be in slight flexion (Fig. 4–155).

Extensor indicis: Patient extends the MP joint of the index finger.

Extensor digiti minimi: Patient extends the MP joint of the 5th digit.

Instructions to Patient: "Bend your knuckles back as far as they will go." Demonstrate motion to patient and instruct to copy.

Grading

Grade 5 (Normal): Completes active extension range of motion with appropriate level of strong resistance.

Grade 4 (Good): Completes active range with some resistance.

Grade 3 (Fair): Completes active range with no resistance.

Grade 2 (Poor), Grade 1 (Trace), and Grade 0 (Zero)

Procedures: Test is the same as that for Grades 5, 4, and 3 except that the forearm is in the midposition.

The tendons of the Extensor digitorum (n = 4), the Extensor indicis (n = 1), and the Extensor digiti minimi (n = 1) are readily apparent on the dorsum of the hand as they course in the direction of each finger.

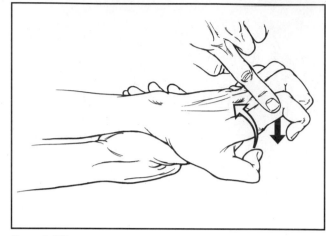

Figure 4–155

FINGER MP EXTENSION
(Extensor digitorum, Extensor indicis, Extensor digiti minimi)

Grading

Grade 2 (Poor): Completes range.

Grade 1 (Trace): Visible tendon activity but no joint motion.

Grade 0 (Zero): No contractile activity.

SUBSTITUTION

Flexion of the wrist will produce IP extension through a tenodesis action.

HELPFUL HINTS

1. MP extension of the fingers is not a strong motion, and only slight resistance is required to "break" the end position.

2. It is usual for the active range of motion to be considerably less than the available passive range. In this test, therefore, the "full available range" is not used, and the active range is accepted.

3. Another way to check whether there is functional extensor strength in the fingers is to "flick" the proximal phalanx of each finger downward; if the finger rebounds, it is usable.

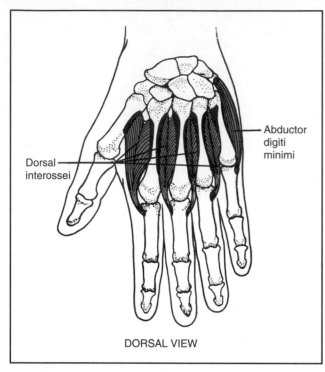

Figure 4–156

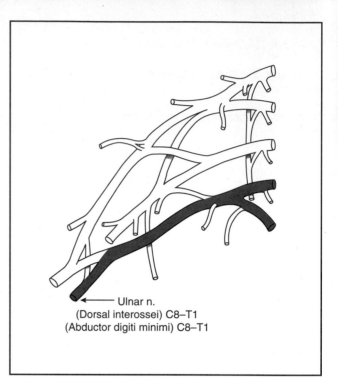

Figure 4–157

Table 4–23 FINGER ABDUCTION		
Muscle	**Origin**	**Insertion**
164. Dorsal interossei	Metacarpals:	Extensor expansion and proximal phalanges:
1st	1st and 2nd	Index finger (radial side)
2nd	2nd and 3rd	Long finger (radial side)
3rd	3rd and 4th	Long finger (ulnar side)
4th	4th and 5th	Ring finger (ulnar side)
159. Abductor digiti minimi	Pisiform Tendon of Flexor carpi ulnaris	5th digit (base of proximal phalanx)

Range of Motion

0° to 20°

Grade 5 (Normal) and Grade 4 (Good)

Position of Patient: Forearm pronated, wrist in neutral. Fingers start in extension and adduction. MP joints in neutral and avoid hyperextension.

Position of Therapist: Support the wrist in neutral. The fingers of the other hand are used to give resistance on the distal phalanx, on the radial side of one finger and the ulnar side of the adjacent finger (i.e., they are squeezed together). The direction of resistance will cause any pair of fingers to approximate (Fig. 4–158).

Test: Abduction of fingers (individual tests):

Dorsal Interossei:
Abduction of ring finger toward little finger.
Abduction of middle finger toward ring finger.
Abduction of middle finger toward index finger.
Abduction of index finger toward thumb.

The long finger (digit 3, finger 2) will move one way when tested with the index finger and the opposite way when tested with the ring finger (see Fig. 4–156, which shows a dorsal Interosseus on either side). When testing the little finger with the ring finger, the Abductor digiti minimi is being tested along with the 4th dorsal Interosseus.

Abductor digiti minimi:
Patient abducts 5th digit away from ring finger.

Instructions to Patient: "Spread your fingers. Hold them. Don't let me push them together."

Grading

Grade 5 (Normal) and Grade 4 (Good):
Neither the dorsal Interossei nor the Abductor digiti minimi will tolerate much resistance. Grading between a 5 and a 4 muscle is a judgment call based on any possible comparison with the contralateral side as well as clinical experience. Figure 4–159 illustrates the test for 2nd and 4th dorsal Interossei.

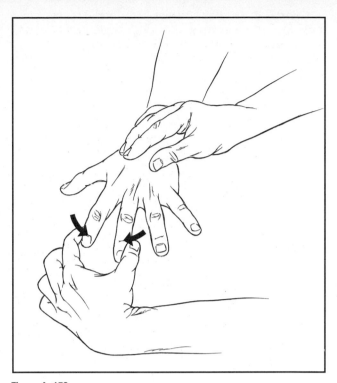

Figure 4–158

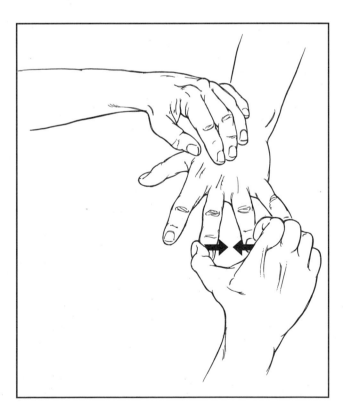

Figure 4–159

Grade 3 (Fair): Patient can abduct any given finger. Remember that the long finger has two dorsal Interossei and therefore must be tested as it moves away from the midline in both directions (Fig. 4–160).

Grade 2 (Poor), Grade 1 (Trace), and Grade 0 (Zero)

Procedures and Grading: Same as for higher grades in this test. A Grade 2 should be assigned if the patient can complete only a partial range of abduction for any given finger. The only dorsal Interosseus that is readily palpable is the 1st at the base of the proximal phalanx (Fig. 4–161).

The Abductor digiti minimi is palpable on the ulnar border of the hand.

ⓒ HELPFUL HINTS

Provide resistance for a Grade 5 test by flicking each finger toward adduction; if the finger tested rebounds, the grade will be Normal.

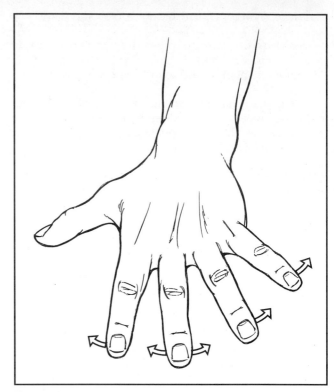

Figure 4–160

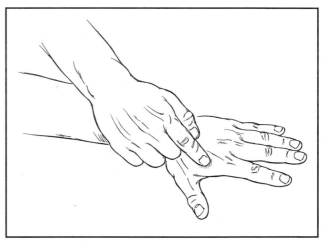

Figure 4–161

FINGER ADDUCTION
(Palmar Interossei)

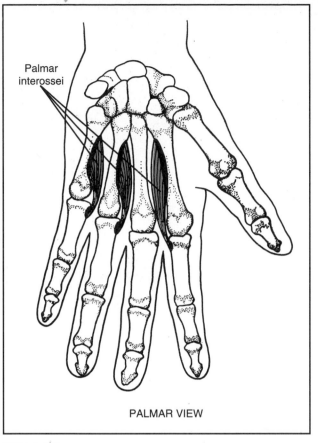

Palmar interossei

PALMAR VIEW

Figure 4–162

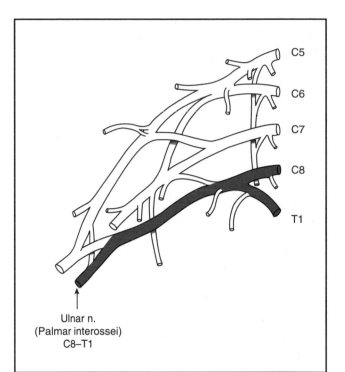

C5
C6
C7
C8
T1

Ulnar n.
(Palmar interossei)
C8–T1

Figure 4–163

Table 4–24 **FINGER ADDUCTION**		
Muscle	**Origin**	**Insertion**
165. Palmar interossei	Metacarpals:	Dorsal expansion and proximal phalanges:
1st	2nd	Index finger (ulnar side)
2nd	4th	Ring finger (radial side)
3rd	5th	Little finger (radial side)

Range of Motion
20° to 0°

Grade 5 (Normal) and Grade 4 (Good)

Position of Patient: Forearm pronated (palm down), wrist in neutral, and fingers extended and adducted. MP joints are neutral; avoid flexion.

Position of Therapist: Examiner grasps the middle phalanx on each of two adjoining fingers (Fig. 4–164). Resistance is given in the direction of abduction for each finger tested. The examiner is trying to "pull" the fingers apart. Each finger should be resisted separately.

Test: Adduction of fingers (individual tests):
Adduction of little finger toward ring finger.
Adduction of ring finger toward middle finger.
Adduction of index finger toward middle finger.
Adduction of thumb toward index finger.

Occasionally there is a 4th palmar Interosseus (not illustrated in Fig. 4–161) that some consider a separate muscle from the Adductor pollicis. In any event, the two muscles cannot be clinically separated.
Because the middle finger (also called the long finger, digit 3, and finger 2) has no palmar Interosseus, it is not tested in adduction.

Instructions to Patient: "Hold your fingers together. Don't let me pull them apart."

Grading

Grade 5 (Normal) and Grade 4 (Good): These muscles are notoriously weak in the sense of not tolerating much resistance. Distinguishing between Grades 5 and 4 is an exercise in futility, and the grade awarded will depend on the amount of the examiner's experience with normal hands.

Grade 3 (Fair): Patient can adduct fingers toward middle finger but cannot hold against resistance (Fig. 4–165).

Grade 2 (Poor), Grade 1 (Trace), and Grade 0 (Zero)

Procedures: Same as for Grades 5, 4, and 3.
For Grade 2, the patient can adduct each of the fingers tested through a partial range of motion. The test for Grade 2 is begun with the fingers abducted.
Palpation of the palmar Interossei is rarely feasible. By placing the examiner's finger against the side of a finger to be tested the therapist may detect a slight outward motion for a muscle less than Grade 2.

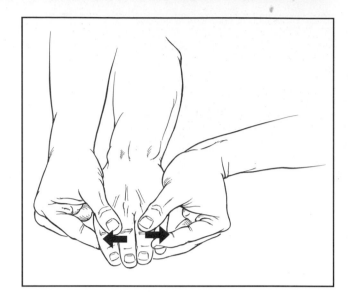

Figure 4–164

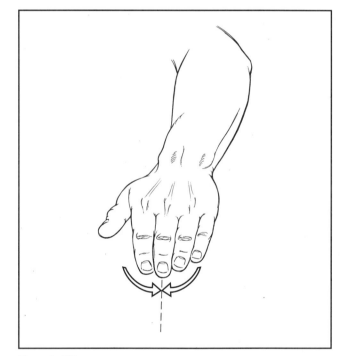

Figure 4–165

FINGER ADDUCTION
(Palmar Interossei)

SUBSTITUTION

Caution must be used to ensure that finger flexion does not occur because the long finger flexors can contribute to adduction.

HELPFUL HINTS

The fingers can be judged quickly by grasping the distal phalanx and flicking the finger in the direction of abduction. If the finger rebounds or snaps back, it is functional.

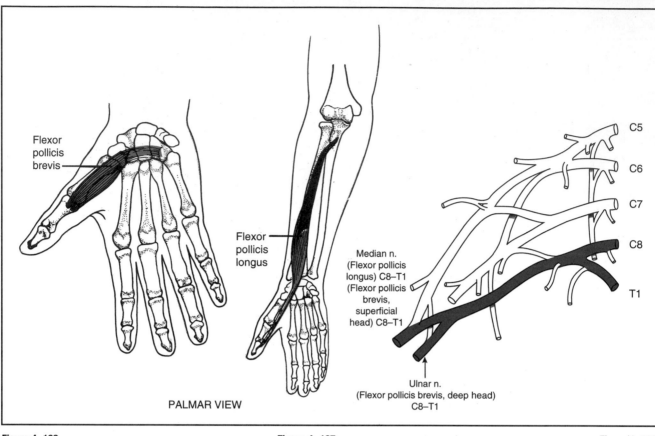

Flexor pollicis brevis

Flexor pollicis longus

PALMAR VIEW

Median n. (Flexor pollicis longus) C8–T1 (Flexor pollicis brevis, superficial head) C8–T1

Ulnar n. (Flexor pollicis brevis, deep head) C8–T1

C5
C6
C7
C8
T1

Figure 4–166

Figure 4–167

Figure 4–168

Table 4–25	THUMB MP AND IP FLEXION	
Muscle	**Origin**	**Insertion**
MP Flexion		
170. Flexor pollicis brevis		
Superficial head	Flexor retinaculum Trapezium	Thumb (base of proximal phalanx)
Deep head	Trapezoid Capitate	
IP Flexion		
169. Flexor pollicis longus	Radius (volar side of middle 1/2)	Thumb (base of distal phalanx)

Range of Motion

MP flexion: 0° to 50°
IP flexion: 0° to 80°

THUMB MP AND IP FLEXION
(Flexor pollicis brevis and Flexor pollicis longus)

THUMB MP FLEXION TESTS
(Flexor pollicis brevis)

Grade 5 (Normal) to Grade 0 (Zero)

Position of Patient: Forearm in supination, wrist in neutral. Carpometacarpal (CMC) joint is at 0°; IP joint is at 0°. Thumb in adduction, lying relaxed and adjacent to the 2nd metacarpal (Fig. 4–169).

Position of Therapist: Stabilize the 1st metacarpal firmly to avoid any wrist or CMC motion. The other hand gives one-finger resistance to MP flexion on the proximal phalanx in the direction of extension (Fig. 4–170).

Test: Patient flexes the MP joint of the thumb, keeping the IP joints straight (Fig. 4–170).

Instructions to Patient: "Bring your thumb across the palm of your hand. Keep the thumb in touch with your palm. Don't bend the end joint. Hold it. Don't let me pull it back."

Demonstrate thumb flexion and have patient practice the motion.

Grading

Grade 5 (Normal): Completes range of motion against maximal thumb resistance.

Grade 4 (Good): Tolerates strong to moderate resistance.

Grade 3 (Fair): Completes full range of motion with perhaps a slight amount of resistance because gravity is eliminated.

Grade 2 (Poor): Completes partial range of motion.

Grade 1 (Trace): Palpate the muscle by initially locating the tendon of the Flexor pollicis longus in the thenar eminence (Fig. 4–171). Then palpate the muscle belly of the Flexor pollicis brevis on the ulnar side of the longus tendon in the thenar eminence.

Grade 0 (Zero): No visible or palpable contractile activity.

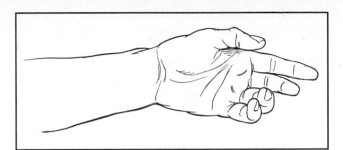

Figure 4–169

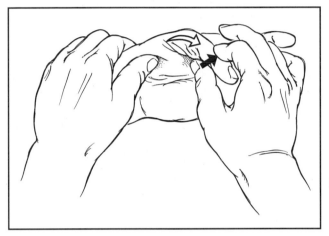

Figure 4–170

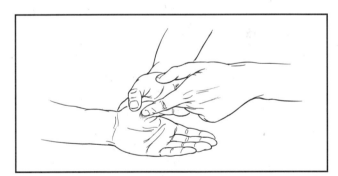

Figure 4–171

SUBSTITUTION BY FLEXOR POLLICIS LONGUS

The long thumb flexor can substitute but only after flexion of the IP joint begins. To avoid this substitution, do not allow flexion of the distal joint of the thumb.

THUMB IP FLEXION TESTS
(Flexor pollicis longus)

Grade 5 (Normal) to Grade 0 (Zero)

Position of Patient: Forearm supinated with wrist in neutral and MP joint of thumb in extension.

Position of Therapist: Stabilize the MP joint of the thumb firmly in extension by grasping the patient's thumb across that joint. Give resistance with the other hand against the palmar surface of the distal phalanx of the thumb in the direction of extension (Fig. 4–172).

Test: Patient flexes the IP joint of the thumb.

Instructions to Patient: "Bend the end of your thumb. Hold it. Don't let me straighten it."

Grading

Grade 5 (Normal) and Grade 4 (Good): Patient tolerates maximal finger resistance from examiner for Grade 5. This muscle is very strong, and a Grade 4 muscle will tolerate strong resistance. Full range always should be completed.

Grade 3 (Fair): Completes a full range of motion with minimal resistance because gravity is eliminated.

Grade 2 (Poor): Completes only a partial range of motion.

Grade 1 (Trace) and Grade 0 (Zero): Palpate the tendon of the Flexor pollicis longus on the palmar surface of the proximal phalanx of the thumb. Palpable activity is graded 1; no activity is graded 0.

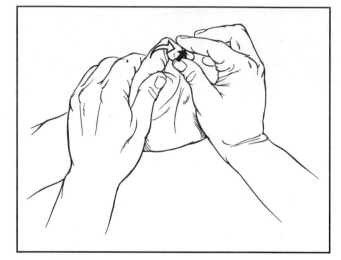

Figure 4–172

SUBSTITUTION

Do not allow the distal phalanx of the thumb to extend at the beginning of the test. If the distal phalanx is extended and then relaxes, the examiner may think active flexion has occurred.

PLATE 3

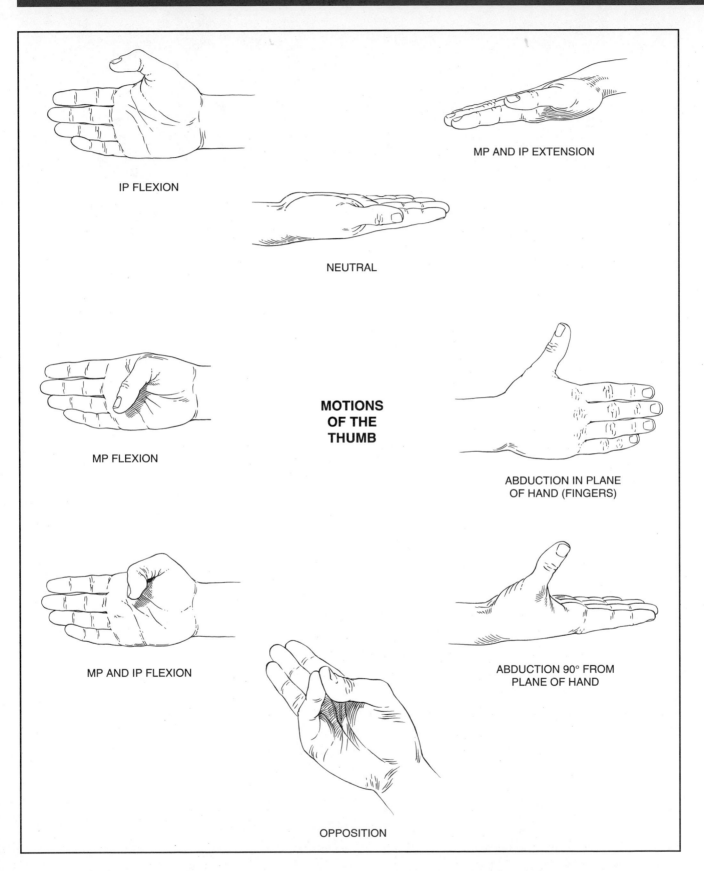

IP FLEXION

MP AND IP EXTENSION

NEUTRAL

MP FLEXION

MOTIONS
OF THE
THUMB

ABDUCTION IN PLANE
OF HAND (FINGERS)

MP AND IP FLEXION

ABDUCTION 90° FROM
PLANE OF HAND

OPPOSITION

THUMB MP AND IP EXTENSION
(Extensor pollicis brevis and Extensor pollicis longus)

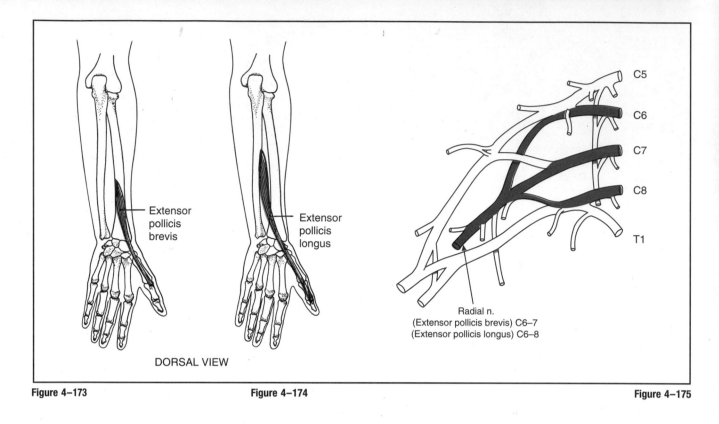

Extensor pollicis brevis

Extensor pollicis longus

DORSAL VIEW

C5
C6
C7
C8
T1

Radial n.
(Extensor pollicis brevis) C6–7
(Extensor pollicis longus) C6–8

Figure 4–173

Figure 4–174

Figure 4–175

Table 4–26	THUMB MP AND IP EXTENSION	
Muscle	Origin	Insertion
MP Extension		
168. Extensor pollicis brevis	Radius (posterior surface)	Thumb (proximal phalanx, dorsal)
IP Extension		
167. Extensor pollicis longus	Ulna (posterior-lateral surface of middle shaft)	Thumb (base of distal phalanx)

Range of Motion

MP extension: 50° to 0°
IP extension: 80° to 0°

The Extensor pollicis brevis is an inconstant muscle that often blends with the Extensor pollicis longus, in which event it is not possible to separate the brevis from the longus by clinical tests, and the test for the longus prevails.

THUMB MP EXTENSION TESTS
(Extensor pollicis brevis)

Grade 5 (Normal) to Grade 0 (Zero)

Position of Patient: Forearm in midposition and wrist in neutral; CMC and IP joints of the thumb are relaxed and in slight flexion. The MP joint of the thumb is in abduction and flexion.

Position of Therapist: Stabilize the first metacarpal firmly, allowing motion to occur only at the MP joint (Fig. 4–176). Resistance is provided with the other hand on the dorsal surface of the proximal phalanx in the direction of flexion. This normally is not a strong muscle.

Test: Patient extends the MP joint of the thumb while keeping the IP joint slightly flexed.

Instructions to Patient: "Bring your thumb up so it points toward the ceiling; don't move the end joint. Hold it. Don't let me push it down."

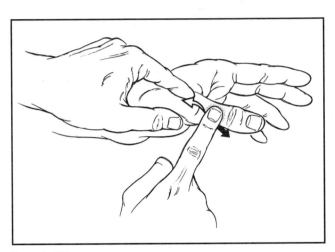

Figure 4–176

Grading

Grade 5 (Normal) and Grade 4 (Good): Only the experienced examiner can accurately distinguish between Grades 5 and 4. Resistance should be applied carefully and slowly because this usually is a weak muscle.

Grade 3 (Fair): Patient moves proximal phalanx of the thumb through full range of extension with little or no resistance.

Grade 2 (Poor): Patient moves proximal phalanx through partial range of motion.

Grade 1 (Trace): The tendon of the Flexor pollicis brevis is palpated (Fig. 4–177) at the base of the 1st metacarpal where it lies between the tendons of the Abductor pollicis and the Extensor pollicis longus.

Grade 0 (Zero): No contractile activity.

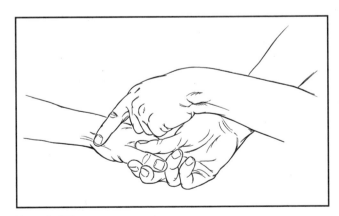

Figure 4–177

SUBSTITUTION

Extension of the IP joint of the thumb with CMC adduction in addition to extension of the MP joint indicates substitution by the Extensor pollicis longus.

THUMB IP EXTENSION TESTS
(Extensor pollicis longus)

Grade 5 (Normal), Grade 4 (Good), and Grade 3 (Fair)

Position of Patient: Forearm in midposition, wrist in neutral with ulnar side of hand resting on the table. Thumb relaxed in a flexion posture.

Position of Therapist: Use the table to support the ulnar side of the hand and stabilize the proximal phalanx of the thumb (Fig. 4–178). Apply resistance over the dorsal surface of the distal phalanx of the thumb in the direction of flexion.

Test: Patient extends the IP joint of the thumb.

Instructions to Patient: "Straighten the end of your thumb. Hold it. Don't let me push it down."

Grading

Grade 5 (Normal) and Grade 4 (Good): Completes full range of motion. This is not a strong muscle, so resistance must be applied accordingly. The distinction between Grades 5 and 4 is based on comparison with the contralateral normal hand and, barring that, extensive experience in testing the hand.

Grade 3 (Fair): Completes full range of motion with no resistance.

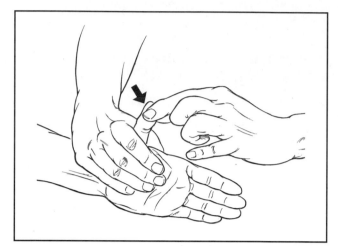

Figure 4–178

Grade 2 (Poor), Grade 1 (Trace), and Grade 0 (Zero)

Position of Patient: Forearm in pronation with wrist in neutral and thumb in relaxed flexion posture to start.

Position of Therapist: Stabilize the wrist over its dorsal surface. Stabilize the fingers by gently placing the other hand across the fingers just below the MP joints (Fig. 4–179).

Test: Patient extends distal joint of the thumb (Fig. 4–179).

Instructions to Patient: "Straighten the end of your thumb."

Grading

Grade 2 (Poor): Thumb completes partial range of motion.

Grade 1 (Trace): Palpate the tendon of the Extensor pollicis longus on the ulnar side of the anatomic snuff box or, alternatively, on the dorsal surface of the proximal phalanx (Fig. 4–180).

Grade 0 (Zero): No contractile activity.

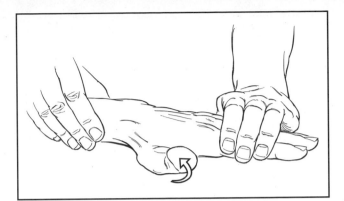

Figure 4–179

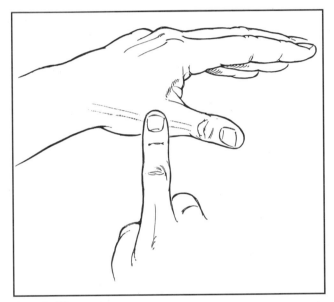

Figure 4–180

SUBSTITUTION

The muscles of the thenar eminence (Abductor pollicis brevis, Flexor pollicis brevis, and Adductor pollicis) can extend the IP joint by flexing the CMC joint (an extensor tenodesis).

HELPFUL HINTS

1. Continued action by the Extensor pollicis longus will extend the MP and CMC joints.

2. A quick way to assess the functional status of the long thumb extensor is to flick the distal phalanx into flexion; if the finger rebounds or snaps back, it is a useful muscle.

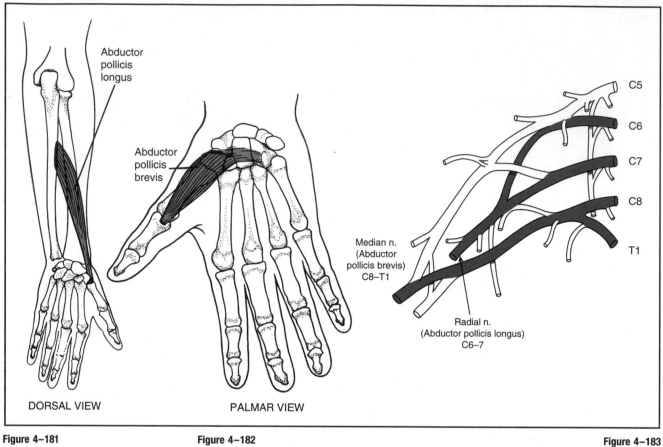

Abductor pollicis longus

Abductor pollicis brevis

Median n. (Abductor pollicis brevis) C8–T1

Radial n. (Abductor pollicis longus) C6–7

C5
C6
C7
C8
T1

DORSAL VIEW

PALMAR VIEW

Figure 4–181

Figure 4–182

Figure 4–183

Table 4–27 THUMB ABDUCTION		
Muscle	**Origin**	**Insertion**
166. Abductor pollicis longus	Ulna (posterior surface laterally) Radius (middle 1/3 of posterior shaft)	1st metacarpal (radial side of base) Trapezium
167. Abductor pollicis brevis	Scaphoid Trapezium Flexor retinaculum	Thumb (base of proximal phalanx, radial side)
Others **152.** Palmaris longus		

Range of Motion
0° to 70°

THUMB ABDUCTION
(Abductor pollicis longus and Abductor pollicis brevis)

ABDUCTOR POLLICIS LONGUS TEST

Grade 5 (Normal) to Grade 0 (Zero)

Position of Patient: Forearm and wrist in neutral; thumb relaxed in adduction.

Position of Therapist: Stabilize the metacarpals of the four fingers and the wrist (Fig. 4–184). Resistance is given on the distal end of the 1st metacarpal in the direction of adduction.

Test: Patient abducts the thumb away from the hand in a plane parallel to the finger metacarpals.

Instructions to Patient: "Lift your thumb straight up." Demonstrate motion to the patient.

Grading

Grade 5 (Normal) and Grade 4 (Good): Completes full range of motion against resistance. Distinguishing Grades 5 and 4 may be difficult.

Grade 3 (Fair): Completes full range of motion with no resistance.

Grade 2 (Poor): Completes partial range of motion.

Grade 1 (Trace): Palpate tendon of the Abductor pollicis longus at the base of the 1st metacarpal on the radial side of the Extensor pollicis brevis (Fig. 4–185). It is the most lateral tendon at the wrist.

Grade 0 (Zero): No contractile activity.

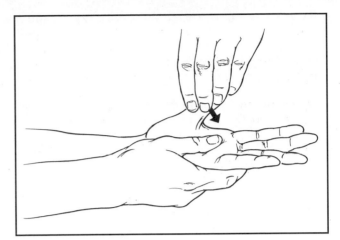

Figure 4–184

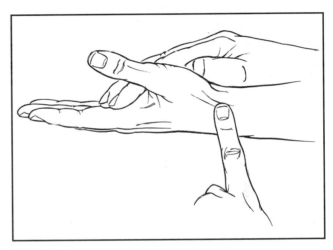

Figure 4–185

SUBSTITUTION

The Extensor pollicis brevis can substitute for the Abductor pollicis longus. If the line of pull is toward the dorsal surface of the forearm (Extensor pollicis brevis), substitution is occurring.

ABDUCTOR POLLICIS BREVIS TEST

Grade 5 (Normal), Grade 4 (Good), and Grade 3 (Fair)

Position of Patient: Forearm in supination, wrist in neutral, and thumb relaxed in adduction.

Position of Therapist: Stabilize the metacarpals (Fig. 4–186) by placing the examiner's hand across the patient's palm with the thumb on the dorsal surface of the patient's hand (somewhat like a handshake but maintaining the patient's wrist in neutral). Apply resistance to the lateral aspect of the proximal phalanx of the thumb in the direction of adduction.

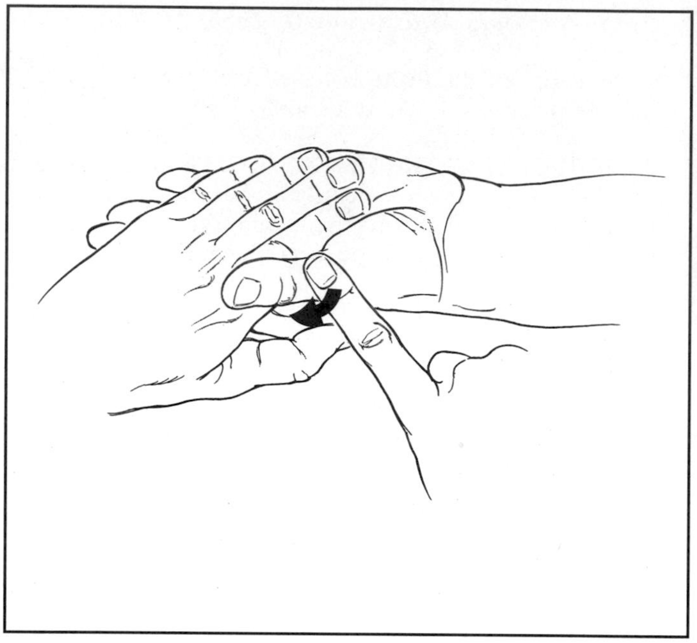

Test: Patient abducts the thumb in a plane perpendicular to the palm. Observe wrinkling of the skin over the thenar eminence and watch for the tendon of the accessory Palmaris longus to "pop out."

Instructions to Patient: "Lift your thumb vertically until it points to the ceiling." Demonstrate motion to the patient.

Grading

Grade 5 (Normal): Completes full range of motion with maximal finger resistance.

Grade 4 (Good): Tolerates moderate resistance.

Grade 3 (Fair): Completes full range of motion with no resistance.

Figure 4–186

Grade 2 (Poor), Grade 1 (Trace), and Grade 0 (Zero)

Position of Patient: Forearm in midposition, wrist in neutral, and thumb relaxed in adduction.

Position of Therapist: Stabilize wrist in neutral.

Test: Patient abducts thumb in a plane perpendicular to the palm.

Instructions to Patient: "Try to lift your thumb so it points at the ceiling."

Grading

Grade 2 (Poor): Completes partial range of motion.

Grade 1 (Trace): Palpate the belly of the Abductor pollicis brevis in the center of the thenar eminence, medial to the Opponens pollicis (Fig. 4–187).

Grade 0 (Zero): No contractile activity.

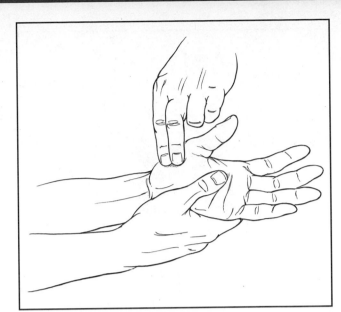

Figure 4–187

SUBSTITUTION

If the plane of motion is not perpendicular, the substitution may come from the Abductor pollicis longus.

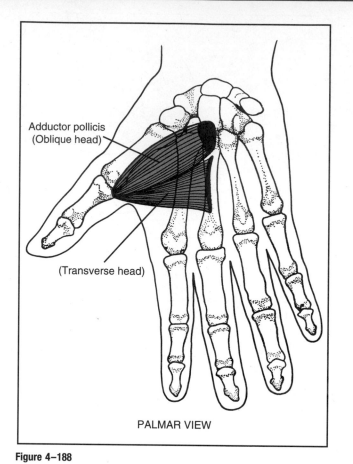

Adductor pollicis
(Oblique head)

(Transverse head)

PALMAR VIEW

Figure 4–188

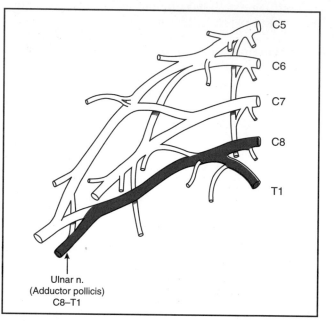

C5

C6

C7

C8

T1

Ulnar n.
(Adductor pollicis)
C8–T1

Figure 4–189

Table 4–28 THUMB ADDUCTION		
Muscle	**Origin**	**Insertion**
173. Adductor pollicis		
Oblique head	Capitate 2nd and 3rd metacarpals Intercarpal ligaments	Thumb (proximal phalanx, base)
Transverse head	3rd metacarpal (palmar surface of distal 2/3)	
Others		
164. 1st dorsal Interosseus		

Range of Motion
70° to 0°

THUMB ADDUCTION
(Adductor pollicis)

Grade 5 (Normal), Grade 4 (Good), and Grade 3 (Fair)

Position of Patient: Forearm in pronation, wrist in neutral, and thumb relaxed and (hanging) down in abduction.

Position of Therapist: Stabilize the metacarpals of the four fingers by grasping the patient's hand around the ulnar side (Fig. 4–190). Resistance is given on the medial side of the proximal phalanx of the thumb in the direction of abduction.

Test: Patient adducts the thumb by bringing the 1st metacarpal up to the 2nd metacarpal. Alternatively, place a sheet of paper between the thumb and the 2nd metacarpal (palmar pinch) and ask the patient to hold while you try to pull the paper away.

Instructions to Patient: "Bring your thumb up to your index finger." Demonstrate motion to the patient.

Grading

Grade 5 (Normal) and Grade 4 (Good): Completes full range of motion and holds against maximal resistance. Patient can resist rigidly (Grade 5), or the muscle yields (Grade 4).

Grade 3 (Fair): Completes full range of motion with no resistance.

Grade 2 (Poor), Grade 1 (Trace), and Grade 0 (Zero)

Position of Patient: Forearm in midposition, wrist in neutral resting on table, and thumb in abduction.

Position of Therapist: Stabilize wrist on the table, and use a hand to stabilize the finger metacarpals (Fig. 4–191).

Test: Patient moves thumb horizontally in adduction. (The end position is shown in Fig. 4–191.)

Instructions to Patient: "Return your thumb to its place next to your index finger." Demonstrate motion to patient.

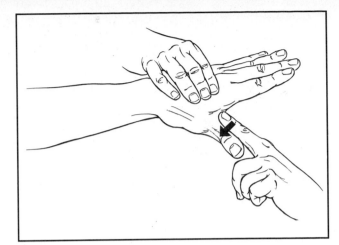

Figure 4–190

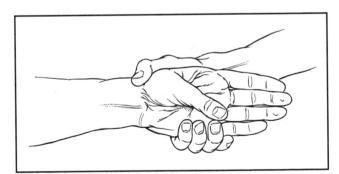

Figure 4–191

Grading

Grade 2 (Poor): Completes partial or full range of motion.

Grade 1 (Trace): Palpate the Adductor pollicis on the palmar side of the web space of the thumb by grasping the web between the index finger and thumb (Fig. 4–192). The Adductor lies between the 1st dorsal Interosseus and the 1st metacarpal bone. This muscle is difficult to palpate, and the therapist may have to ask the patient to perform a palmar pinch to assist in its location.

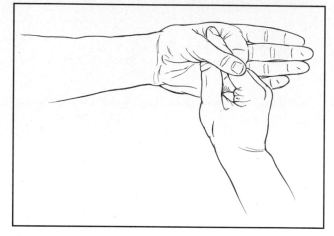

Figure 4–192

SUBSTITUTIONS

1. The Flexor pollicis longus and the Flexor pollicis brevis will flex the thumb, drawing it across the palm. These muscles should be kept inactive during the adduction test.

2. The Extensor pollicis longus may attempt to substitute for the thumb adductor, in which case the CMC joint will extend.

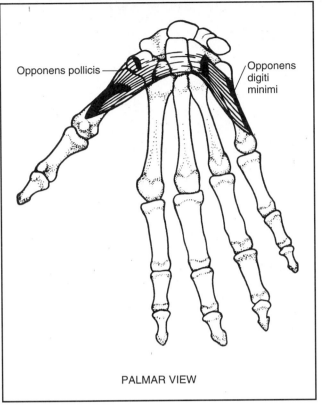

PALMAR VIEW

Figure 4–193

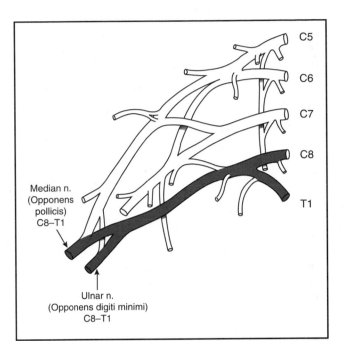

Figure 4–194

Table 4–29 THUMB OPPOSITION		
Muscle	**Origin**	**Insertion**
172. Opponens pollicis	Trapezium Flexor retinaculum	1st metacarpal
161. Opponens digiti minimi	Hamate (hamulus) Flexor retinaculum	5th metacarpal (ulnar margin)
Others **171.** Abductor pollicis brevis		

Range of Motion
Pad of thumb to pad of 5th digit

This motion is a combination of abduction, flexion, and medial rotation of the thumb (Fig. 4–195).

The two muscles in thumb-to-5th digit opposition (Opponens pollicis and Opponens digiti minimi) should not be tested together and also should be graded separately.

Grade 5 (Normal) to Grade 0 (Zero)

Position of Patient: Forearm is supinated, wrist in neutral, and thumb in adduction with MP and IP flexion.

Position of Therapist: Stabilize the hand by holding the wrist on the dorsal surface. The examiner may prefer the hand to be stabilized on the table.

Opponens pollicis: Apply resistance for the Opponens pollicis at the head of the 1st metacarpal in the direction of lateral rotation, extension, and adduction (See Fig. 4–195).

Opponens digiti minimi: Give resistance for the Opponens digiti minimi on the palmar surface of the 5th metacarpal in the direction of medial rotation (flattening the palm) (Fig. 4–196).

Test: Patient raises the thumb away from the palm and rotates it so that its distal phalanx apposes the distal phalanx of the little finger. Such apposition must be pad to pad and not tip to tip. Opposition also can be evaluated by asking the patient to hold an object between the thumb and little finger (in opposition), which the examiner tries to pull away.

Instructions to Patient: "Bring your thumb to your little finger and touch the two pads, forming the letter 'O' with your thumb and little finger." Demonstrate motion to the patient and require practice.

Grading

Grade 5 (Normal): Completes the full motion correctly against maximal thumb resistance.

Grade 4 (Good): Completes the range against moderate resistance.

Grade 3 (Fair): Moves thumb and 5th digit through full range of opposition with no resistance.

Grade 2 (Poor): Moves through partial range of opposition. (The two Opponens muscles are evaluated separately.)

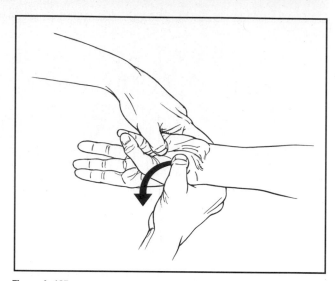

Figure 4–195

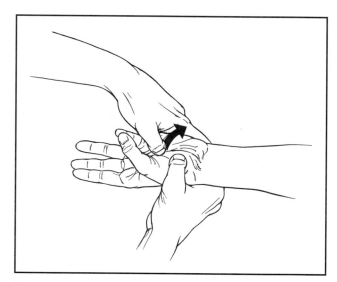

Figure 4–196

OPPOSITION (THUMB TO LITTLE FINGER)
(Opponens pollicis and Opponens digiti minimi)

Grade 1 (Trace): Palpate the Opponens pollicis along the radial shaft of the 1st metacarpal (Fig. 4–197). It lies lateral to the Abductor pollicis brevis. During Grade 5 and Grade 4 contractions the examiner will have difficulty in palpating the metacarpal because of the muscle mass. In Grade 3 muscles and below, the weaker contractions do not obscure palpation of the metacarpal.

Palpate the Opponens digiti minimi on the hypothenar eminence on the radial side of the 5th metacarpal (Fig. 4–198). Be careful not to cover the muscle with the finger or thumb used for palpation lest any contractile activity be missed.

Grade 0 (Zero): No contractile activity.

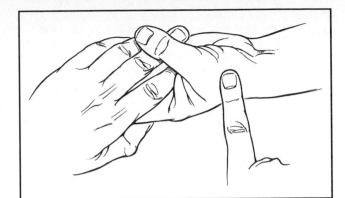

Figure 4–197

SUBSTITUTIONS

1. The Flexor pollicis longus and the Flexor pollicis brevis can draw the thumb across the palm toward the little finger. If such motion occurs in the plane of the palm it is not opposition; contact will be at the tips, not the pads, of the digits.

2. The Abductor pollicis brevis may substitute, but the rotation component of the motion will not be present.

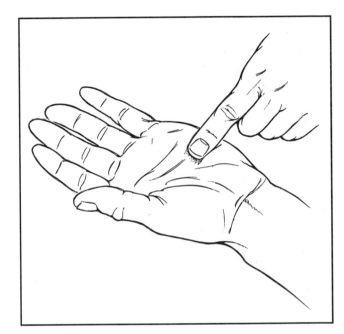

Figure 4–198

REFERENCES

1. Kendall FP, McCreary EK, Provance PG. *Muscles: Testing and Function,* 4th ed. Baltimore: Williams & Wilkins, 1993.
2. Perry J. Shoulder function for the activities of daily living. *In* Matsen FA, Fu FH, Hawkins RJ. *The Shoulder: A Balance of Mobility and Stability.* Chap. 10. Rosemont, IL: American Academy of Orthopedic Surgeons, 1993.

Testing the Muscles of the Lower Extremity

Hip Flexion
Hip Flexion, Abduction, and External Rotation with Knee
    Flexion
Hip Extension
Hip Abduction
Hip Abduction from Flexed Position
Hip Adduction
Hip External Rotation
Hip Internal Rotation
Knee Flexion
Knee Extension
Ankle Plantar Flexion
Foot Dorsiflexion and Inversion
Foot Inversion
Foot Eversion with Plantar Flexion or Dorsiflexion
Hallux and Toe MP Flexion
Hallux and Toe DIP and PIP Flexion
Hallux and Toe MP and IP Extension
Extension of the Metacarpophalangeal and Interphalangeal
    Joints of the Hallux and All Joints of the Four Lateral
    Toes

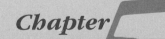

Chapter 5

Introduction to Testing the Hip

Knowledge of the ranges of motion of the hip is imperative before manual tests of hip strength are conducted. If the examiner does not have a clear idea of hip joint ranges, especially tightness in the hip flexor muscles, test results will be contaminated. For example, in the presence of a hip flexion contracture, the patient must be standing and leaning over the edge of the table. This position (described on page 181) will decrease the influence of the flexion contracture and will allow the patient to move against gravity through the available range.

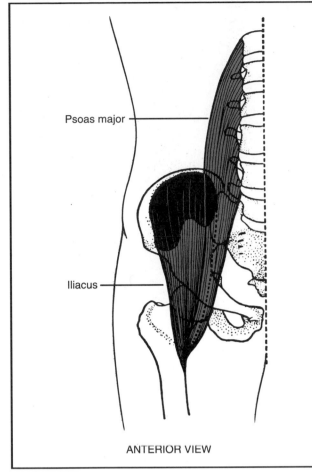

ANTERIOR VIEW

Figure 5–1

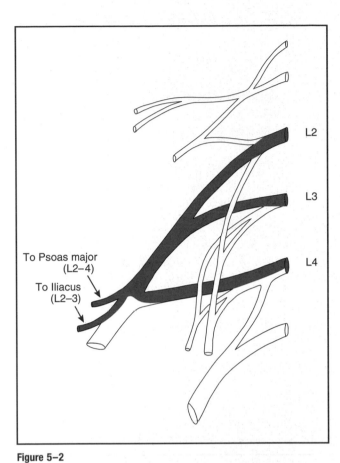

L2

L3

L4

To Psoas major
(L2–4)

To Iliacus
(L2–3)

Figure 5–2

Table 5–1 HIP FLEXION		
Muscle	**Origin**	**Insertion**
174. Psoas major	L1–L5 vertebrae (transverse processes) T12–L5 vertebral bodies	Femur (lesser trochanter)
176. Iliacus	Iliac fossa (anterior 2/3)	Femur (lesser trochanter)

Others
196. Rectus femoris
195. Sartorius
185. Tensor fasciae latae
177. Pectineus
180. Adductor brevis
179. Adductor longus
181. Adductor magnus

Range of Motion
0° to 120°

Grade 5 (Normal), Grade 4 (Good), and Grade 3 (Fair)

Position of Patient: Short sitting with thighs fully supported on table and legs hanging over the edge. Patient may use arms to provide trunk stability by grasping table edge or with hands on table at each side (Fig. 5–3).

Position of Therapist: Standing next to limb to be tested. Contoured hand to give resistance over distal thigh just proximal to the knee joint (Fig. 5–3).

Test: Patient flexes hip, clearing the table and maintaining neutral rotation; holds that position against the examiner's resistance, which is given in a downward direction toward floor.

Instructions to Patient: "Lift your leg off the table and don't let me push it down."
For Grade 3: "Lift your leg straight up off the table."

Grading

Grade 5 (Normal): Thigh clears table. Patient tolerates maximal resistance.

Grade 4 (Good): Hip flexion holds against strong to moderate resistance. There may be some "give" at the end position.

Grade 3 (Fair): Patient completes test range and holds the position without resistance (Fig. 5–4).

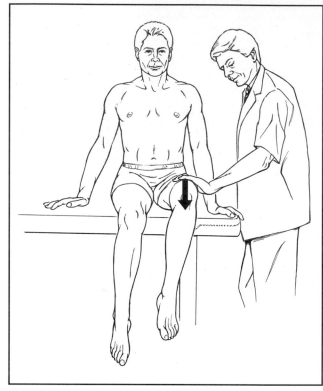

Figure 5–3

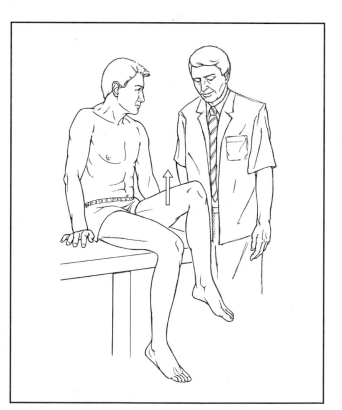

Figure 5–4

Grade 2 (Poor)

Position of Patient: Side-lying with limb to be tested uppermost and supported by examiner (Fig. 5–5). Trunk in neutral alignment. Lowermost limb may be flexed for stability.

Position of Therapist: Standing behind patient. Cradle test limb in one arm with hand support under the knee. Opposite hand maintains trunk alignment at hip (Fig. 5–5).

Test: Patient flexes supported hip. Knee is permitted to flex to prevent hamstring tension.

Instructions to Patient: "Bring your knee up toward your chest."

Grading

Grade 2 (Poor): Patient completes the range of motion in side-lying position.

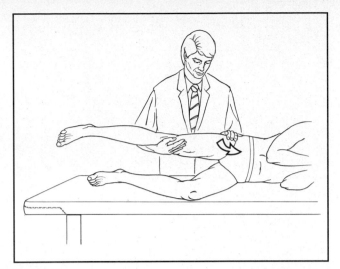

Figure 5-5

Grade 1 (Trace) and Grade 0 (Zero)

Position of Patient: Supine. Test limb supported by examiner under calf with hand behind knee (Fig. 5–6).

Position of Therapist: Standing at side of limb to be tested. Test limb is supported under calf with hand behind knee. Free hand palpates the Psoas major muscle just distal to the inguinal ligament on the medial side of the Sartorius (see Fig. 5–6).

Test: Patient attempts to flex hip.

Instructions to Patient: "Try to bring your knee up to your nose."

Grading

Grade 1 (Trace): Palpable contraction but no visible movement.

Grade 0 (Zero): No palpable contraction of muscle.

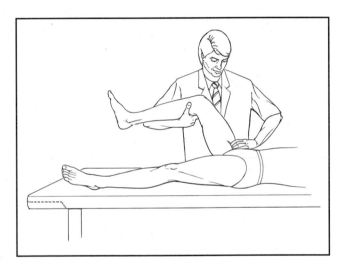

Figure 5–6

SUBSTITUTIONS

1. Use of the Sartorius will result in external rotation and abduction of the hip. The Sartorius, because it is superficial, will be seen and can be palpated in most limbs (Fig. 5–7).
2. If the Tensor fascia lata substitutes for the hip flexors, internal rotation and abduction of the hip will result. If, however, the patient is tested in the supine position, gravity will cause the limb to externally rotate. The Tensor may be seen and palpated at its origin on the anterior superior iliac spine (ASIS).

HELPFUL HINTS

1. When the trunk is weak the test will be more accurate from a supine position.

2. Hip flexion is not a strong motion, so experience is necessary to appreciate what constitutes a normal level of resistance.

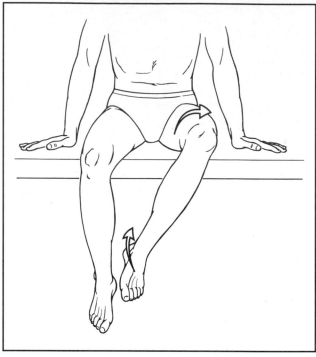

Figure 5–7

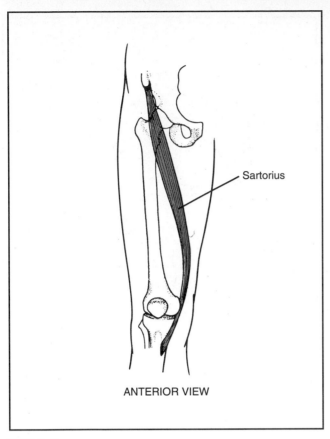

Figure 5–8

ANTERIOR VIEW

Sartorius

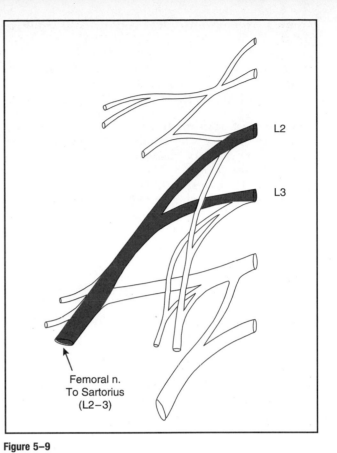

Figure 5–9

L2

L3

Femoral n.
To Sartorius
(L2–3)

Table 5–2 HIP FLEXION, ABDUCTION, AND EXTERNAL ROTATION		
Muscle	**Origin**	**Insertion**
195. Sartorius	Ilium (anterior superior spine)	Tibia (medial surface)

Others
Hip and knee flexors
Hip external rotators
Hip abductors

Range of Motion
Two-joint motion.
Incomplete range in both joints

Grade 5 (Normal), Grade 4 (Good), and Grade 3 (Fair)

Position of Patient: Short sitting with thighs supported on table and legs hanging over side. Arms may be used for support.

Position of Therapist: Standing lateral to the leg to be tested. Place one hand on the lateral side of knee; the other hand grasps the medial-anterior surface of distal leg (Fig. 5–10).

Hand at knee resists hip flexion and abduction (down and inward direction) in the Grade 5 and 4 tests. Hand at the ankle resists hip external rotation and knee flexion (up and outward) in Grade 5 and 4 tests. No resistance for Grade 3 test.

Test: Patient flexes, abducts, and externally rotates the hip and flexes the knee (Fig. 5–10).

Instructions to Patient: Therapist may demonstrate the required motion passively and then ask the patient to repeat the motion, or the therapist may place the limb in the desired end position.

"Hold it! Don't let me move your leg or straighten your knee."

Alternate instruction: "Slide your heel up the shin of your other leg."

Grading

Grade 5 (Normal): Holds end point against maximal resistance.

Grade 4 (Good): Tolerates moderate to heavy resistance.

Grade 3 (Fair): Completes movement and holds end position but takes no resistance (Fig. 5–11).

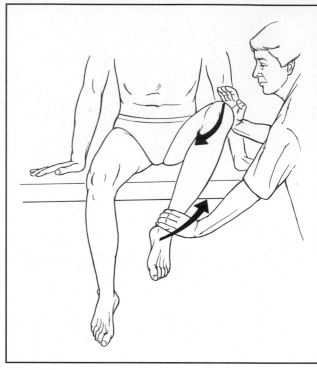

Figure 5–10

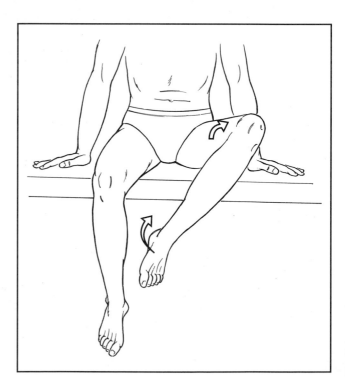

Figure 5–11

Grade 2 (Poor)

Position of Patient: Supine. Heel of limb to be tested is placed on contralateral shin (Fig. 5–12).

Position of Therapist: Standing at side of limb to be tested. Support limb as necessary to maintain alignment.

Test: Patient slides test heel upward along shin to knee.

Instructions to Patient: "Slide your heel up to your knee."

Grading

Grade 2 (Poor): Completes desired movement.

Grade 1 (Trace) and Grade 0 (Zero)

Position of Patient: Supine.

Position of Therapist: Standing on side to be tested. Cradle test limb under calf with hand supporting limb behind knee. Opposite hand palpates Sartorius on medial side of thigh where the muscle crosses the femur (Fig. 5–13). Examiner may prefer to palpate near the muscle origin just below the anterior spine of the ilium (ASIS).

Test: Patient attempts to slide heel up shin toward knee.

Instructions to Patient: "Try to slide your heel up to your knee."

Grading

Grade 1 (Trace): Therapist can detect slight contraction of muscle; no visible movement.

Grade 0 (Zero): No palpable contraction.

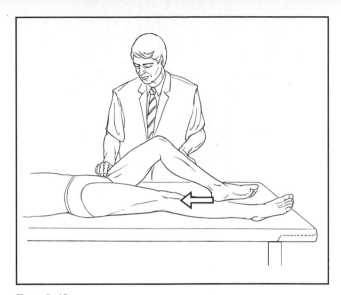

Figure 5–12

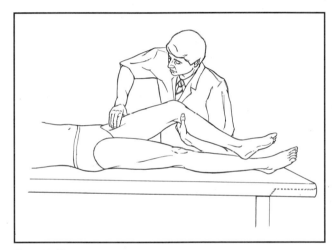

Figure 5–13

HELPFUL HINTS

1. Substitution by the Iliopsoas or the Rectus femoris results in straight hip flexion without abduction and external rotation.

2. *Never* grasp the belly of a muscle—the calf, in this instance.

3. Therapist is reminded that failure of the patient to complete the full range of motion (ROM) in the Grade 3 (Fair) sitting test is not an automatic Grade 2. Patient should be tested in supine to confirm the Grade 2 score.

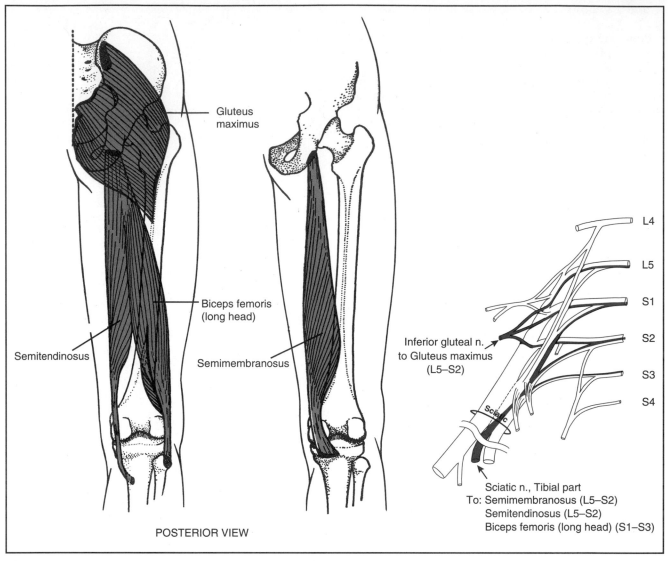

POSTERIOR VIEW

Gluteus maximus

Biceps femoris (long head)

Semitendinosus

Semimembranosus

Inferior gluteal n. to Gluteus maximus (L5–S2)

Sciatic

Sciatic n., Tibial part
To: Semimembranosus (L5–S2)
Semitendinosus (L5–S2)
Biceps femoris (long head) (S1–S3)

L4
L5
S1
S2
S3
S4

Figure 5–14   Figure 5–15   Figure 5–16

Table 5–3 HIP EXTENSION		
Muscle	**Origin**	**Insertion**
182. Gluteus maximus	Ilium (posterior gluteal line) Sacrum (posterior) Coccyx (posterior) Sacrotuberous ligament	Femur (gluteal tuberosity) Iliotibial band
193. Semitendinosus	Ischial tuberosity	Tibia (proximal shaft)
194. Semimembranosus	Ischial tuberosity	Tibia (medial condyle) Femur (lateral condyle)
192. Biceps femoris (long head)	Ischial tuberosity	Fibula (head) Tibia (lateral condyle)

Range of Motion

0° to 120°

HIP EXTENSION
(Gluteus maximus and Hamstrings)

Grade 5 (Normal), Grade 4 (Good), and Grade 3 (Fair) (Aggregate of All Hip Extensor Muscles)

Position of Patient: Prone. (*Note:* If there is a hip flexion contracture, immediately go to the test described for hip extension modified for hip flexion tightness [page 181]). Arms may be overhead or abducted to hold sides of table.

Position of Therapist: Standing at side of limb to be tested at level of pelvis. (*Note:* Illustration (Fig. 5–17) shows examiner on opposite side to avoid obscuring activity.)

The hand providing resistance is placed on the posterior leg just above the ankle. The opposite hand may be used to stabilize or maintain pelvis alignment in the area of the posterior superior spine of the ilium (Fig. 5–17). This is the most demanding test because the lever arm is longest.

Alternate Position: The hand that gives resistance is placed on the posterior thigh just above the knee (Fig. 5–18). This is a less demanding test.

Test (For Both Positions): Patient extends hip through entire available range of motion. Resistance is given straight downward toward the floor. (No resistance is given in the Grade 3 test.)

Instructions to Patient: "Lift your leg off the table as high as you can without bending your knee."

Grading

Grade 5 (Normal): Patient completes available range and holds test position against maximal resistance.

Grade 4 (Good): Patient completes available range against strong to moderate resistance.

Grade 3 (Fair): Completes range and holds the position without resistance (Fig. 5–19).

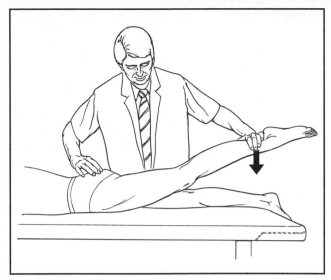
Figure 5–17

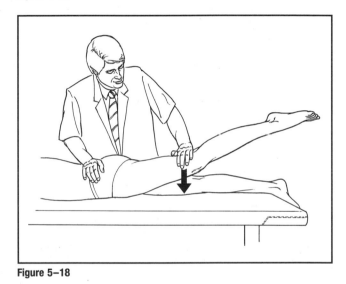
Figure 5–18

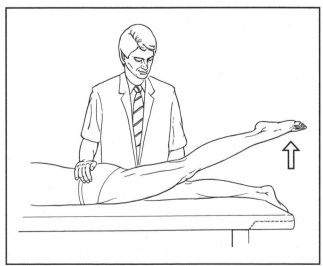
Figure 5–19

Grade 2 (Poor)

Position of Patient: Side-lying with test limb uppermost. Knee straight and supported by examiner. Lowermost limb is flexed for stability.

Position of Therapist: Standing behind patient at thigh level. Therapist supports test limb just below the knee, cradling the leg (Fig. 5–20). Opposite hand over the pelvic crest to maintain pelvic and hip alignment.

Test: Patient extends hip through full range of motion.

Instructions to Patient: "Bring your leg back toward me. Keep your knee straight."

Grading

Grade 2 (Poor): Completes range of extension motion in side-lying position.

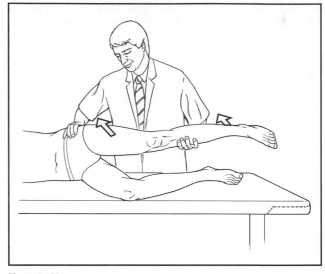

Figure 5–20

Grade 1 (Trace) and Grade 0 (Zero)

Position of Patient: Prone.

Position of Therapist: Standing on side to be tested at level of hips. Palpate hamstrings (deep into tissue with fingers) at the ischial tuberosity (Fig. 5–21). Palpate the Gluteus maximus with deep finger pressure over the center of the buttocks and also over the upper and lower fibers (not illustrated).

Test: Patient attempts to extend hip in prone position or tries to squeeze buttocks together.

Instructions to Patient: "Try to lift your leg from the table" OR "Squeeze your buttocks together."

Grading

Grade 1 (Trace): Palpable contraction of either hamstrings or Gluteus maximus but no visible joint movement. Contraction of Gluteus maximus will result in narrowing of the gluteal crease.

Grade 0 (Zero): No palpable contraction.

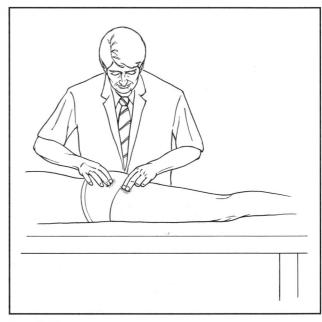

Figure 5–21

◎ HELPFUL HINTS

Therapist should be aware that the hip extensors are among the most powerful muscles in the body, and most therapists will not be able to "break" a Grade 5 hip extension. Care should be taken not to overgrade a Grade 4 muscle.

HIP EXTENSION TEST TO ISOLATE GLUTEUS MAXIMUS

Grade 5 (Normal), Grade 4 (Good), and Grade 3 (Fair)

Position of Patient: Prone with knee flexed to 90°. (*Note:* In the presence of a hip flexion contracture, do not use this test but refer to the test for hip extension modified for hip flexion tightness (see page 181).

Position of Therapist: Standing on side to be tested at level of pelvis. (*Note:* The therapist in the illustration deliberately is shown on the wrong side to avoid obscuring test positions.) Hand for resistance is contoured over posterior thigh just above the knee. The opposite hand may stabilize or maintain the alignment of the pelvis (Fig. 5–22).

For the Grade 3 test the knee may need to be supported in flexion (by cradling at the ankle).

Test: Patient extends hip through available range, maintaining knee flexion. Resistance is given in a straight downward direction (toward floor).

Instructions to Patient: "Lift your foot to the ceiling" OR "Lift your leg keeping your knee bent."

Grading

Grade 5 (Normal): Completes available range of motion and holds end position against maximum resistance.

Grade 4 (Good): Limb position can be held against heavy to moderate resistance.

Grade 3 (Fair): Completes available range of motion and holds end position but takes no resistance (Fig. 5–23).

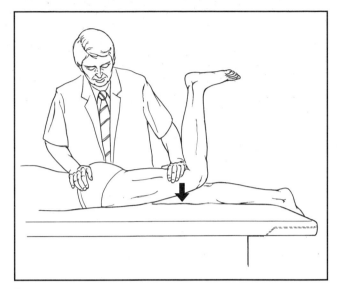

Figure 5–22

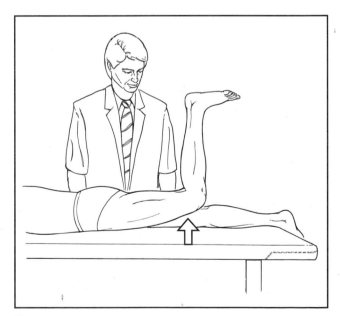

Figure 5–23

Grade 2 (Poor)

Position of Patient: Side-lying with test limb uppermost. Knee is flexed and supported by examiner. Lowermost hip and knee should be flexed for stability (Fig. 5–24).

Position of Therapist: Standing behind the patient at thigh level. Therapist cradles uppermost leg with forearm and hand under the flexed knee. Other hand is on pelvis to maintain postural alignment.

Test: Patient extends hip with supported knee flexed.

Instructions to Patient: "Move your leg back toward me."

Grading

Grade 2 (Poor): Completes available range of motion in side-lying position.

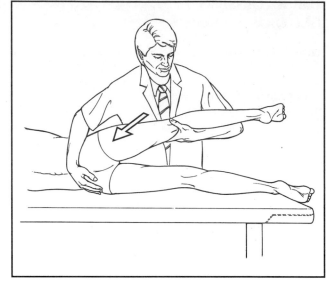

Figure 5–24

Grade 1 (Trace) and Grade 0 (Zero)

This test is identical to the Grade 1 and 0 tests for aggregate hip extension (see Fig. 5–21). The patient is prone and attempts to extend the hip or squeeze the buttocks together while the therapist palpates the Gluteus maximus.

ⓒ HELPFUL HINTS

Hip extension range is less when the knee is flexed because of tension in the Rectus femoris. A diminished range may be observed, therefore, in tests that isolate the Gluteus maximus.

HIP EXTENSION TESTS MODIFIED FOR HIP FLEXION TIGHTNESS

Grade 5 (Normal), Grade 4 (Good), and Grade 3 (Fair)

Position of Patient: Patient stands with hips flexed and places torso prone on the table (Fig. 5–25). The arms are used to "hug" the table for support. The knee of the nontest limb should be flexed to allow test limb to rest on the floor at start of the test.

Position of Therapist: Standing at side of limb to be tested. (*Note:* The illustration [Fig. 5–25] shows the examiner on the opposite side to avoid obscuring test positions.) The hand used to provide resistance is contoured over the posterior thigh just above the knee. The opposite hand stabilizes the pelvis laterally to maintain hip and pelvis posture (see Fig. 5–22).

Test: Patient extends hip through available range, but hip extension range is less when the knee is flexed (see page 180). Keeping the knee in extension will test all hip extensor muscles; with the knee flexed, the isolated Gluteus maximus will be evaluated.

Resistance is applied downward (toward floor) and forward.

Instructions to Patient: "Lift your foot off the floor as high as you can."

Grading

Grade 5 (Normal): Completes available range of hip extension. Holds end position against maximal resistance.

Grade 4 (Good): Completes available range of hip extension. (*Note:* Because of the intrinsic strength of these muscles, weakened extensor muscles frequently are overgraded.) Limb position can be held against heavy to moderate resistance.

Grade 3 (Fair): Completes available range and holds end position without resistance.

Grade 2 (Poor), Grade 1 (Trace), and Grade 0 (Zero)

Do not test the patient with hip flexion contractures and weak extensors (less than Grade 3) in the standing position. Position the patient side-lying on the table. Conduct the test as described for the aggregate of extensor muscles (see page 177) or for the isolated Gluteus maximus (see page 179).

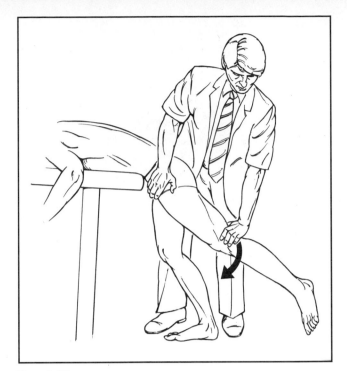

Figure 5–25

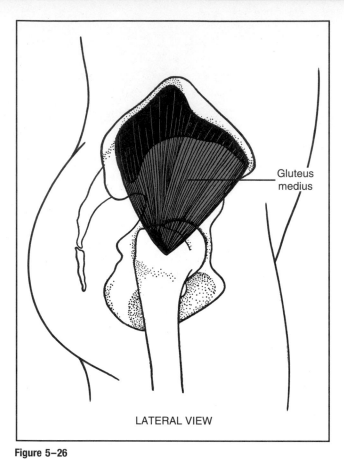

Gluteus medius

LATERAL VIEW

Figure 5–26

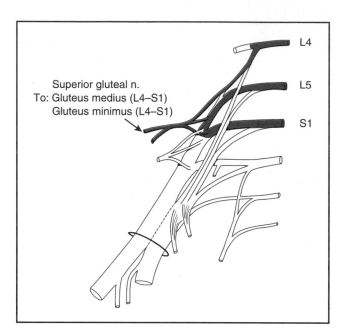

Superior gluteal n.
To: Gluteus medius (L4–S1)
Gluteus minimus (L4–S1)

L4

L5

S1

Figure 5–27

Table 5–4	HIP ABDUCTION	
Muscle	**Origin**	**Insertion**
183. Gluteus medius	Ilium (outer surface between crest and posterior gluteal line)	Femur (greater trochanter)
184. Gluteus minimus	Ilium (outer surface between anterior and inferior gluteal lines) Greater sciatic notch	Femur (greater trochanter)

Others
185. Tensor fasciae latae
182. Gluteus maximus (upper fibers)

Range of Motion
0° to 45°

Grade 5 (Normal), Grade 4 (Good), and Grade 3 (Fair)

Position of Patient: Side-lying with test leg uppermost. Start test with the limb slightly extended beyond the midline and the pelvis rotated slightly forward. Lowermost leg is flexed for stability.

Position of Therapist: Standing behind patient. Hand used to give resistance is contoured across the lateral surface of the knee. The hand used to palpate the Gluteus medius is just proximal to the greater trochanter of the femur (Fig. 5–28). (No resistance is used in a Grade 3 test.)

Alternatively, resistance may be applied at the ankle, which gives a longer lever arm and requires greater strength on the part of the patient to achieve a grade of 5 or 4. Examiner is reminded always to use the same lever in a given test sequence and in subsequent comparison tests.

To distinguish a Grade 5 from a Grade 4 result, first apply resistance at the ankle and then at the knee.

Test: Patient abducts hip through the complete available range of motion without flexing the hip or rotating it in either direction. Resistance is given in a straight downward direction.

Instructions to Patient: "Lift your leg up in the air. Hold it. Don't let me push it down."

Grading

Grade 5 (Normal): Completes available range and holds end position against maximal resistance.

Grade 4 (Good): Completes available range and holds against heavy to moderate resistance.

Grade 3 (Fair): Completes range of motion and holds end position without resistance (Fig. 5–29).

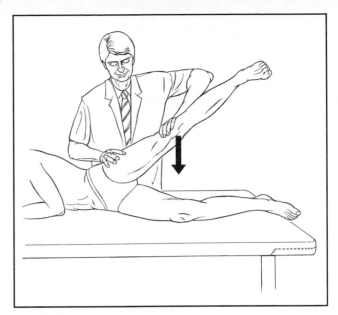

Figure 5–28

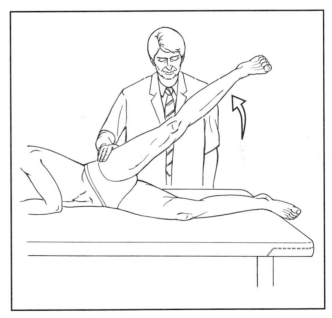

Figure 5–29

Grade 2 (Poor)

Position of Patient: Supine.

Position of Therapist: Standing on side of limb being tested. One hand supports and lifts the limb by holding it under the ankle to raise limb just enough to decrease friction. This hand offers no resistance, nor should it be used to offer assistance to the movement. On some smooth surfaces such support may not be necessary (Fig. 5–30).

The other hand palpates the Gluteus medius just proximal to the greater trochanter of the femur.

Test: Patient abducts hip through available range.

Instructions to Patient: "Bring your leg out to the side. Keep your kneecap pointing to the ceiling."

Grading

Grade 2 (Poor): Completes range of motion supine with no resistance and minimal to zero friction.

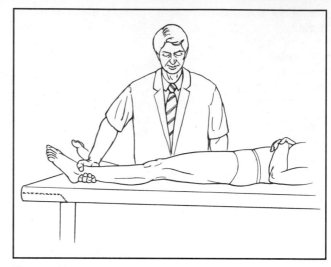

Figure 5–30

Grade 1 (Trace) and Grade 0 (Zero)

Position of Patient: Supine.

Position of Therapist: Standing at side of limb being tested at level of thigh. (*Note:* Illustration [Fig. 5–31] shows therapist on opposite side to avoid obscuring test positions.) One hand supports the limb under the ankle just above the malleoli. The hand should provide neither resistance nor assistance to movement (Fig. 5–31). Palpate the Gluteus medius on the lateral aspect of the hip just above the greater trochanter.

Test: Patient attempts to abduct hip.

Instructions to Patient: "Try to bring your leg out to the side."

Grading

Grade 1 (Trace): Palpable contraction of Gluteus medius but no movement of the part.

Grade 0 (Zero): No palpable contraction.

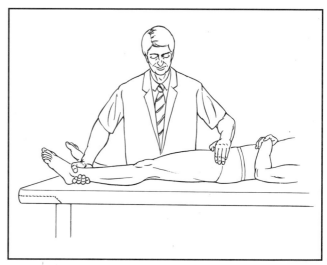

Figure 5–31

HIP ABDUCTION
(Glutei medius and minimus)

SUBSTITUTIONS IN HIP ABDUCTION TESTING

1. Hip-hike substitution:
    Patient may "hike hip" by approximating pelvis to thorax using the lateral trunk muscles, which moves the limb through partial abduction range (Fig. 5–32). This movement may be detected by observing the lateral trunk and hip (move clothing aside) and palpating the Gluteus medius above the trochanter.
2. External rotation and flexion substitution:
    The patient may try to externally rotate during the motion of abduction (Fig. 5–33). This could allow the oblique action of the hip flexors to substitute for the Gluteus medius.
3. Tensor fasciae latae substitution:
    If the test is allowed to begin with active hip flexion or with the hip positioned in flexion, there is an opportunity for the Tensor fasciae latae to abduct the hip.

◖ HELPFUL HINTS

1. The examiner should not be able to "break" a Grade 5 muscle and most therapists will not be able to "break" a Grade 4 muscle. A grade of 4 often masks significant weakness because of the intrinsic great strength of these muscles. Giving resistance at the ankle rather than at the knee assists in overcoming this problem.

2. It is difficult to impossible to palpate minimal contractile activity of muscle through clothing. (This is one of the cardinal principles of manual muscle testing.)

3. When the patient is supine, the weight of the opposite limb stabilizes the pelvis. It is not necessary, therefore, to use a hand to stabilize the contralateral limb.

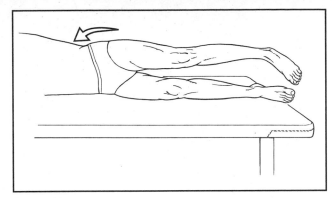

Figure 5–32

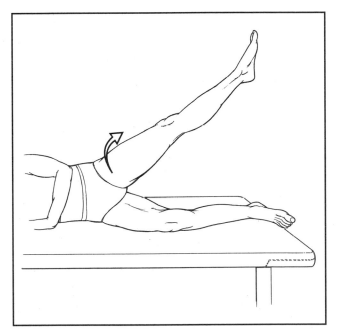

Figure 5–33

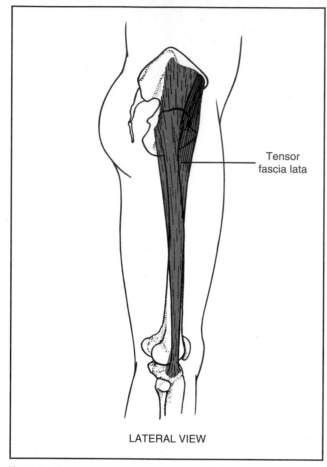

Tensor
fascia lata

LATERAL VIEW

Figure 5–34

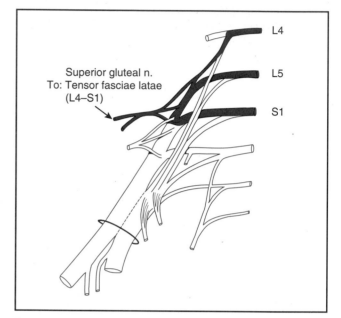

Superior gluteal n.
To: Tensor fasciae latae
(L4–S1)

L4

L5

S1

Figure 5–35

Table 5–5 HIP ABDUCTION FROM FLEXION		
Muscle	**Origin**	**Insertion**
185. Tensor fasciae latae	Iliac crest Anterior superior iliac spine	Iliotibial band
Others		
183. Gluteus medius		
184. Gluteus minimus		

Range of Motion

Two-joint muscle. No specific range of motion can be assigned solely to the Tensor

HIP ABDUCTION FROM FLEXED POSITION
(Tensor fasciae latae)

Grade 5 (Normal), Grade 4 (Good), and Grade 3 (Fair)

Position of Patient: Side-lying. Uppermost limb (test limb) is flexed to 45° and lies across the lowermost limb with the foot resting on the table (Fig. 5–36).

Position of Therapist: Standing behind patient at level of pelvis. Hand for resistance is placed on lateral surface of the thigh just above the knee. Hand providing stabilization is placed on the crest of the ilium (Fig. 5–37).

Test: Patient abducts hip through approximately 30° of motion. Resistance is given downward (toward floor) from the lateral surface of the distal femur. No resistance is given for the Grade 3 test.

Instructions to Patient: "Lift your leg and hold it. Don't let me push it down."

Grading

Grade 5 (Normal): Completes available range; holds end position against maximal resistance.

Grade 4 (Good): Completes available range and holds against strong to moderate resistance.

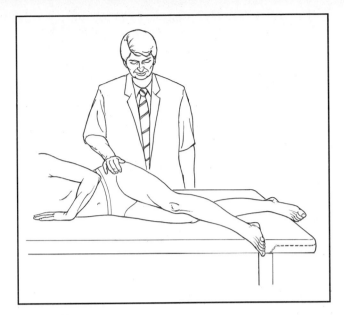

Figure 5–36

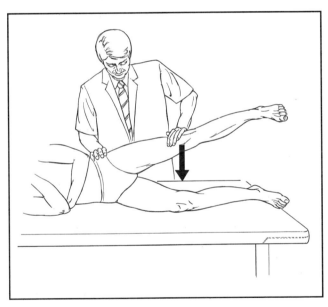

Figure 5–37

Grade 3 (Fair): Completes movement; holds end position but takes no resistance (Fig. 5–38).

Grade 2 (Poor)

Position of Patient: Patient is in long-sitting position supporting trunk with hands placed behind body on table. Trunk may lean backward up to 45° from vertical (Fig. 5–39).

Position of Therapist: Standing at side of limb to be tested. (*Note:* Illustration [Fig. 5–39] deliberately shows therapist on wrong side to avoid obscuring test positions.) One hand supports the limb under the ankle; this hand will be used to reduce friction with the surface as the patient moves but should neither resist nor assist motion.

The other hand palpates the Tensor fascia lata on the proximal anterolateral thigh where it inserts into the iliotibial band.

Test: Patient abducts hip through 30° of range.

Instructions to Patient: "Bring your leg out to the side."

Grading

Grade 2 (Poor): Completes hip abduction motion to 30°.

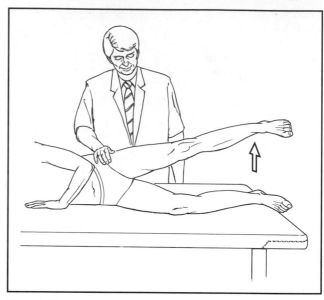

Figure 5–38

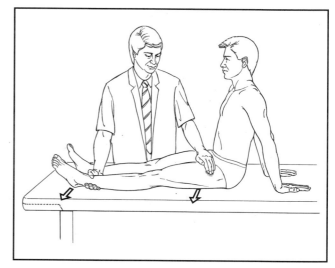

Figure 5–39

HIP ABDUCTION FROM FLEXED POSITION
(Tensor fasciae latae)

Grade 1 (Trace) and Grade 0 (Zero)

Position of Patient: Long sitting.

Position of Therapist: One hand palpates the insertion of the Tensor at the lateral aspect of the knee. The other hand palpates the Tensor on anterolateral thigh (Fig. 5–40).

Test: Patient attempts to abduct hip.

Instructions to Patient: "Try to move your leg out to the side."

Grading

Grade 1 (Trace): Palpable contraction of Tensor fibers but no limb movement.

Grade 0 (Zero): No palpable contractile activity.

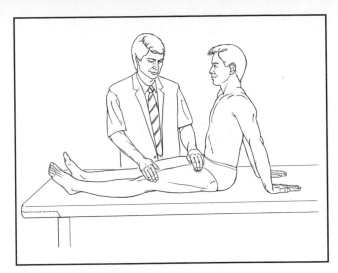

Figure 5–40

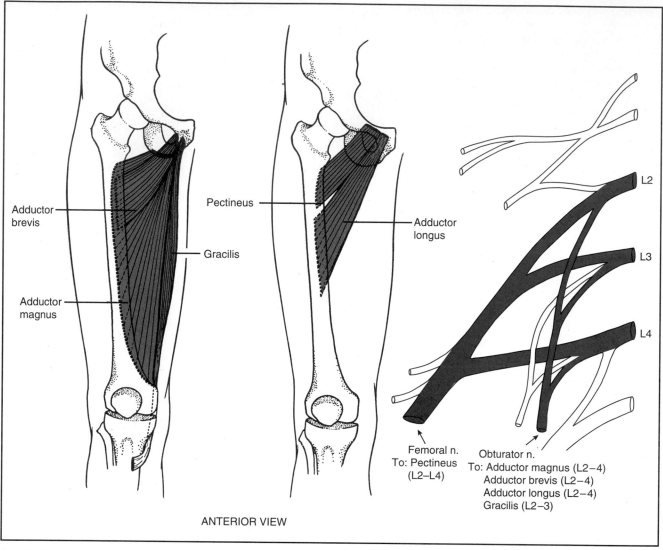

ANTERIOR VIEW

Figure 5–41

Figure 5–42

Figure 5–43

Table 5–6	HIP ADDUCTION		
Muscle	**Origin**	**Insertion**	
181. Adductor magnus	Ischial tuberosity Pubis (inferior ramus)	Femur (linea aspera and adductor tubercle on medial condyle)	
180. Adductor brevis	Pubis (body and inferior ramus)	Femur (linea aspera)	
179. Adductor longus	Pubis (anterior crest)	Femur (linea aspera)	
177. Pectineus	Pubis (pectineal line)	Femur (posterior)	
178. Gracilis	Pubis (body and inferior ramus)	Tibia (shaft distal to condyle)	

Range of Motion
0° to 15–20°

HIP ADDUCTION
(Adductors magnus, brevis and longus; Pectineus and Gracilis)

Grade 5 (Normal), Grade 4 (Good), and Grade 3 (Fair)

Position of Patient: Side-lying with test limb (lowermost) resting on the table. Uppermost limb in 25° of abduction, supported by the examiner. The therapist cradles the leg with the forearm, the hand supporting the limb on the medial surface of the knee (Fig. 5–44).

Position of Therapist: Standing behind patient at knee level. The hand giving resistance to the test limb (lowermost limb) is placed on the medial surface of the distal femur, just proximal to the knee joint. Resistance is directed straight downward toward the table (Fig. 5–45).

Test: Patient adducts hip until the lower limb contacts the upper one.

Instructions to Patient: "Lift your bottom leg up to your top one. Hold it. Don't let me push it down."

For Grade 3: "Lift your bottom leg up to your top one. Don't let it drop!"

Grading

Grade 5 (Normal): Completes full range; holds end position against maximal resistance.

Grade 4 (Good): Completes full movement, but tolerates strong to moderate resistance.

Grade 3 (Fair): Completes full movement; holds end position but takes no resistance (Fig. 5–46).

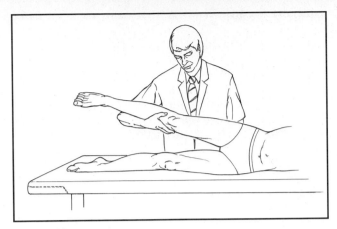

Figure 5–44

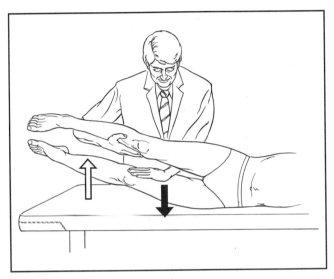

Figure 5–45

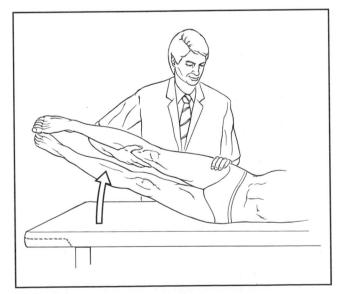

Figure 5–46

Grade 2 (Poor)

Position of Patient: Supine. The nontest limb is positioned in some abduction to prevent interference with motion of the test limb.

Position of Therapist: Standing at side of test limb at knee level. One hand supports the ankle and elevates it slightly from the table surface to decrease friction as the limb moves across (Fig. 5–47). The examiner uses this hand neither to assist nor resist motion. The opposite hand palpates the adductor mass on the inner aspect of the proximal thigh.

Test: Patient adducts hip without rotation.

Instructions to Patient: "Bring your leg in toward the other one."

Grading

Grade 2 (Poor): Patient adducts limb through full range.

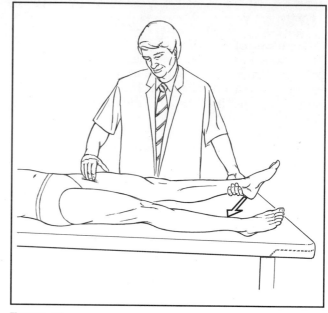

Figure 5–47

Grade 1 (Trace) and Grade 0 (Zero)

Position of Patient: Supine.

Position of Therapist: Standing on side of test limb. One hand supports the limb under the ankle. The other hand palpates the adductor mass on the proximal medial thigh (Fig. 5–48).

Test: Patient attempts to adduct hip.

Instructions to Patient: "Try to bring your leg in."

Grading

Grade 1 (Trace): Palpable contraction, no limb movement.

Grade 0 (Zero): No palpable contraction.

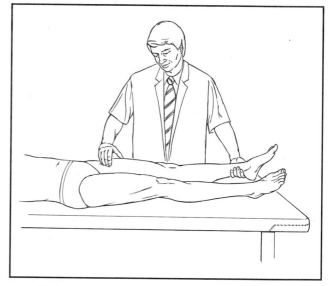

Figure 5–48

HIP ADDUCTION
(Adductors magnus, brevis and longus; Pectineus and Gracilis)

SUBSTITUTIONS FOR HIP ADDUCTION

1. Hip flexor substitution:
   The patient may attempt to substitute the hip flexors for the ad-ductors by internally rotating the hip using a posterior pelvic tilt (Fig. 5–49). The patient will appear to be trying to turn supine from side-lying. Maintenance of true side-lying is necessary for an accurate test.
2. Hamstring substitution:
   Patient may attempt to substitute the hamstrings for the adduc-tors by externally rotating the test hip with an anterior pelvic tilt. Patient will appear to move toward prone. Again, true side-lying is important.

HELPFUL HINTS

In the supine test position for Grades 2, 1, and 0 the weight of the opposite leg stabilizes the pelvis, so there is no need for manual stabilization of the nontest hip.

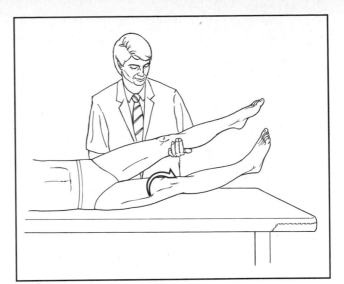

Figure 5–49

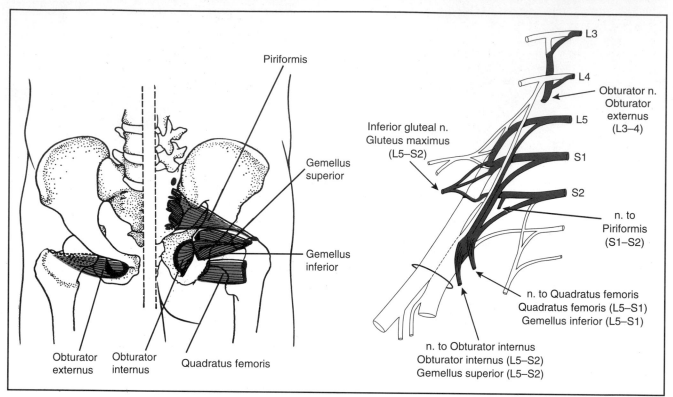

Figure 5–50

Figure 5–51

Figure 5–52

Table 5–7 HIP EXTERNAL ROTATION		
Muscle	**Origin**	**Insertion**
188. Obturator externus	Ischium and pubis (Obturator foramen, medial side)	Femur (trochanteric fossa)
187. Obturator internus	Pubis (inferior ramus) Ischium (inferior ramus) Ischium and pubis (Obturator foramen, internal side and membrane)	Femur (greater trochanter)
191. Quadratus femoris	Ischial tuberosity	Femur (quadrate tubercle)
186. Piriformis	Sacrum (anterior) Ilium (sciatic notch) Sacrotuberous ligament	Femur (greater trochanter)
189. Gemellus superior	Ischium (spine)	Femur (greater trochanter)
190. Gemellus inferior	Ischial tuberosity	Femur (greater trochanter)
182. Gluteus maximus	Ilium (posterior gluteal line) Sacrum (posterior) Coccyx (posterior) Sacrotuberous lig.	Femur (gluteal tuberosity) Iliotibial band

Range of Motion
0° to 45°

HIP EXTERNAL ROTATION
(Obturators internus and externus, Gemellae superior and inferior. Piriformis, Quadratus femoris, Gluteus maximus [posterior])

Grade 5 (Normal), Grade 4 (Good), and Grade 3 (Fair)

Position of Patient: Short sitting. Trunk may be supported by placing hands flat or fisted at sides (Fig. 5–53).

Position of Therapist: Sits on a low stool or kneels beside limb to be tested. The hand that gives resistance grasps the ankle just above the malleolus. Resistance is applied as a laterally directed force at the ankle (Fig. 5–53).

The other hand, which will offer counterpressure, is contoured over the lateral aspect of the distal thigh just above the knee. Resistance is given as a medially directed force at the knee. The two forces are applied in counter directions for this rotary motion (Fig. 5–53).

Test: Patient externally rotates the hip. This is a test where it is preferable for the examiner to place the limb in the test end position rather than to ask the patient to perform the movement.

Instructions to Patient: "Don't let me turn your leg out."

Grading

Grade 5 (Normal): Holds at end of range against maximal resistance.

Grade 4 (Good): Holds at end of range against strong to moderate resistance.

Grade 3 (Fair): Holds end position but tolerates no resistance (Fig. 5–54).

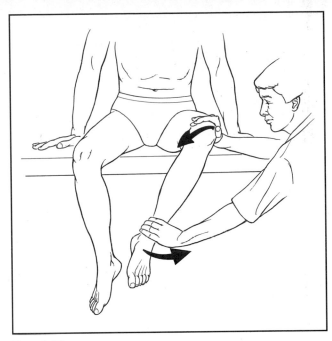

Figure 5–53

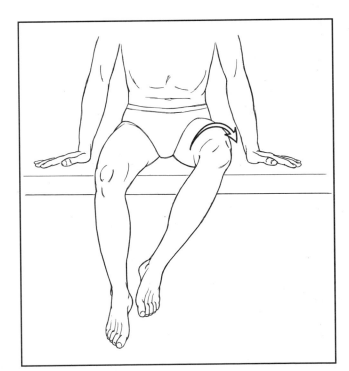

Figure 5–54

Grade 2 (Poor)

Position of Patient: Supine. Test limb is in internal rotation.

Position of Therapist: Standing at side of limb to be tested.

Test: Patient externally rotates hip in available range of motion (Fig. 5–55). One hand may be used to maintain pelvic alignment at lateral hip.

Instructions to Patient: "Roll your leg out."

Grading

Grade 2 (Poor): Completes external rotation range of motion.

Alternate Test for Grade 2: With the patient short-sitting, the therapist places the test limb in maximal internal rotation. The patient then is instructed to return the limb actively to the midline (neutral) position against slight resistance. Care needs to be taken to ensure that gravity is not the predominant force. If this motion is performed satisfactorily, the test may be assessed as a Grade 2.

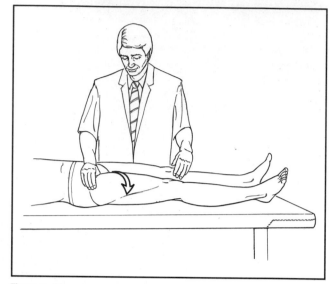

Figure 5–55

HIP EXTERNAL ROTATION
(Obturators internus and externus, Gemellae superior and inferior. Piriformis, Quadratus femoris, Gluteus maximus [posterior])

Grade 1 (Trace) and Grade 0 (Zero)

Position of Patient: Supine with test limb placed in internal rotation.

Position of Therapist: Standing at side of limb to be tested.

Test: Patient attempts to externally rotate hip.

Instructions to Patient: "Try to roll your leg out."

Grading

Grades 1 (Trace) and 0 (Zero): The external rotator muscles, except for the Gluteus maximus, are not palpable. If there is any discernible movement (contractile activity) a grade of 1 should be given; otherwise, a grade of 0 is assigned on the principle that whenever uncertainty exists, the lesser grade should be awarded.

⬤ HELPFUL HINTS

1. There is wide variation in the amount of hip external rotation ROM that can be considered normal. It is imperative, therefore, that an individual's accurate range (in each test position) be known before manual muscle testing takes place.

2. There is greater range of rotation at the hip when the hip is flexed than when it is extended, probably secondary to laxity of joint structures.

3. In short-sitting tests, the patient should not be allowed to use the following motions, lest they add visual distortion and contaminate the test results:
   (a) lift the contralateral buttock off the table or lean in any direction to lift the pelvis;
   (b) increase flexion of the test knee;
   (c) abduct the test hip.

4. In the supine test for a Grade 2 muscle, as the hip rolls past the midline, minimal resistance can be offered to offset the assistance of gravity.

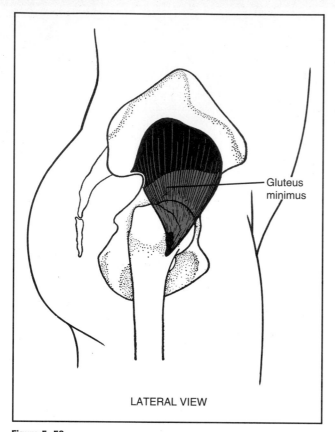

LATERAL VIEW

Gluteus minimus

Figure 5–56

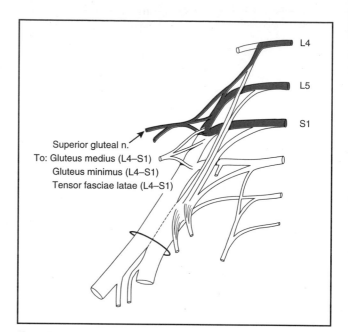

L4

L5

S1

Superior gluteal n.
To: Gluteus medius (L4–S1)
Gluteus minimus (L4–S1)
Tensor fasciae latae (L4–S1)

Figure 5–57

Table 5–8 HIP INTERNAL ROTATION		
Muscle	**Origin**	**Insertion**
184. Gluteus minimus	Ilium (outer surface between anterior and inferior gluteal lines) Greater sciatic notch	Femur (greater trochanter)
183. Gluteus medius	Ilium (outer surface between crest and posterior gluteal line)	Femur (greater trochanter)
185. Tensor fasciae latae	Iliac crest ASIS	Iliotibial band

Range of Motion

0° to 45°

Grade 5 (Normal), Grade 4 (Good), and Grade 3 (Fair)

Position of Patient: Short sitting. Arms may be used for trunk support at sides or may be crossed over chest.

Position of Therapist: Sitting or kneeling in front of patient. One hand grasps the lateral surface of the ankle just above the malleolus (Fig. 5–58). Resistance is given (Grades 5 and 4 only) as a medially directed force at the ankle.

The opposite hand, which offers counterpressure, is contoured over the medial surface of the distal thigh just above the knee. Resistance is applied as a laterally directed force at the knee. Note the counter directions of the force applied.

Test: The limb should be placed in the end position of full internal rotation by the examiner for best test results (Fig. 5–58).

Grading

Grade 5 (Normal): Holds end position against maximal resistance.

Grade 4 (Good): Holds end position against strong to moderate resistance.

Grade 3 (Fair): Holds end position but takes no resistance (Fig. 5–59).

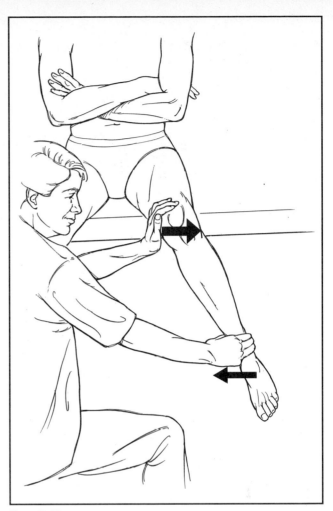

Figure 5–58

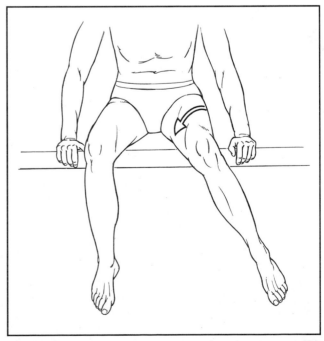

Figure 5–59

Grade 2 (Poor)

Position of Patient: Supine. Test limb in partial external rotation.

Position of Therapist: Standing next to test leg. Palpate the Gluteus medius proximal to the greater trochanter and the Tensor fascia latae (Fig. 5–60) over the anterolateral hip below the anterior superior iliac spine.

Test: Patient internally rotates hip through available range.

Instructions to Patient: "Roll your leg in toward the other one."

Grading

Grade 2 (Poor): Completes the range of motion.

Alternate Test for Grade 2: With patient short-sitting, the examiner places the test limb in maximal external rotation. The patient then is instructed to return the limb actively to the midline (neutral) position against slight resistance. Care needs to be taken to ensure that gravity is not the predominant force. If this motion is performed satisfactorily, the test may be assessed a Grade 2.

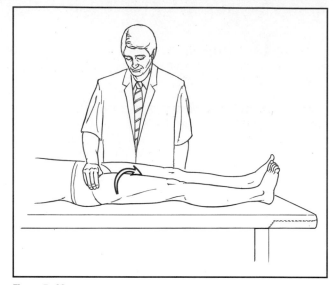

Figure 5–60

Grade 1 (Trace) and Grade 0 (Zero)

Position of Patient: Patient supine with test limb placed in external rotation.

Position of Therapist: Standing next to test leg.

Test: Patient attempts to internally rotate hip. One hand is used to palpate the Gluteus medius (over the posterolateral surface of the hip above the greater trochanter). The other hand is used to palpate the Tensor fascia latae (on the anterolateral surface of the hip below the ASIS).

Instructions to Patient: "Try to roll your leg in."

Grading

Grade 1 (Trace): Palpable contractile activity in either or both muscles.

Grade 0 (Zero): No palpable contractile activity.

HELPFUL HINTS

1. In the short-sitting tests, do not allow the patient to assist internal rotation by lifting the pelvis on the side of the limb being tested.

2. Neither should the patient be allowed to extend the knee or adduct and extend the hip during performance of the test. These motions contaminate the test by offering visual distortion to the therapist.

3. The reader is referred to Helpful Hints 1, 2, and 4 for the external rotation test (page 197), which apply here as well.

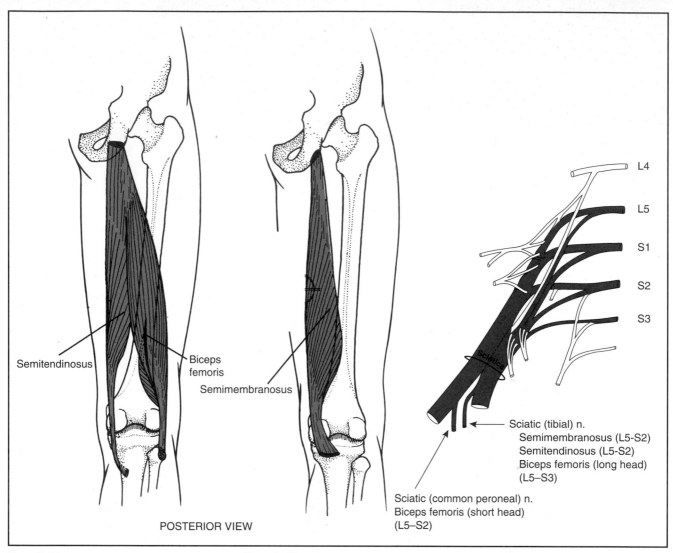

Semitendinosus

Biceps femoris

Semimembranosus

POSTERIOR VIEW

L4

L5

S1

S2

S3

Sciatic

Sciatic (tibial) n.
Semimembranosus (L5-S2)
Semitendinosus (L5-S2)
Biceps femoris (long head)
(L5–S3)

Sciatic (common peroneal) n.
Biceps femoris (short head)
(L5–S2)

Figure 5–61   Figure 5–62   Figure 5–63

Table 5–9 KNEE FLEXION		
Muscle	**Origin**	**Insertion**
192. Biceps femoris		
Long head	Ischium (tuberosity)	Fibula (lateral head)
	Sacrotuberous ligament	Tibia (lateral condyle)
Short head	Femur (linea aspera and lateral condyle)	
193. Semitendinosus	Ischial tuberosity	Tibia (proximal shaft) Pes anserina
194. Semimembranosus	Ischial tuberosity	Tibia (medial condyle) Femur (lateral condyle)

Range of Motion

0° to 135°

Grade 5 (Normal), Grade 4 (Good), and Grade 3 (Fair)

There are three basic muscle tests for the hamstrings at Grades 5 and 4. The examiner should test first for the aggregate of the three hamstring muscles (with the foot in midline). Only if there is deviation (or asymmetry) in the movement or a question in the examiner's mind is there a need to test the medial and lateral hamstrings separately.

Hamstring Muscles in Aggregate

Position of Patient: Prone with limbs straight and toes hanging over the edge of the table. Test may start in about 45° of knee flexion (Fig. 5–64).

Position of Therapist: Standing next to limb to be tested. (Illustration is deliberately incorrect to avoid obscuring test activity.) Hand giving resistance is contoured around the posterior surface of the leg just above the ankle (Fig. 5–64). Resistance is applied in the direction of knee extension for Grades 5 and 4.

The other hand is placed over the hamstring tendons on the posterior thigh (optional).

Test: Patient flexes knee while maintaining leg in neutral rotation.

Instructions to Patient: "Bend your knee. Hold it! Don't let me straighten it."

Medial Hamstring Test (Semitendinosus and Semimembranosus)

Position of Patient: Prone with knee flexed to less than 90°. Leg in internal rotation (toes pointing toward midline).

Position of Therapist: Hand giving resistance grasps the leg at the ankle. Resistance is applied in an oblique direction (down and out) toward knee extension (Fig. 5–65).

Test: Patient flexes knee, maintaining the leg in internal rotation (heel toward examiner, toes pointing toward midline).

Figure 5–64

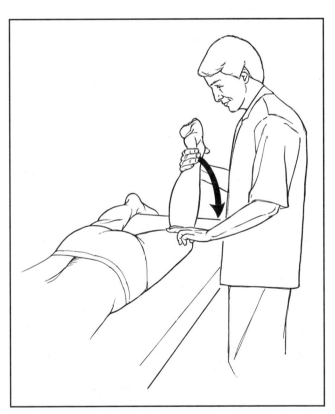

Figure 5–65

Lateral Hamstring Test (Biceps femoris)

Position of Patient: Prone with knee flexed to less than 90°. Leg is in external rotation (toes pointing laterally).

Position of Therapist: Therapist resists knee flexion at the ankle using a downward and inward force (Fig. 5–66).

Test: Patient flexes knee, maintaining leg in external rotation (heel away from examiner, toes pointing toward examiner) (Fig. 5–66).

Grading the Hamstring Muscles (Grades 5 to 3)

Grade 5 (Normal) for All Three Tests: Resistance will be maximal, and the end knee flexion position (approximately 90°) cannot be broken.

Grade 4 (Good) for All Three Tests: End knee flexion position is held against strong to moderate resistance.

Grade 3 (Fair) for All Three Tests: Holds end range position but tolerates no resistance (Fig. 5–67).

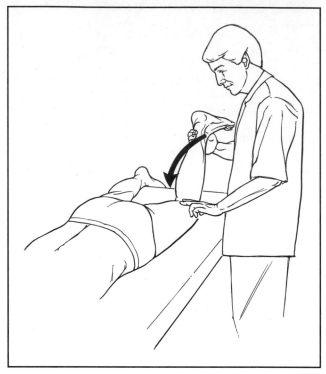

Figure 5–66

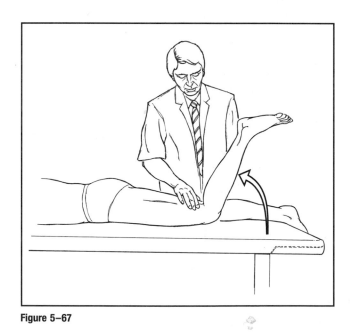

Figure 5–67

Grade 2 (Poor)

Position of Patient: Side-lying with test limb (uppermost limb) supported by examiner. Lower limb flexed for stability.

Position of Therapist: Standing behind patient at knee level. One arm is used to cradle thigh, providing hand support at medial side of knee. Other hand supports the leg at the ankle just above the malleolus (Fig. 5–68).

Test: Patient flexes knee through available range of motion.

Instructions to Patient: "Bend your knee."

Grading

Grade 2 (Poor): Completes available range of motion in side-lying.

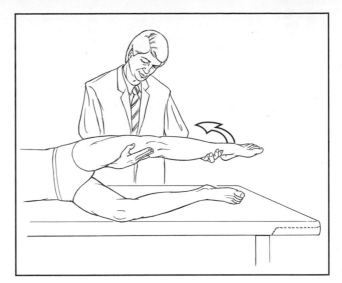

Figure 5–68

Grade 1 (Trace) and Grade 0 (Zero)

Position of Patient: Prone. Limbs are straight with toes extending over end of table. Knee is partially flexed and supported at ankle by examiner.

Position of Therapist: Standing next to test limb at knee level. One hand supports the flexed limb at the ankle (Fig. 5–69). The opposite hand palpates both the medial and lateral hamstring tendons just above the posterior knee.

Test: Patient attempts to flex knee.

Instructions to Patient: "Try to bend your knee."

Grading

Grade 1 (Trace): Tendons become prominent, but no visible movement occurs.

Grade 0 (Zero): No palpable contraction of the muscles, and tendons do not stand out.

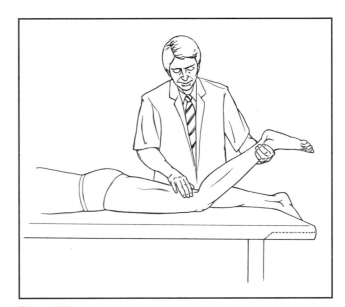

Figure 5–69

SUBSTITUTIONS FOR KNEE FLEXION
(WHEN THE HAMSTRINGS ARE VERY WEAK)

1. Hip flexion substitution:
   The prone patient may flex the hip to start knee flexion. The buttock on the test side will rise as the hip flexes, and the patient may appear to roll slightly toward supine (Fig. 5–70).
2. Sartorius substitution:
   The Sartorius may try to assist with knee flexion, but this also causes flexion and external rotation of the hip. Knee flexion when the hip is externally rotated is less difficult because the leg is not raised vertically against gravity.
3. Gracilis substitution:
   Action of the Gracilis contributes a hip adduction motion.
4. Gastrocnemius substitution:
   Do not permit the patient to use strong plantar flexion in an attempt to substitute with the Gastrocnemius.

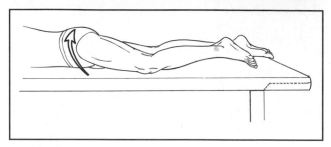

Figure 5–70

HELPFUL HINTS

1. If the Biceps femoris is stronger than the medial hamstrings, the leg will externally rotate during knee flexion. Similarly, if the Semitendinosus and Semimembranosus are the stronger components, the leg will internally rotate during knee flexion. This is the situation that, when observed, indicates asymmetry and the need to test the medial and lateral hamstrings separately.

2. In tests for Grades 3 and 2, the knee may be placed in a 10° flexed position to start the test when Gastrocnemius weakness is present (the Gastrocnemius assists in knee flexion).

3. If the hip flexes at the end of the knee flexion range of motion, check for a tight Rectus femoris muscle because this tightness will limit the range of knee flexion.

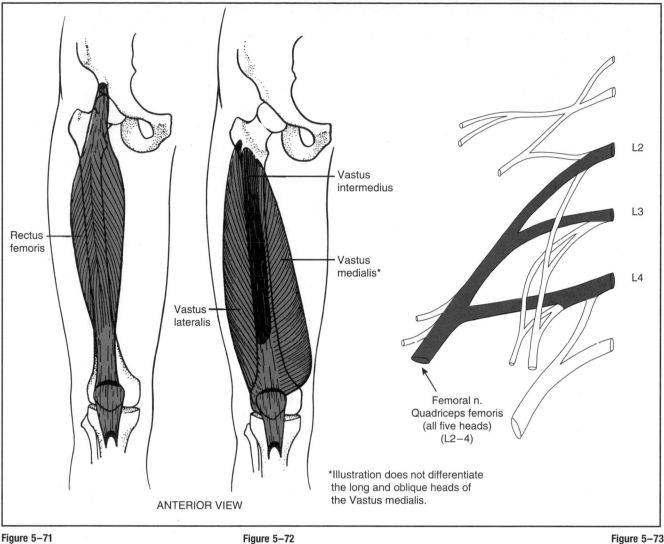

Rectus
femoris

Vastus
lateralis

Vastus
intermedius

Vastus
medialis*

Femoral n.
Quadriceps femoris
(all five heads)
(L2–4)

L2

L3

L4

*Illustration does not differentiate
the long and oblique heads of
the Vastus medialis.

ANTERIOR VIEW

Figure 5–71   Figure 5–72   Figure 5–73

Table 5–10 KNEE EXTENSION		
Muscle	**Origin**	**Insertion**
196. Rectus femoris	Ilium (anterior spine) Acetabulum (posterior)	Patella (base)
198. Vastus intermedius	Femur (upper 2/3 shaft)	Patella (base)
197. Vastus lateralis	Femur (linea aspera; greater trochanter; intertrochanteric line)	Patella (lateral)
199. Vastus medialis longus	Femur (linea aspera; intertrochanteric line) Tendons of Adductor magnus and longus	Patella (medial)
200. Vastus medialis oblique	Femur (linea aspera; supracondylar line) Tendon of Adductor magnus	Patella

Range of Motion

135° to 0°
May extend 10° beyond 0° in those with hyperextension

The Quadriceps femoris muscles are tested together as a functional group. Any given head cannot be separated from any other by manual muscle testing. The Rectus femoris is isolated from the other Quadriceps during a hip flexion test.

Grade 5 (Normal), Grade 4 (Good), and Grade 3 (Fair)

Position of Patient: Short sitting. Place wedge or pad under the distal thigh to maintain the femur in the horizontal position. An experienced examiner may replace the padding under the thigh with his hand (Fig. 5–74). Hands rest on the table on either side of the body for stability, or may prefer to grasp the table edge. The patient should be allowed to lean backward to relieve hamstring muscle tension.

Do not allow the patient to hyperextend the knee because this may lock it into position.

Position of Therapist: Stand at side of limb to be tested. The hand giving resistance is contoured over the anterior surface of the distal leg just above the ankle. For Grades 5 and 4, resistance is applied in a downward direction (toward the floor) in the direction of knee flexion.

Test: Patient extends knee through available range of motion but not beyond 0°.

Instructions to Patient: "Straighten your knee. Hold it! Don't let me bend it."

Grading

Grade 5 (Normal): Holds end position against maximal resistance. Most physical therapists will not be able to break the Normal knee extensors.

Grade 4 (Good): Holds end position against strong to moderate resistance.

Grade 3 (Fair): Completes available range and holds the position without resistance (Fig. 5–75).

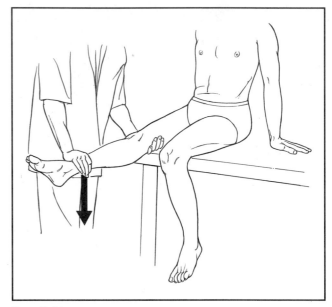

Figure 5–74

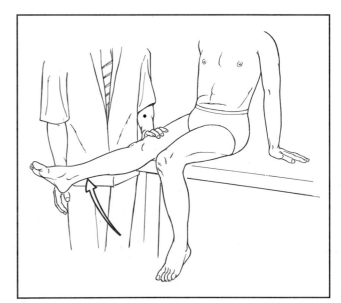

Figure 5–75

Grade 2 (Poor)

Position of Patient: Side-lying with test limb uppermost. Lowermost limb may be flexed for stability. Limb to be tested is held in about 90° of knee flexion. The hip should be in full extension.

Position of Therapist: Standing behind patient at knee level. One arm cradles the test limb around the thigh with the hand supporting the underside of the knee (Fig. 5–76). The other hand holds the leg just above the malleolus.

Test: Patient extends knee through the available range of motion. The therapist supporting the limb provides neither assistance nor resistance to the patient's voluntary movement. This is part of the art of muscle testing that must be acquired.

Instructions to Patient: "Straighten your knee."

Grading

Grade 2 (Poor): Completes available range of motion.

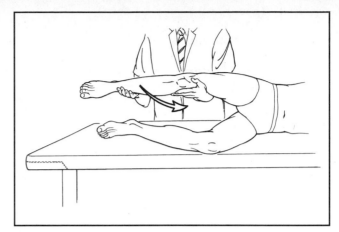

Figure 5–76

Grade 1 (Trace) and Grade 0 (Zero)

Position of Patient: Supine.

Position of Therapist: Standing next to limb to be tested at knee level. Hand used for palpation should be on the Quadriceps tendon just above the knee with the tendon "held" gently between the thumb and fingers. The examiner also may want to palpate the patellar tendon with two to four fingers just below the knee (Fig. 5–77).

Test: Patient attempts to extend knee.

As an alternate test, the therapist may place one hand under the slightly flexed knee; palpate either the Quadriceps or the patellar tendon while the patient tries to extend the knee.

Instructions to Patient: "Push the back of your knee down into the table" OR "Tighten your knee cap" (Quadriceps setting).

For alternate test: "Push the back of your knee down into my hand."

Grading

Grade 1 (Trace): Contractile activity can be palpated in muscle through the tendon. No joint movement occurs.

Grade 0 (Zero): No palpable contractile activity.

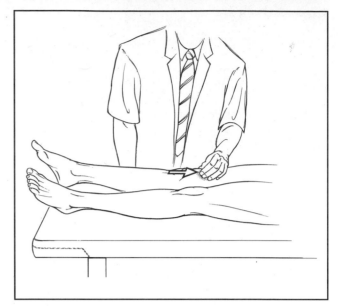

Figure 5–77

SUBSTITUTION

The patient is side-lying (as in the Grade 2 test). May use the hip internal rotators to substitute for the Quadriceps, thereby allowing the knee to fall into extension.

◑ HELPFUL HINTS

Knowledge of the patient's hamstring range of motion is imperative before conducting tests for knee extension strength. Straight leg raising (SLR) range dictates the optimal position for the knee extension test in the sitting position. In short-sitting for Grades 5, 4, and 3, the less the range of straight leg raising, the greater the backward trunk lean. Range of SLR also informs the examiner of the "available range" within the patient's comfort zone for side-lying tests.

ANKLE PLANTAR FLEXION
(Gastrocnemius and Soleus)

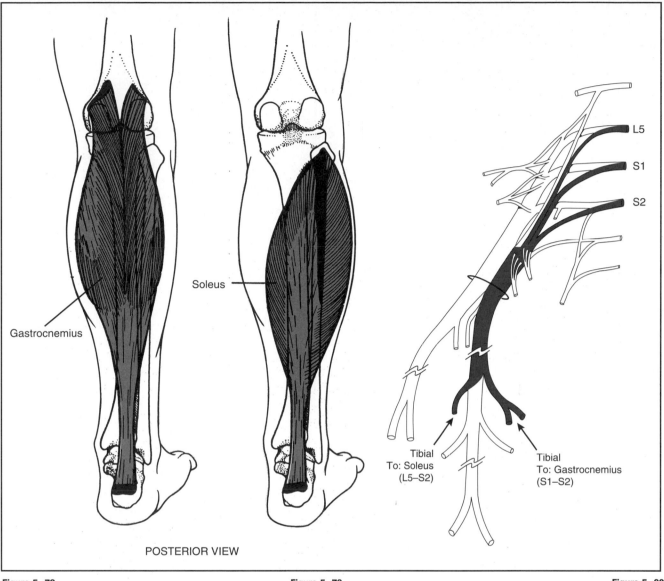

Gastrocnemius

Soleus

L5

S1

S2

Tibial
To: Soleus
(L5–S2)

Tibial
To: Gastrocnemius
(S1–S2)

POSTERIOR VIEW

Figure 5–78   **Figure 5–79**   **Figure 5–80**

Table 5–11	PLANTAR FLEXION	
Muscle	**Origin**	**Insertion**
205. Gastrocnemius		
Medial head	Femur (medial condyle; popliteal surface)	Tendo calcaneus
Lateral head	Femur (lateral condyle)	Tendo calcaneus
206. Soleus	Fibula (head and proximal 1/3 of shaft) Tibia (popliteal line)	Tendo calcaneus

Range of Motion
0° to 45°

GASTROCNEMIUS AND SOLEUS TEST

Grade 5 (Normal), Grade 4 (Good), and Grade 3 (Fair)

Position of Patient: Patient stands on limb to be tested with knee extended. Patient is likely to need external support; no more than one or two fingers should be used on a table (or other surface) for balance assist only (Fig. 5–81).

Position of Therapist: Standing or sitting with a lateral view of test limb.

Test: Patient raises heel from floor consecutively through full range of plantar flexion.

Instructions to Patient: Therapist demonstrates correct heel rise to patient. "Stand on your right leg; go up on your tiptoes; now down. Repeat this 20 times." Repeat test for left limb.

Grading

Grade 5 (Normal): Patient successfully completes a minimum of 20 heel rises through full range of motion without rest between rises and without fatigue. (Twenty heel rises represent over 60 percent of maximum electromyographic activity of the plantar flexors.)[1]

Grade 4 (Good): A grade 4 is conferred when the patient completes any number of correct heel rises between 19 and 10 with no rest between repetitions and without fatigue. Grade 4 is conferred only if the patient uses correct form in all repetitions. Any failure to complete the full range in any given repetition automatically drops the grade to at least the next lower level.

Grade 3 (Fair): Patient completes between nine and one heel rises correctly with no rest or fatigue.
If the patient cannot complete at least one correct full-range heel rise in the standing position, the grade must be less than 3 (Fair). Regardless of any resistance in a nonstanding position for any reason, the patient must be given a grade of less than 3.

Figure 5–81

Grade 2 (Poor)

Standing Test

Position of Patient: Standing on limb to be tested with knee extended, with a two-finger balance assist.

Position of Therapist: Standing or sitting with a clear lateral view of test limb.

Test: Patient attempts to raise heel from the floor through the full range of plantar flexion (Fig. 5–82).

Instructions to Patient: "Stand on your right leg. Try to go up on your tiptoes."
Repeat test for left leg.

Grading

Grade 2+ (Poor +): The patient can just clear the heel from the floor and cannot get up on the toes for the end test position.
Note: This is a rare exception for the use of a 2+ (Poor +) grade. There is no Grade 2 from the standing position.

Prone Test

Position of Patient: Prone with feet off end of table.

Position of Therapist: Standing at end of table in front of foot to be tested. One hand is contoured under and around the test leg just above the ankle (Fig. 5–83). Heel and palm of hand giving resistance are placed against the plantar surface at the level of the metatarsal heads.

Test: Patient plantar flexes ankle through the available range of motion. Manual resistance is down and forward toward dorsiflexion.

Grading

Grade 2+ (Poor +): Completes plantar flexion range and holds against maximum resistance.

Grade 2 (Poor): Patient completes plantar flexion range but tolerates no resistance.

Grade 2 – (Poor –): Patient completes only a partial range of motion.

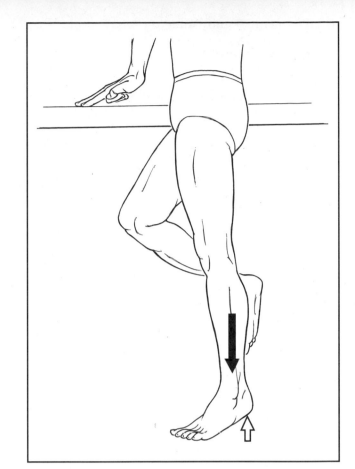

Figure 5–82

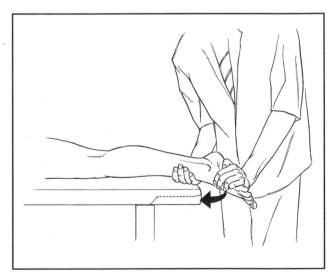

Figure 5–83

Grade 1 (Trace) and Grade 0 (Zero)

Position of Patient: Prone with feet off end of table.

Position of Therapist: Standing at end of table in front of foot to be tested. One hand palpates Gastrocnemius-Soleus activity by monitoring tension in the Achilles tendon just above the calcaneus (Fig. 5–84). The muscle bellies of the two muscles also may be palpated (not illustrated).

Test: Patient attempts to plantar flex the ankle.

Instructions to Patient: "Point your toes down, like a toe dancer."

Grading

Grade 1 (Trace): Tendon reflects some contractile activity in muscle, but no joint motion occurs. Contractile activity may be palpated in muscle bellies.

The best location to palpate the Gastrocnemius is at midcalf with thumb and fingers on either side of the midline but above the Soleus. Palpation of the Soleus is best done on the posterolateral surface of the distal calf. In most people with calf strength of Grade 3 or better, the two muscles can be observed and differentiated during plantar flexion testing because their definition is clear.

Grade 0 (Zero): No palpable contraction.

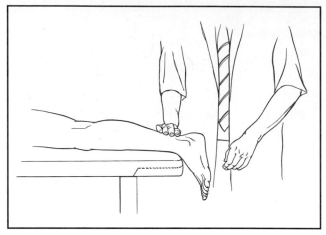

Figure 5–84

PLANTAR FLEXION, SOLEUS ONLY

All plantar flexor muscles are active in all positions of plantar flexion testing; therefore, no true isolation of the Soleus is possible. Testing during standing with the test leg flexed results in a 70 percent decrease in Gastrocnemius activity.[2] The test performed to "isolate" the Soleus should be interpreted with this caveat in mind.

Grade 5 (Normal), Grade 4 (Good), and Grade 3 (Fair)

Position of Patient: Standing on limb to be tested with knee slightly flexed (Fig. 5–85). Use one or two fingers for balance assist.

Position of Therapist: Standing or sitting with clear lateral view of test limb.

Test: Patient raises heel from floor through full range of plantar flexion, maintaining flexed position of knee (Fig. 5–85). Twenty (20) correct heel raises must be done consecutively without rest and without great fatigue.

Instructions to Patient: Therapist demonstrates test position and motion. "Stand on your right leg with your knee bent. Keep your knee bent and go up and down on your tiptoes at least 20 times."
Repeat test for left leg.

Grading

Grade 5 (Normal): Patient completes 20 consecutive heel rises to full range without rest or complaint of fatigue.

Grade 4 (Good): Patient completes between 19 and 10 correct heel rises without rest.

Grade 3 (Fair): Patient completes between nine and one correct heel rises with the knee flexed.

Note: If patient cannot complete all heel rises through a full range, the grade must be lower than 3. If the patient is unable to stand for the Grade 3 test for any reason, the grade awarded may not exceed a 2.

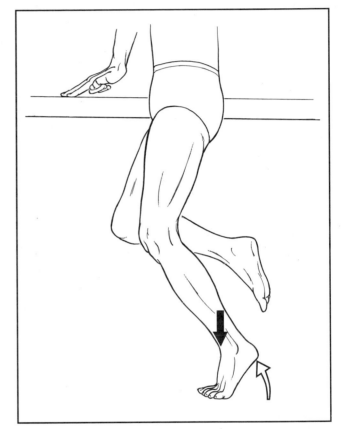

Figure 5–85

Grade 2 (Poor), Grade 1 (Trace), and Grade 0 (Zero)

Position of Patient: Prone with knee flexed to 90°.

Position of Therapist: Standing next to patient. Resistance is given with the heel of the hand placed under plantar surface of forefoot in the direction of dorsiflexion.

Test: Patient attempts to plantar flex the ankle while the knee is maintained in flexion.

Instructions to Patient: "Point your toes toward the ceiling."

Grading

Grade 2+ (Poor +): Completes full plantar flexion range against maximal resistance.

Grade 2 (Poor): Completes full plantar flexion range with no resistance.

Grade 2 − (Poor −): Completes only a partial range of motion with knee flexed.

Grades 1 and 0: Palpable contraction or Achilles tendon tightening is Grade 1. No contractile activity is Grade 0.

SUBSTITUTIONS IN PLANTAR FLEXION TESTING

1. By Flexor hallucis longus and Flexor digitorum longus:
   When substitution by the toe flexors occurs, their motions will be accompanied by plantar flexion of the forefoot and incomplete movement of the calcaneus (Fig. 5–86).
2. By Peroneus longus and Peroneus brevis:
   These muscles substituting for the Gastrocnemius and Soleus will pull the foot into eversion.
3. By Tibialis posterior:
   The foot will move into inversion during plantar flexion testing if the Tibialis posterior substitutes for the primary plantar flexors.
4. By Tibialis posterior, Peroneus longus, and Peroneus brevis:
   Substitution by these three muscles will plantar flex the forefoot instead of the ankle.

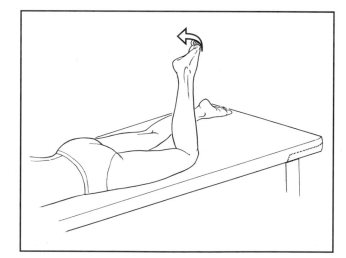

Figure 5–86

ANKLE PLANTAR FLEXION
(Gastrocnemius and Soleus)

HELPFUL HINTS

1. If for any reason the patient cannot lie prone, an alternative is to use the supine position for nonweight-bearing testing. The highest grade awarded in this case may not exceed a 2+.

2. If the patient is unable to perform a standing plantar flexion test but has a stable forefoot, a different application of resistance may be used with the patient supine. The resistance is applied against the sole of the foot with the forearm while the heel is cupped with the hand of the same arm and the ankle is forced into dorsiflexion. The highest grade that may be awarded in this case is a 2+.

3. During standing plantar flexion tests, the Tibialis posterior and the Peroneus longus and brevis muscles must be Grade 5 or 4 to stabilize the forefoot to attain and hold the tiptoe position.

4. During standing heel rise testing, it is important to be sure that the patient maintains a fully erect posture. If the subject leans forward, such posture can bring the heel off the ground, creating a testing artifact.

5. In the test that isolates the Soleus action, the knee is placed in flexion to put slack on the Gastrocnemius head which crosses the knee joint.

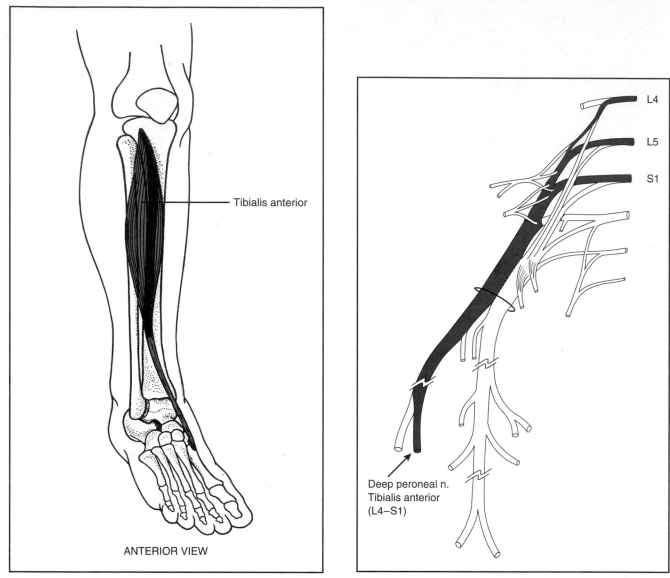

Tibialis anterior

ANTERIOR VIEW

Figure 5–87

Deep peroneal n.
Tibialis anterior
(L4–S1)

L4

L5

S1

Figure 5–88

Table 5–12 FOOT DORSIFLEXION AND INVERSION			Range of Motion
Muscle	**Origin**	**Insertion**	**0° to 20°**
203. Tibialis anterior	Tibia (lateral condyle and proximal 2/3 of lateral shaft)	1st cuneiform 1st metatarsal	

All Grades 5 (Normal) to 0 (Zero)

Position of Patient: Short sitting. Alternatively, patient may be supine.

Position of Therapist: Sitting on stool in front of patient with patient's heel resting on thigh. One hand is contoured around the posterior leg just above the malleoli for Grades 5 and 4 (Fig. 5–89). The hand providing resistance for the same grades is cupped over the dorsomedial aspect of the foot (Fig. 5–89).

Test: Patient dorsiflexes ankle and inverts foot, keeping toes relaxed.

Instructions to Patient: "Bring your foot up and in. Hold it! Don't let me push it down."

Grading

Grade 5 (Normal): Completes full range and holds against maximal resistance.

Grade 4 (Good): Completes available range against strong to moderate resistance.

Grade 3 (Fair): Completes available range of motion and holds end position without resistance (Fig. 5–90).

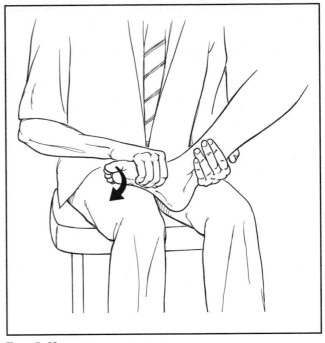

Figure 5–89

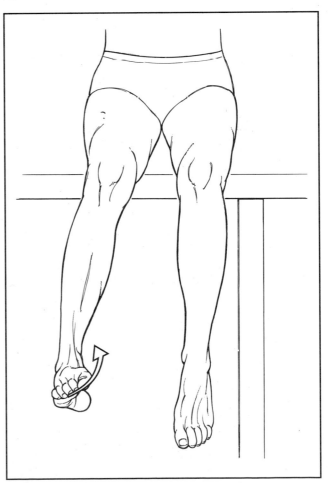

Figure 5–90

Grade 2 (Poor): Completes only a partial range of motion.

Grade 1 (Trace): Therapist will be able to detect some contractile activity in the muscle, or the tendon will "stand out." There is no joint movement.

Palpate the tendon of the Tibialis anterior on the anterior-medial aspect of the ankle at about the level of the malleoli (Fig. 5–91, lower hand). Palpate the muscle for contractile activity over its belly just lateral to the "shin" (Fig. 5–91, upper hand).

Grade 0 (Zero): No palpable contraction.

SUBSTITUTION

Substitution by the Extensor digitorum longus and the Extensor hallucis longus muscles results also in toe extension. Instruct the patient, therefore, to keep the toes relaxed so that they are not part of the test movement.

HELPFUL HINTS

If the supine position is used in lieu of the short-sitting position for the Grade 3 test, the therapist should add a degree of difficulty to the test to compensate for the lack of gravity. For example, give mild resistance in the supine position but award no more than a Grade 3.

In the supine position, to earn a Grade 2 the patient must complete a full range of motion.

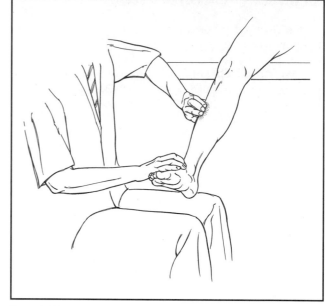

Figure 5–91

FOOT INVERSION
(Posterior tibialis)

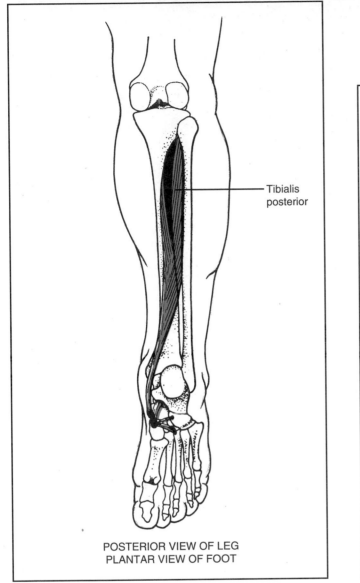

Tibialis
posterior

POSTERIOR VIEW OF LEG
PLANTAR VIEW OF FOOT

Figure 5–92

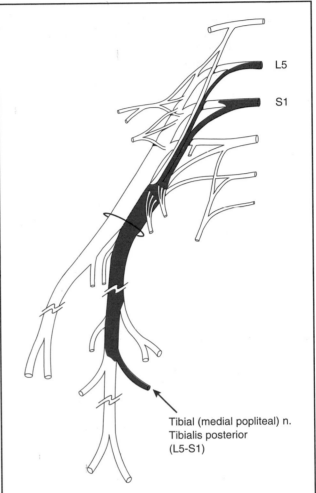

L5

S1

Tibial (medial popliteal) n.
Tibialis posterior
(L5-S1)

Figure 5–93

Table 5–13	FOOT INVERSION	
Muscle	**Origin**	**Insertion**
204. Tibialis posterior	Tibia (proximal 2/3 of shaft and distal condyle)	Navicular (tuberosity)
	Fibula (proximal 2/3 and posterior head)	Expansions to calcaneus Cuneiform bones (three)

Others
213. Flexor digitorum longus
222. Flexor hallucis longus
205. Gastrocnemius (medial head)

Range of Motion
0° to 35°

Grades 5 (Normal) to 2 (Poor)

Position of Patient: Short sitting with ankle in slight plantar flexion.

Position of Therapist: Sitting on low stool in front of patient or on side of test limb. One hand is used to stabilize the ankle just above the malleoli (Fig. 5–94). Hand providing resistance is contoured over the dorsum and medial side of the foot at the level of the metatarsal heads. Resistance is directed toward eversion and slight dorsiflexion.

Test: Patient inverts foot through available range of motion.

Instructions to Patient: Therapist may need to demonstrate motion. "Turn your foot down and in. Hold it."

Grading

Grade 5 (Normal): The patient completes the full range and holds against maximal resistance.

Grade 4 (Good): The patient completes available range against strong to moderate resistance.

Grade 3 (Fair): The patient will be able to invert the foot through the full available range of motion (Fig. 5–95).

Grade 2 (Poor): The patient will be able to complete only a partial range of motion.

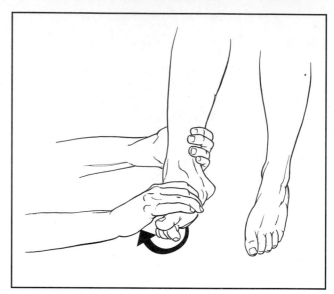

Figure 5–94

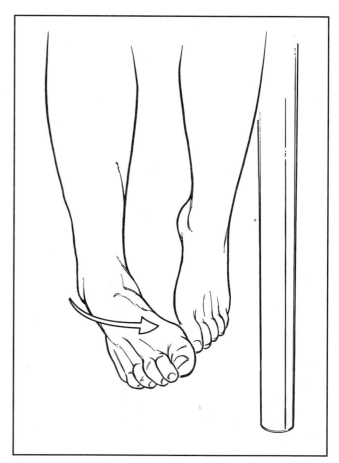

Figure 5–95

Grade 1 (Trace) and Grade 0 (Zero)

Position of Patient: Short sitting or supine.

Position of Therapist: Sitting on low stool or standing in front of patient. Palpate tendon of the Tibialis posterior between the medial malleolus and the navicular bone (Fig. 5–96). Alternatively, palpate tendon above the malleolus.

Test: Patient attempts to invert foot.

Instructions to Patient: "Try to turn your foot down and in."

Grading

Grade 1 (Trace): The tendon will stand out if there is contractile activity in the muscle. If palpable activity occurs in the absence of movement, the Grade is 1.

Grade 0 (Zero): No palpable contraction.

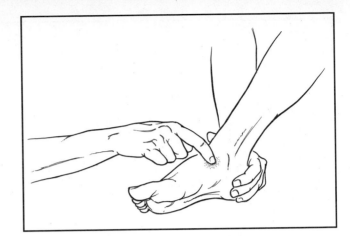

Figure 5–96

HELPFUL HINT

Flexors of the toes should remain relaxed to prevent substitution by the Flexor digitorum longus and Flexor hallucis longus.

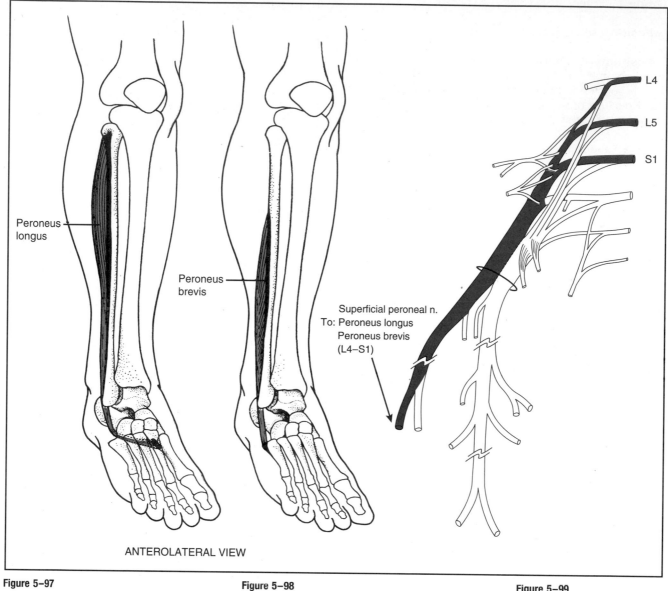

Peroneus longus

Peroneus brevis

Superficial peroneal n.
To: Peroneus longus
     Peroneus brevis
     (L4–S1)

L4

L5

S1

ANTEROLATERAL VIEW

Figure 5–97

Figure 5–98

Figure 5–99

Table 5–14 FOOT EVERSION		
Muscle	**Origin**	**Insertion**
208. Peroneus longus	Fibula (head and upper 2/3 of shaft) Tibia (lateral condyle)	1st metatarsal 1st cuneiform
209. Peroneus brevis	Fibula (distal of 2/3 shaft)	5th metatarsal
Others		
211. Extensor digitorum longus		
210. Peroneus tertius		

Range of Motion
0° to 25°

Grades 5 (Normal) to 0 (Zero)

Position of Patient: Short sitting with ankle in neutral position (midway between dorsiflexion and plantar flexion) (Fig. 5–100). Test also may be performed with patient supine.

Position of Therapist: Sitting on low stool in front of patient or standing at end of table if patient is supine.

One hand grips the ankle just above the malleoli for stabilization. Hand giving resistance is contoured around the dorsum and lateral border of the forefoot (Fig. 5–100). Resistance is directed toward inversion and slight dorsiflexion.

Test: Patient everts foot with depression of first metatarsal head and some plantar flexion.

Instructions to Patient: "Turn your foot down and out. Hold it! Don't let me move it in."

Grading

Grade 5 (Normal): Patient completes full range and holds end position against maximal resistance.

Grade 4 (Good): Patient completes available range of motion against strong to moderate resistance.

Grade 3 (Fair): Patient completes available range of eversion but tolerates no resistance (Fig. 5–101).

Grade 2 (Poor): The patient will be able to complete only a partial range of eversion motion.

Grade 1 (Trace) and Grade 0 (Zero)

Position of Patient: Short sitting or supine.

Position of Therapist: Sitting on low stool or standing at end of table.

To palpate the Peroneus longus, place fingers on the lateral leg over the upper one-third just below the head of the fibula. The tendon of the muscle can be felt posterior to the lateral malleolus but behind the tendon of the Peroneus brevis.

To palpate the tendon of the Peroneus brevis, place index finger over the tendon as it comes forward from behind the lateral malleolus, proximal to the base of the 5th metatarsal (Fig. 5–102). The belly of the Peroneus brevis can be palpated on the lateral surface of the distal leg over the fibula.

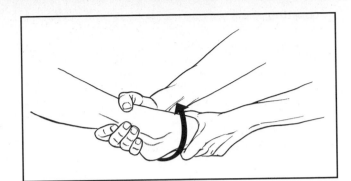

Figure 5–100

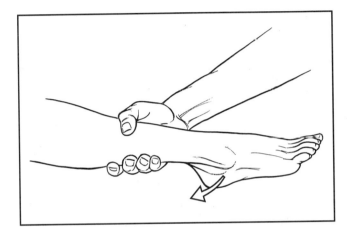

Figure 5–101

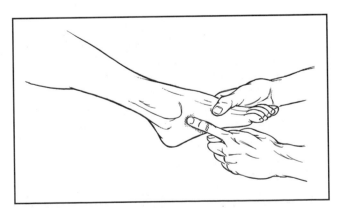

Figure 5–102

Grading

Grade 1 (Trace): Palpation will reveal contractile activity in either or both muscles, which may cause the tendon to stand out. No motion occurs.

Grade 0 (Zero): No palpable contractile activity.

Isolation of Peroneus Longus

Give resistance against the plantar surface of the head of the 1st metatarsal in a direction toward inversion and dorsiflexion.

Foot Eversion with Dorsiflexion

If the Peroneus tertius is present, it can be tested by asking the patient to evert and dorsiflex the foot. In this motion, however, the Extensor digitorum longus participates.

The tendon of the Peroneus tertius can be palpated on the lateral aspect of the dorsum of the foot, where it lies lateral to the tendon of the Extensor digitorum longus slip to the little toe.

HELPFUL HINTS

1. Foot eversion is accompanied by either dorsiflexion or plantar flexion. The toe extensors are the primary dorsiflexors accompanying eversion because the Peroneus tertius is an inconstant muscle.

2. The primary motion of eversion with plantar flexion is accomplished by the Peroneus brevis because the Peroneus longus is primarily a depressor of the 1st metatarsal head rather than an evertor.

3. The Peroneus brevis cannot be isolated if both peronei are innervated and active.

4. If there is a difference in strength between the Peroneus longus and the Peroneus brevis, the stronger of the two can be ascertained by the relative amount of resistance taken in eversion versus the resistance taken at the 1st metatarsal head. If greater resistance is taken at the 1st metatarsal head, the Peroneus longus is the stronger muscle.

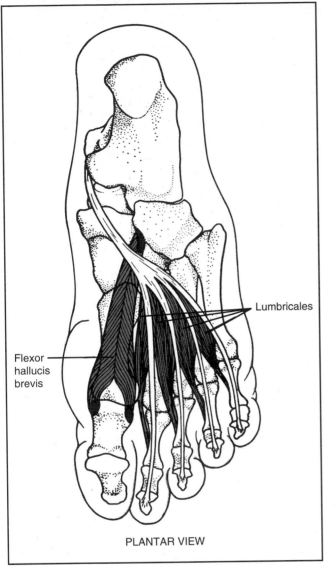

PLANTAR VIEW

Lumbricales

Flexor hallucis brevis

Figure 5–103

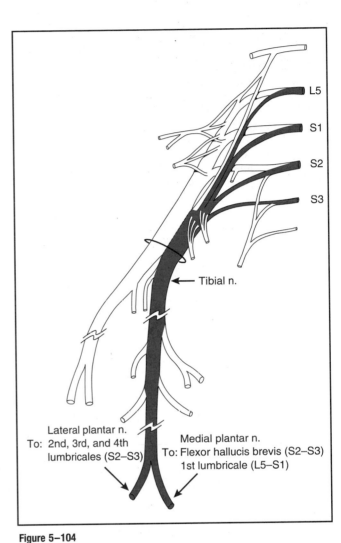

L5
S1
S2
S3

← Tibial n.

Lateral plantar n.
To: 2nd, 3rd, and 4th lumbricales (S2–S3)

Medial plantar n.
To: Flexor hallucis brevis (S2–S3)
1st lumbricale (L5–S1)

Figure 5–104

Table 5–15	FLEXION OF MP JOINTS OF TOES AND HALLUX	
Muscle	**Origin**	**Insertion**
Toes		
218. Lumbricales	Tendons of Flexor digitorum longus	Toes 2–5 (via tendons of Extensor digitorum longus)
Hallux		
223. Flexor hallucis brevis	Cuboid (plantar surface) Cuneiform (lateral)	Hallux (proximal phalanx)
Others		
219/220. Interossei (dorsal and plantar)		
216. Flexor digiti minimi brevis		
213. Flexor digitorum longus		
214. Flexor digitorum brevis		

Range of Motion
Great toe, 0° to 45°
Lateral 4 toes, 0° to 40°

HALLUX MP FLEXION
(Flexor hallucis brevis)

All Grades: 5 (Normal) to 0 (Zero)

Position of Patient: Short sitting (alternate position: supine) with legs hanging over edge of table. Ankle is in neutral position (midway between dorsiflexion and plantar flexion).

Position of Therapist: Sitting on low stool in front of patient. Alternate position: standing at side of table near patient's foot.

Test foot rests on examiner's lap. One hand is contoured over the dorsum of the foot just below the ankle for stabilization (Fig. 5–105). The index finger of the other hand is placed beneath the proximal phalanx of the great toe. Alternatively, the tip of the finger (with very short fingernails) is placed up under the proximal phalanx.

Test: Patient flexes great toe.

Instructions to Patient: "Bend your big toe over my finger. Hold it. Don't let me straighten it."

Grading

Grade 5 (Normal): Patient completes available range and tolerates strong resistance.

Grade 4 (Good): Patient completes available range and tolerates moderate to mild resistance.

Grade 3 (Fair): Patient completes available range of metatarsophalangeal (MP) flexion of the great toe but is unable to hold against any resistance.

Grade 2 (Poor): Patient completes only partial range of motion.

Grade 1 (Trace): Therapist may note contractile activity but no toe motion.

Grade 0 (Zero): No contractile activity.

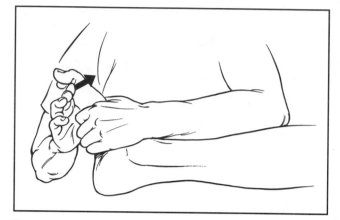

Figure 5–105

HELPFUL HINTS

1. The muscle and tendon of the Flexor hallucis brevis cannot be palpated.

2. When the Flexor hallucis longus is not functional, the Flexor hallucis brevis will flex the MP joint but with no flexion of the IP joint. In the opposite condition, when the Flexor hallucis brevis is not functional, the IP joint flexes and the MP joint may hyperextend. (When this condition is chronic, the posture is called hammer toe.)

TOE MP FLEXION
(Lumbricales)

All Grades: 5 (Normal) to 0 (Zero)

Position of Patient: Short sitting with foot on examiner's lap. Alternate position: supine. Ankle is in neutral (midway between dorsiflexion and plantar flexion).

Position of Therapist: Sitting on low stool in front of patient. Alternate position: standing next to table beside test foot.

One hand grasps the dorsum of the foot just below the ankle to provide stabilization (as in test for flexion of the hallux) (Fig. 5–106). The index finger of the other hand is placed under the MP joints of the four lateral toes to provide resistance to flexion.

Test: Patient flexes lateral four toes at the MP joints, keeping the IP joints neutral.

Instructions to Patient: "Bend your toes over my finger."

Grading: Grading is the same as that used for the great toe.

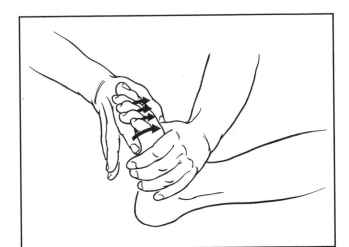

Figure 5–106

HELPFUL HINTS

In actual practice, the great toe and the lateral toes are rarely tested independently. Many patients cannot separate hallux motion from motion of the lateral toes, nor can they separate MP and IP motions.

Purists will ask the examiner to test each toe separately because the Lumbricales are notoriously uneven in strength. This may not, however, be practicable.

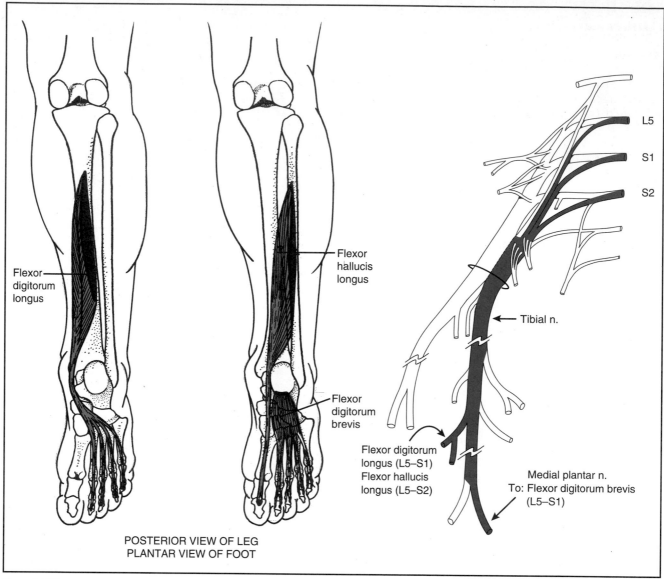

Flexor digitorum longus

Flexor hallucis longus

Flexor digitorum brevis

Flexor digitorum longus (L5–S1)
Flexor hallucis longus (L5–S2)

Tibial n.

L5

S1

S2

Medial plantar n.
To: Flexor digitorum brevis
(L5–S1)

POSTERIOR VIEW OF LEG
PLANTAR VIEW OF FOOT

Figure 5–107   Figure 5–108   Figure 5–109

Table 5–16 FLEXION OF IP JOINTS OF HALLUX AND TOES		
Muscle	**Origin**	**Insertion**
DIP, Toes		
213. Flexor digitorum longus	Tibia (posterior, middle 2/3 of shaft)	Distal phalanges (base of lateral four toes)
PIP, Toes		
214. Flexor digitorum brevis	Calcaneus (tuberosity)	Toes 2–5 (middle phalanges)
IP, Hallux		
222. Flexor hallucis longus	Fibula (inferior 2/3 of shaft)	Great toe (base of distal phalanx)

Range of Motion
PIP flexion, four lateral toes: 0° to 35°
DIP flexion, four lateral toes: 0° to 60°
IP flexion of hallux: 0° to 90°

All Grades: 5 (Normal) to 0 (Zero)

Position of Patient: Short sitting with foot on examiner's lap, or supine.

Position of Therapist: Sitting on short stool in front of patient or standing at side of table near patient's foot.

One hand grasps the anterior foot with the fingers placed across the dorsum of the foot and the thumb under the proximal phalanges (PIP) or distal phalanges (DIP) or under the IP of the hallux for stabilization (Figs. 5–110 to 5–112).

The other hand applies resistance using the examiner's four fingers or thumbs under the middle phalanges (for IP test) (Fig. 5–110); under the distal phalanges for the distal phalanges test (Fig. 5–111); and with the index finger under the distal phalanx of the hallux (Fig. 5–112).

Test: Patient flexes the toes or hallux.

Instructions to Patient: "Curl your toes; hold it. Curl your big toe and hold it."

Grading

Grades 5 (Normal) and 4 (Good): Patient completes range of motion of toes and then hallux; resistance in both tests may be minimal.

Grades 3 (Fair) and 2 (Poor): Patient completes range of motion with no resistance (Grade 3) or completes only a partial range (Grade 2).

Grades 1 (Trace) and 0 (Zero): Minimal to no palpable contractile activity occurs. Tendon of the Flexor hallucis longus may be palpated on the plantar surface of the proximal phalanx of the great toe.

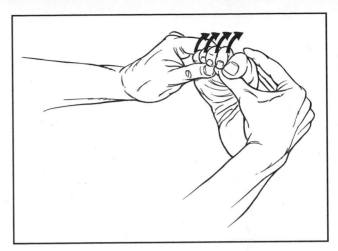
Figure 5–110

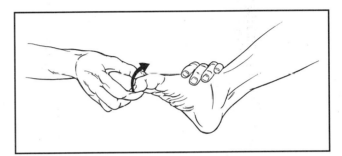
Figure 5–111

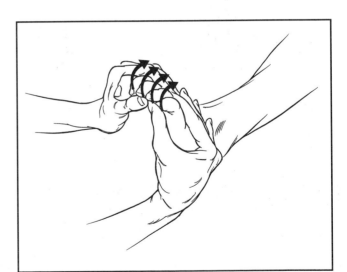
Figure 5–112

◑ HELPFUL HINTS

1. As with all toe motions, the patient may not be able to move one toe separately from another, nor to separate MP from IP activity among individual toes.

2. Some persons can separate hallux activity from toe motions, but fewer can separate MP from IP hallux activity.

3. Many people can "pinch" with their great toe (Adductor hallucis), but this is not a common clinical test.

4. The Abductor hallucis is not commonly tested because it is only rarely isolated. Its activity can be observed by resisting adduction of the forefoot, which will bring the great toe into abduction, but the lateral toes commonly extend at the same time.

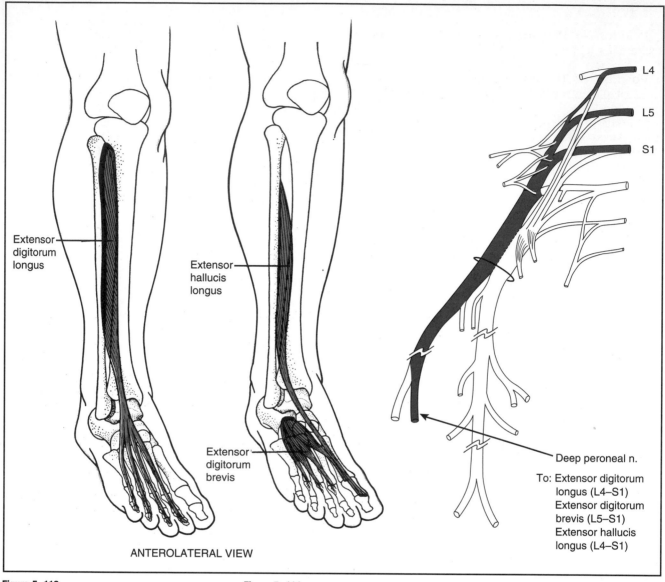

Extensor
digitorum
longus

Extensor
hallucis
longus

Extensor
digitorum
brevis

ANTEROLATERAL VIEW

L4

L5

S1

Deep peroneal n.

To: Extensor digitorum
longus (L4–S1)
Extensor digitorum
brevis (L5–S1)
Extensor hallucis
longus (L4–S1)

Figure 5–113   **Figure 5–114**   **Figure 5–115**

Table 5–17	EXTENSION OF MP JOINTS OF TOES AND IP JOINT OF HALLUX	
Muscle	**Origin**	**Insertion**
211. Extensor digitorum longus	Tibia (lateral condyle) Fibula (shaft, proximal 3/4 on anterior)	Toes 2–5 (to each proximal and each distal phalanx)
212. Extensor digitorum brevis	Calcaneus	Ends in four tendons: 1. Hallux (proximal phalanx) 2–4. Join Extensor digitorum longus of toes 2–4
221. Extensor hallucis longus	Fibula (shaft, 2/4 anterior surface)	Hallux (base of distal phalanx)

Range of Motion

0° to 75–80°

All Grades: 5 (Normal) to 0 (Zero)

Position of Patient: Short sitting with foot on examiner's lap. Alternate position: supine. Ankle in neutral (midway between plantar and dorsiflexion).

Position of Therapist: Sitting on low stool in front of patient, or standing beside table near the patient's foot.

Lateral Toes: One hand stabilizes the metatarsals with the fingers on the plantar surface and the thumb on the dorsum of the foot (Fig. 5–116). The other hand is used to give resistance with the thumb placed over the dorsal surface of the proximal phalanges of the toes.

Hallux: Stabilize the metatarsal area by contouring hand around the plantar surface of the foot with the thumb curving around to the base of the hallux (Fig. 5–117). The other hand stabilizes the foot at the heel. For resistance, place thumb over the MP joint (Fig. 5–117) or over the IP joint (Fig. 5–118).

Test: Patient extends lateral four toes or extends hallux.

Instructions to Patient: "Straighten your big toe. Hold it." "Straighten your toes and hold it."

Grading

Grades 5 (Normal) and 4 (Good): Patient can extend the toes fully against variable resistance (which may be small).

Grades 3 (Fair) and 2 (Poor): Patient can complete range of motion with no resistance (Grade 3) or can complete a partial range of motion (Grade 2).

Grades 1 (Trace) and 0 (Zero): Tendons of the Extensor digitorum longus can be palpated or observed over dorsum of metatarsals. Tendon of the Extensor digitorum brevis often can be palpated on the lateral side of the dorsum of the foot just in front of the malleolus.
Palpable contractile activity is a Grade 1; no contractile activity is a Grade 0.

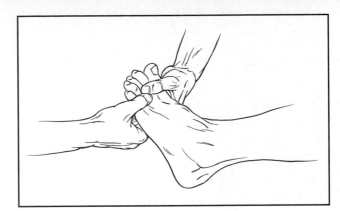

Figure 5–116

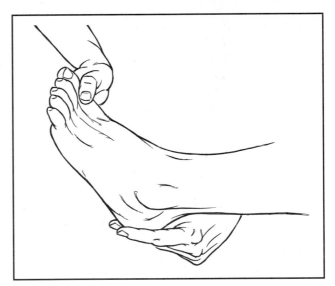

Figure 5–117

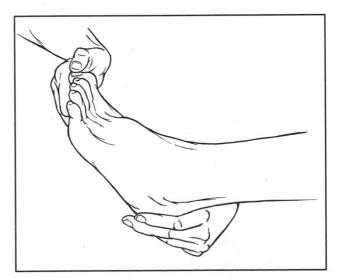

Figure 5–118

C **HELPFUL HINTS**

1. Many (if not most) patients cannot separate great toe extension from extension of the four lateral toes. Nor can most separate MP from IP activity.

2. The test is used not so much to ascertain strength as to determine whether the toe muscles are active.

REFERENCES

1. Mulroy S. Functions of the triceps surae during strength testing and gait. Dissertation (Ph.D.), Department of Biokinesiology and Physical Therapy, University of Southern California, Los Angeles, 1994.
2. Perry J, Easterday CS, Antonelli DJ. Surface versus intramuscular electrodes for electromyography for superficial and deep muscles. Phys Ther 61:6–15, 1981.

Testing in Infants and Children

Barbara Connolly, Ed.D., P.T.

Neck Extension
Neck Flexion
Trunk Flexion
Trunk Rotation
Trunk Extension
Hip and Knee Flexion
Hip and Knee Extension
Hip Abduction
Hip Adduction
Ankle Plantarflexion
Ankle Dorsiflexion
Scapular Abduction
Scapular Adduction and Shoulder Abduction
Shoulder Flexion
Elbow Extension
Elbow Flexion
Elbow Supination

Chapter 6

Manual muscle testing in children below the age of 5 years presents a challenge to physical therapists. Either because of their inability to understand instructions or their unwillingness to cooperate with the therapist, a true manual muscle test may not be possible.[1] Some children between the ages of 2 and 5 years may be able to participate in a true manual muscle test if they are cooperative and interested in the activity. However, they may not be able to understand the counterforce that they must exert against the resistance administered by the examiner once they have assumed the test position.[2]

A knowledge of normal movement during these early years (birth to 5 years) is essential for the examiner who wishes to assess muscle strength in young children. The examiner should be familiar with the early motor milestones achieved by children without disabilities and with children's performance in functional activities. The examiner should observe the range, symmetry, rhythm, and smoothness of movement during the child's play as well as during specific muscle test activities.

Grading

Isolated muscles are not graded during functional assessment that involves spontaneous or elicited movements or during play activities. Groups of muscles that perform the desired movement are graded instead. As in muscle assessment of adults, the grading scales of Normal, Good, Fair, Poor, Trace, or Zero can be used with only minor changes in the operational definitions of the terms.

The amount of resistance to be applied to obtain a Normal or Good grade can be assessed either by observing the child (if the child can hold briefly against resistance applied by the examiner) or by assessing the child's ability to move against the resistance of a small weight placed on the child's extremity or body part during the movement. For example, a small wrist weight may be placed around the child's wrist while the child reaches for an object held overhead. This method is also subjective to some degree but allows assessment of improvements made in muscle strength in specific movements during a short period of time.

A grade of Fair is defined as the ability of the muscle or muscle group to move a part through a complete range of motion against gravity (a vertical movement). Poor is defined as either a complete horizontal movement or a partial movement in the vertical plane.

Evidence of the presence or absence of a contraction in a muscle or muscle group yields a grade of Trace (slight contraction without joint movement) or Zero (no detectable contraction).

An alternative method of describing muscle strength is to characterize weakness in a muscle group as minimal, mild, moderate, or severe.[3] However, these are subjective terms of description, and the same examiner who performs the initial assessment of the child should perform all subsequent assessments for interrater reliability.

Neck Extension

Test Position:	Child is suspended by the therapist's hands placed under the chest.[4]	
Response:	2 months	Child raises head to midline and holds it for 2 to 3 seconds (Fig. 6–1).
	3 months	Child lifts head beyond plane of body.

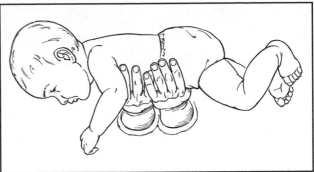

Figure 6–1

Test Position:	Child is placed on stomach, and the therapist shakes a rattle above the child's head.[4]	
Response:	2 months	Child actively extends head to 45 degrees.
	4 months	Child actively extends head to 90 degrees and maintains position (Fig. 6–2).

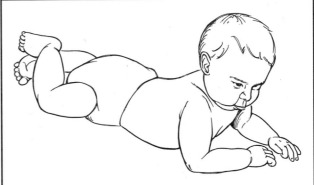

Figure 6–2

Neck Flexion

Test Position: Child is in the supine position. Therapist grasps the wrists and hands and pulls child to a sitting position.[4]

Response:

4 months — Child keeps the chin tucked and maintains the head in line with trunk. (Fig. 6–3).

6 months — Child holds head 15 degrees in front of midline (Fig. 6–4).

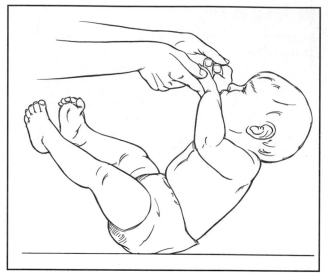

Figure 6–3

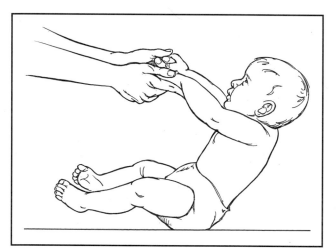

Figure 6–4

Test Position: Child is in the supine position. Therapist's hands are held out to the child ready to pull him to a sitting position.[4]

Response:

6 months — Child lifts head independently without stimulus of hands (Fig. 6–5).

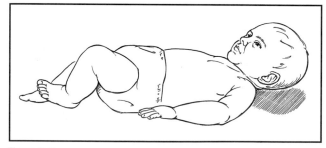

Figure 6–5

Test Position: Child is held vertically under the arms by the therapist and tilted backward 45 degrees or more.[4]

Response: 6 months   Child lifts head and holds it steadily (Fig. 6–6).

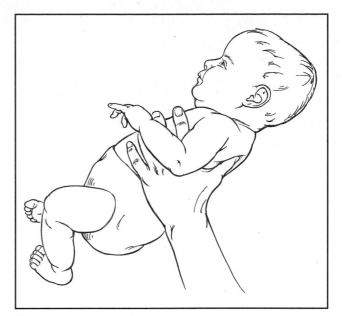

Figure 6–6

Test Position: Child is held vertically under the arms by the therapist and tilted sideways 45 degrees or more.[4]

Response: 2 to 3 months   Child holds head in midline of trunk with lag of no more than 10 degrees (Fig. 6–7).

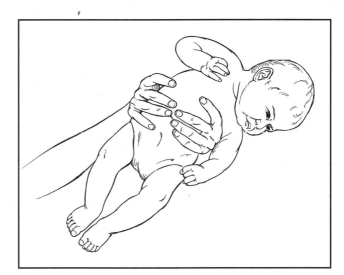

Figure 6–7

Trunk Flexion

Test Position: Child is in the supine position. Therapist grasps the child's wrist or hands and pulls him to a sitting position.[4]

Response: 4 months   Abdominals stabilize the rib cage and hips, and knees flex to assist movement actively.

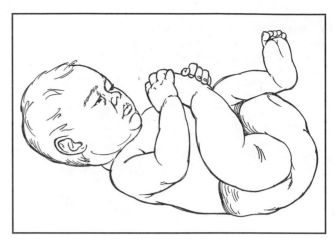

Figure 6–8

Test Position: Child is in the supine position. A toy is placed around the foot, or pegs are placed between toys.[4]

Response:

4 to 5 months   Child lifts legs and brings foot to mouth (Fig. 6–8).

6 months   Child lifts legs straight upright. Demonstrates midrange control in space (Fig. 6–9).

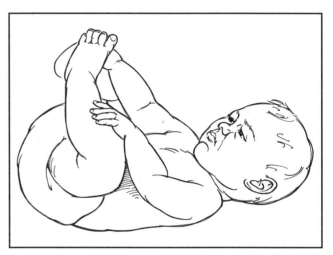

Figure 6–9

Test Position: Child is placed on hands and knees. Therapist observes him or her for the presence of lordosis. Posture without lordosis requires balance between the flexors and extensors—the trunk extensors and abdominal control. Lordosis in this position indicates that the abdominals are not strong enough to lift the pelvis.[4]

Response: 7 months   Lumbar lordosis should not be present, and the back should be straight (Fig. 6–10).

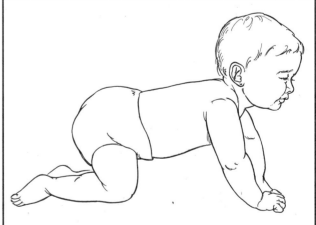

Figure 6–10

Test Position: Child is in the sitting position.[4]

Response: 7 months — Good interplay of flexors and extensors should be evident when the child is sitting, and the pelvic position should be neutral. If the pelvis is tilted anteriorly, poor abdominal control should be suspected (Fig. 6–11).

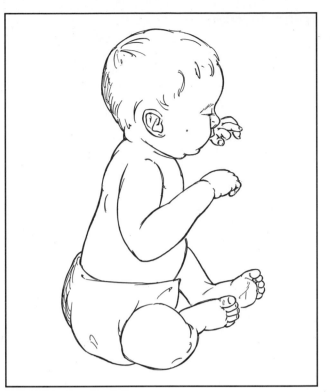

Figure 6–11

Test Position: Child lies supine on the mat with the knees bent to a 90-degree angle and the hands clasped behind the head. The child is asked to sit up and touch the elbows to the knees.[3]

Response: 4 to 4.5 years — Child should be able to perform three or four sit-ups within 30 seconds.

5 to 5.5 years — Child should be able to perform six to eight sit-ups within 30 seconds (Fig. 6–12).

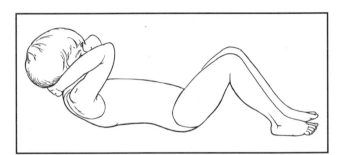

Figure 6–12

Test Position: Child lies supine on the mat. She is asked to assume and hold a "curled" position with the head and knees flexed to the chest.[5]

Response: 8 years — Child should be able to hold this position for 20 to 30 seconds (Fig. 6–13).

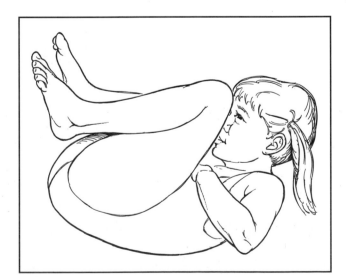

Figure 6–13

Trunk Rotation

Test Position: Child is in the supine position. Therapist holds a toy at the child's side and waves it to attract child's attention.[4]

Response:

5 months	Child rolls to side using lateral head righting, trunk rotation, and lower extremity dissociation (Fig. 6–14).
8 to 9 months	Child rolls to the prone position using counterrotation, the shoulder rotating in one direction and the hip in the other.

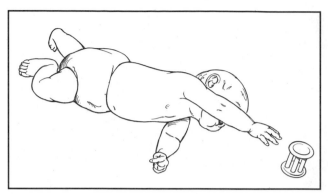

Figure 6–14

Test Position: Child is in the prone position. Therapist holds a toy at the child's side and waves it to attract child's attention.[4]

Response:

7 to 8 months	Child pivots 90 degrees to retrieve the toy placed at both the right and left sides (Fig. 6–15).
8 months	Child pushes his back over his knees and into a sitting position using a rotational pattern (Fig. 6–16).

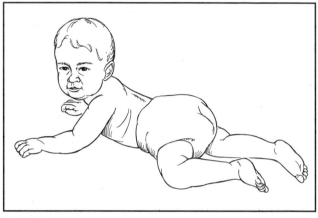

Figure 6–15

Test Position: Child is in a sitting position. Therapist holds a toy at the child's side and waves it to attract child's attention.[4]

Response:

7 months	Child rotates trunk while maintaining hips in contact with surface.
10 to 11 months	Child pivots 180 degrees in a circular pattern on the buttocks.

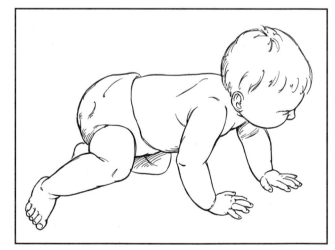

Figure 6–16

Trunk Extension

Test Position: Child is placed on the stomach or is suspended ventrally. Limb actions are observed.[4]

Response:

4 to 5 months	Child extends the arms and legs off the surface (Fig. 6–17).
5 months	Child rocks on his stomach.

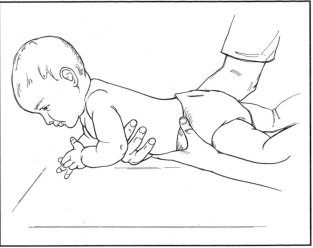

Figure 6–17

Test Position: Child is in a sitting position.[4]

Response:

8 months	Child exhibits good trunk extension with slight lumbar lordosis (see Fig. 6–11).
10 to 11 months	Child leans forward to get toy without losing balance or touching floor to recover.

Test Position: Child is placed on his hands and knees. Therapist observes the child for the presence of lordosis. Posture without lordosis requires balance between flexion and extension—the trunk extensors and abdominal control.[4]

Response:

7 months	Lumbar lordosis should not be present, and the back should be straight (see Fig. 6–10).

Test Position: Child is standing. She is asked to touch the toes and then return to a standing position.[3]

Response:

3 to 4 years	Performance of this activity without using the hands for support demonstrates good strength in the trunk extensors and gluteus maximus (Fig. 6–18).

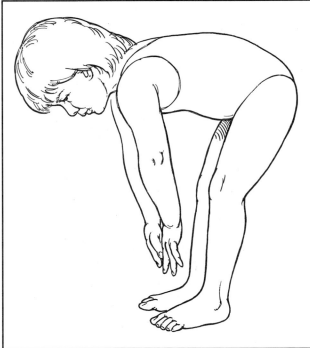

Figure 6–18

Test Position: The child places his or her legs on either side of the therapist's body. Therapist then grasps the child on each side of the thorax so that the trunk is in an arched position. The therapist then stands up and asks the child to "fly."[3]

Response: 5 years Child can maintain a "flying" position for 16 seconds or longer when held at the hip level (Fig. 6–19).

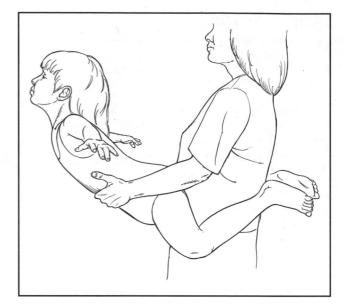

Figure 6–19

Test Position: Child lies on his or her stomach on the floor and is asked to assume an "airplane" position with the head, arms, and legs lifted from the floor.[3,5]

Response: 8 years Child is able to assume and hold the "airplane" position for 20 to 30 seconds[5] (Fig. 6–20).

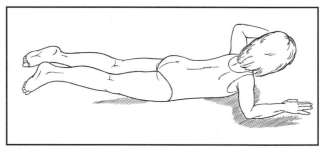

Figure 6–20

Hip and Knee Flexion

Test Position:	Child is in the supine position. Coverings, shoes, and socks are removed.[4]	
Response:	4 to 5 months	Child flexes at hips bilaterally, knees turned outward and feet brought to mouth for play.
	2 to 3 years	On command, child pumps the legs as in riding a bicycle. Flexion of knee to chest demonstrates hip and knee flexor strength (Fig. 6–21).

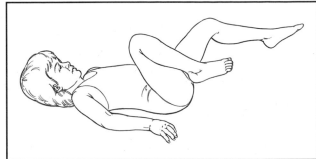

Figure 6–21

Test Position:	Child is in a prone position with a toy placed in front of him or her.[4]	
Response:	7 months	Child moves forward on stomach toward the toy using both arms and legs in contact with the surface.
	9 to 10 months	Child moves forward on hands and knees using separate motions for each leg.

Test Position:	Child is in a sitting position, with a toy placed in front of him or her.[4]	
Response:	7 months	Child lifts legs 1 to 2 inches while sitting.
	8 to 9 months	Child moves toward the toy while maintaining a sitting posture. He or she scoots along the surface using the hands and feet to propel body.

Test Position: Child is in a standing position at a set of steps with a toy placed at the top of the steps. The hip flexors and hamstrings are used to raise the leg up onto the step.[4]

Response:

15 to 17 months	Child climbs four steps holding onto a railing or wall and placing both feet on each step.
18 to 23 months	Child climbs four steps without support, placing both feet on each step.
24 to 29 months	Child climbs four steps holding onto a railing or wall and alternating feet.
36 to 41 months	Child climbs four steps without support and alternating feet.

Test Position: Child lies supine on the mat. He or she is asked to assume and hold a "curled" position with the head and knees flexed to the chest.[5]

Response:

8 years	Child should be able to hold this position for 20 to 30 seconds (see Fig. 6–13).

Hip and Knee Extension

Test Position:	Child lies in a prone position, and limb actions are observed.[4]	
Response:	4 to 5 months	Child extends arms and legs off mattress.
	5 months	Child rocks on stomach.
	6 months	Active kicking of the legs into extension occurs when the child is stimulated.

Test Position:	Child is in a prone position with the chest and pelvis supported by a small bench or table.[3]	
Response:	2 to 5 years	Child is asked to kick one leg up toward the ceiling. The knee remains bent so that the gluteus maximus works without the assistance of the hamstrings (Fig. 6–22).

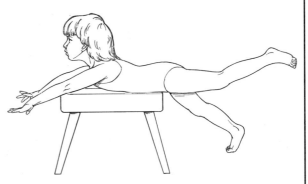

Figure 6–22

Test Position:	Child lies on his or her stomach on the floor and is asked to assume an "airplane" position with the head, arms, and legs lifted from the floor.[3,5]	
Response:	8 years	Child is able to assume and hold the "airplane" position for 20 to 30 seconds[5] (see Fig. 6–20).

Test Position:	Child lies in the supine position, and limb actions are observed.[4]	
Response:	6 months	Child performs half "bridges" by pushing with one leg and extending the other.[4]
	2 to 5 years[3]	Child is asked to make a "bridge" by raising his hips from the floor. This movement demonstrates good strength in the gluteus maximus (Fig. 6–23).

Figure 6–23

Test Position: Child lies in a supine position on a firm surface and is asked to move the legs as if riding a bicycle.[3]

Response: 2 to 5 years   Child attempts to "bicycle" the legs on command. Extensor strength of the hips and knees is demonstrated as the legs are lowered from the chest (see Fig. 6–21).

Test Position: Child is placed in a kneeling position.[4]

Response: 12 to 14 months   Child can maintain a kneeling position with hips aligned under shoulders for 5 seconds (Fig. 6–24).

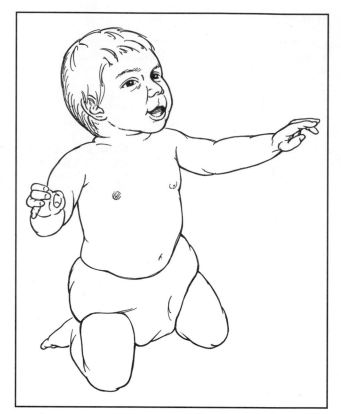

Figure 6–24

Test Position: Child stands without support. A tennis ball or toy is placed on the floor within 1 foot of the child. He is encouraged to pick up the object.[3]

Response: 18 to 23 months   Child squats, recovers ball, and returns to the standing position without falling (Fig. 6–25).

Figure 6–25

Hip Abduction

Test Position:	Child is placed on a small tilt board.[6]	
Response:	7 to 8 months	In the sitting position, as the child is tilted to the side, the opposite arm and leg should abduct (Fig. 6–26).
	9 to 12 months	In the quadruped position as the child is tilted to the side, the opposite arm and leg should abduct (Fig. 6–27).

Test Position:	Child is placed on a sofa or small table and encouraged to walk to a toy placed at the opposite end.[4]	
Response:	9 to 10 months	Child abducts both legs while moving sideways.

Figure 6–26

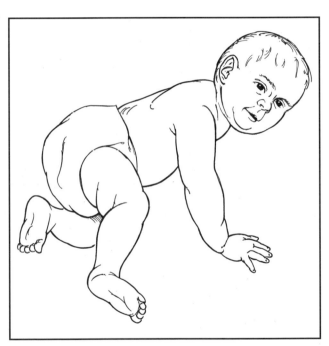

Figure 6–27

Test Position: While standing, the child is asked to lift the right leg while support is given with one hand held by the therapist. After the child performs the activity with the right leg, the left leg should be tested.[3]

Response: 2 to 5 years   As the right leg is lifted, the left hip should remain level. If the hip drops, the right hip abductors are weak (Fig. 6–28).

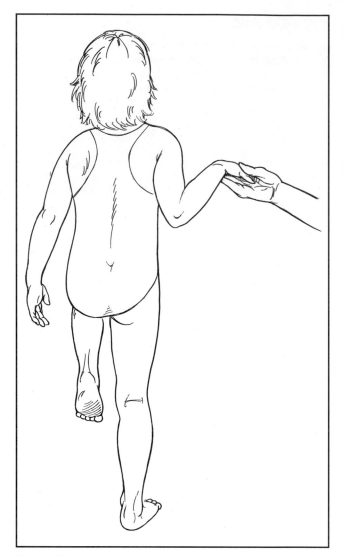

Figure 6–28

Hip Adduction

Test Position: Child is placed on a small tilt board.[6]

Response: 7 to 8 months   In the sitting position, as the child is tilted to the side, the arm and leg on the downhill (same) side should adduct (see Fig. 6–26).

9 to 12 months   In the quadruped position, as the child is tilted to the side, the arm and leg on the downhill (same) side should adduct (see Fig. 6–27).

Test Position: Child is in the sitting position.[4]

Response: 11 to 12 months   Child is able to sit with the legs straight out in front of him rather than in an abducted position.

Ankle Plantarflexion

Test Position: Therapist demonstrates walking on tip-toe, hands on hips.[3]

Response: 24 to 29 months   On verbal request, child imitates examiner and walks for five steps (Fig. 6–29).

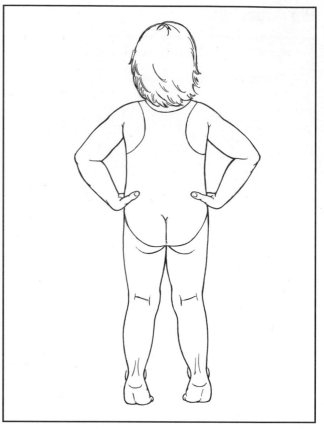

Figure 6–29

Ankle Dorsiflexion

Test Position: Therapist demonstrates walking on heels, hands on hips.[3] .

Response: 3 years   On verbal request, child imitates examiner and walks for five steps (Fig. 6–30).

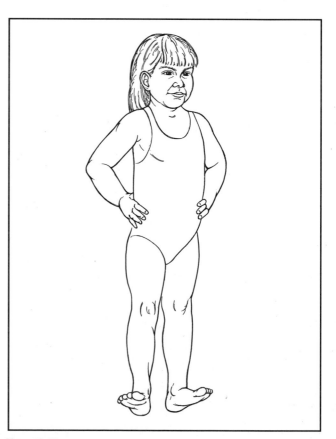

Figure 6–30

Scapular Abduction

Test Position: Child is in the supine position.[4]

Response:	3 months	Child brings both hands to midline to play with toy, clothing, or fingers (Fig. 6–31).
	4 months	Child reaches out with both hands for toy held away from chest (Fig. 6–32).
	5 months	Child reaches out with hands to play with feet and hold onto toes.

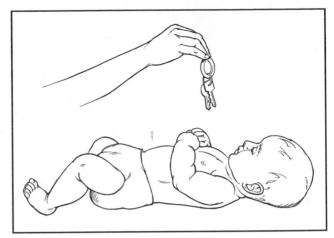

Figure 6–31

Test Position: Child sits in therapist's lap and is given a cube or toy to play with.[4]

| **Response:** | 4 to 5 months | Child grasps cube or toy and brings hands together to play with it. |

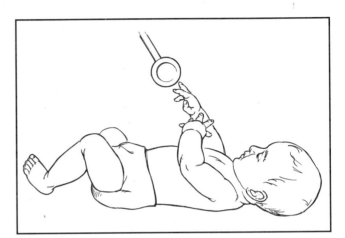

Figure 6–32

Test Position: Child is in the prone position on a supporting surface.

| **Response:** | 4 months | Child brings arms forward to prop them on the forearms in front of shoulder girdle; arms are close to the midline (Fig. 6–33). |

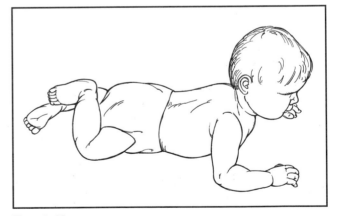

Figure 6–33

Test Position: Child is placed in a wheelbarrow position with the lower extremities supported by the therapist.[3,4,7]

Response: 6 to 7 months — Child should be able to maintain his or her weight on the extended arms without "winging" of the scapula when the therapist holds the child at the pelvis. A winged scapula is seen with weakness of the Serratus anterior.[4]

5 years — Child should be able to perform a "wheelbarrow" walk for a distance of 9 to 10 feet when held at the ankles by the therapist[3] (Fig. 6–34).

Figure 6–34

Test Position: Child is in a standing position facing a wall.[3]

Response: 2 to 5 years — Child is asked to push arms forward with the palms pressed against the wall. Winging of the scapula should not be seen, and the scapula should remain flat against the thoracic wall (Fig. 6–35).

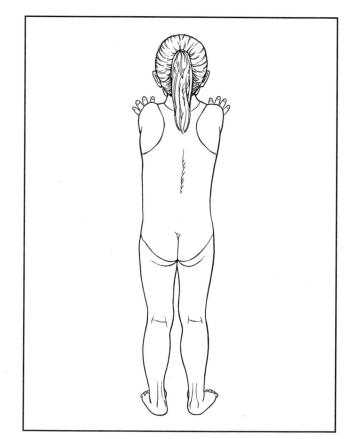

Figure 6–35

Scapular Adduction and Shoulder Abduction

Test Position:	Child is in the prone position on a supporting surface. Toy is held above the child's head by the therapist.[4]

Response:	6 months	Child extends head, trunk, and lower extremities, using bilateral scapular adduction to reinforce extension.

Test Position:	Child is in a sitting position, and toy is held out to the side of the child.[4]

Response:	9 to 10 months	Child lifts his or her arm to the side to reach for the toy.
	Over 10 months	Small weights can be attached to the wrists of the child during the reaching activity to determine ability to move against resistance, OR child may be asked to give the therapist a weighted object (e.g., bean bag, weight, heavy toy).

Test Position:	The child places his or her legs on either side of the therapist's body. The therapist then grasps the child on each side of the thorax so that the trunk is in an arched position. The therapist then stands up and asks the child to "fly."[7]

Response:	5 years	Child can maintain a "flying" position for 16 seconds or longer when held at the hip level. Arms are abducted at shoulders with elbows extended (see Fig. 6–19).

Test Position:	Child lies prone on the floor and is asked to assume an "airplane" position with the head, arms, and legs lifted from the floor.[3,5]

Response:	8 years	Child can assume and hold this "airplane" position for 20 to 30 seconds[5] (see Fig. 6–20).

Shoulder Flexion

Test Position: Child is in the supine position on a flat surface. Therapist holds child's hands and encourages him to pull himself to a sitting position.[4]

Response: 6 months Child pulls himself to a sitting position with the upper extremities using active flexion of the shoulders, arms, and abdominals.

Test Position: Child is in a sitting position, and toy is held above child's head.[4]

Response: 11 months Child should be able to complete the last 20 degrees of overhead reach in a sitting position.

Over 11 months Small weights can be attached to the child's wrists during the reaching activity to determine ability to move against resistance, OR the child may be asked to give the therapist a weighted object (e.g., bean bag, weight, or heavy toy).

Elbow Extension

Test Position: Child lies in a prone position on a supporting surface. A rattle is shaken above the child's head.[4]

Response: 6 months   Child elevates head and stomach by pushing up on the extended arms (Fig. 6–36).

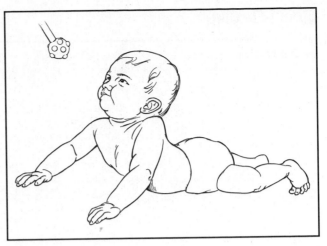

Figure 6–36

Test Position: Child lies in a prone position on a supporting surface. Movement of arms is observed by therapist.[4]

Response: 6 to 7 months   Child pushes himself backward on the stomach by extending the elbows and abducting the scapula.

Test Position: Child is in a sitting position, and toy is offered to the child.[4]

Response: 7 months   Child reaches for toy with extended elbow.

Over 7 months   Small weights can be attached to the child's wrists during reaching activity to determine ability to move against resistance, OR the child can be asked to give the therapist a weighted object (e.g., bean bag, weight, or heavy toy).

Test Position: Child is standing facing a chair with hands on front corners of the chair and hips and legs extended.[3]

Response: 6 to 7 years   Child performs seven to eight push-ups in 20 seconds by extending the arms to raise the chest.

Elbow Flexion

Test Position: Child is in the supine position, and therapist gives child index fingers to grasp.[4]

Response: 4 to 5 months   Child pulls self to a sitting position with the assistance of arms.

Test Position: Child is in the prone position with a toy placed in front of him or her.[4]

Response: 6 to 7 months   Child pulls self forward a distance of 3 feet using arms.

Test Position: Child is in a sitting position on a scooter board inside a hula hoop on a tile or linoleum floor.[7]

Response: 3 to 5 years   Child can push and pull self forward and backward inside the hoop five or more times (Fig. 6–37).

Figure 6–37

Elbow Supination

Test Position: Child is in a sitting position and is offered a toy.[4]

Response: 11 months — Child reaches for the toy with supinated forearm and extended elbow.

Test Position: Child is standing at a door and is told to open the door.[4]

Response: 24 to 29 months — Child opens door using forearm rotation of knob.

Test Position: Child is given a bottle with the cap screwed on and a pellet inside. The child is told to "Get a pellet."[4]

Response: 36 to 41 months — Child removes cap by using forearm rotation.

Test Position: Child is given a wind-up toy with a key and told to "wind up the toy" after the therapist demonstrates the motion.[4]

Response: 36 to 41 months — Child can turn key at least 90 degrees.

REFERENCES

1. Alexander J, Molnar GE. Muscular strength in children: Preliminary report on objective standards. Arch Phys Med Rehabil 54:424, 1973.
2. Molnar GE, Alexander J. Development of quantitative standards for muscle strength in children. Arch Phys Med Rehabil 55:490, 1974.
3. Pact V, Sirothkin-Roses M, Beatuc J. *The Muscle Testing Handbook*. Boston: Little, Brown, 1984.
4. Folio MR, Fewell RR. Peabody Developmental Motor Scales and Activity Cards. Allen, TX: DLM Teaching Resources, 1983.
5. Fisher AG, Murray EA, Bundy AC. *Sensory Integration: Theory and Practice*. Philadelphia: F.A. Davis, 1991.
6. Effgen SK. Developing postural reactions. In Connolly BH, Montgomery PC (eds): *Therapeutic Exercise in Developmental Disabilities,* 2nd ed. Chattanooga, TN: Chattanooga Group, 1993.
7. Berk RA, DeGangi GA. *DeGangi-Berk Test of Sensory Integration*. Los Angeles: Western Psychological Services, 1983.

Assessment of Muscles Innervated by Cranial Nerves

Introduction To Testing and Grading
Extraocular Muscles of the Eye
Muscles of the Face
The Nose Muscles
Muscles of the Mouth
Muscles of Mastication
Muscles of the Tongue
Muscles of the Palate
Muscles of the Pharynx
Muscles of the Larynx
Swallowing

Note: This chapter describes the muscles innervated by motor branches of the cranial nerves and describes test methods of assessing the muscles of the eye, eyelid, face, jaw, tongue, soft palate, posterior pharyngeal wall, and larynx. The tests are appropriate for patients whose neurologic deficits are either central or peripheral. The only requirement for the patient to participate in the test is the ability to follow simple directions.

Chapter 7

Introduction to Testing and Grading

Muscles innervated by the cranial nerves are not amenable to classic methods of manual muscle testing and grading. In many, if not most, cases they do not move a bony lever, so manual resistance as a means of evaluation of their strength and function is not always the primary procedure.

The therapist needs to become familiar with the cranial nerve muscles in normal persons. Their appearance, strength, excursion, and rate of motion are all variables that are unlike the other skeletal muscles, which are more familiar. As for the infant and young child, the best way to assess the gross function of these muscles is to observe the child while crying. In any event, experience with assessment requires considerable practice with both normal persons and a wide variety of patients with suspected and known cranial nerve motor deficits emanating from both upper and lower motor neuron lesions.

An anecdote from the personal experience of one of the authors involves a patient who was being evaluated for bulbar function because of a motor neuron disease. A "strange" structure appeared in the back of the throat when the patient opened wide to say "Ah-h-h." As it turned out, there was no tumor, no foreign object, and no structural deformity. The strange structure was the epiglottis, which is observable in these situations in a sizable number of people.

The issue of symmetry is particularly important in testing the ocular, facial, tongue, jaw, pharyngeal, and palate muscles. The symmetry of these muscles, except for the laryngeal muscles, is visible to the examiner. Asymmetry is more readily detected just by observation in these than in the limb muscles and should always be documented.

In all tests in this section, the movements or instructions may not be entirely familiar to the patient, so each test should be demonstrated and the patient should be allowed to practice. In the presence of unusual or unexpected test results, the examiner should inquire about prior facial reconstructive surgery.

GENERAL GRADING PROCEDURES

The distinction to be made in testing the muscles described in this chapter is to ascertain their relative functional level with respect to their intended activity. The scoring system, therefore, is a functional one, and motions or function are graded as follows:

F: Functional; normal or only slight impairment

WF: Weak functional; moderate impairment that affects the degree of active motion

NF: Nonfunctional; severe impairment

O: Absent

UNIVERSAL PRECAUTIONS IN BULBAR TESTING

In testing the muscles of the head, oral cavity, and throat the examiner frequently encounters body fluids such as saliva, tears, and broncho-tracheal-pharyngeal secretions. The precaution of wearing gloves should always be followed. If the patient has any infectious disease or if there are copious secretions, the examiner should be masked and gowned as well as gloved.

The examiner should be cautious about standing directly in front of a patient who has been instructed to cough. This also is true for the patient who has an open tracheostomy.

When a tongue blade is used, it should be sterile, and care should be used about where it is placed between tests on a given patient.

PATIENT AND EXAMINER POSITIONS FOR ALL TESTS

The short sitting position is preferred. The head and trunk should be supported as necessary to maintain normal alignment or to accommodate deformities. If the patient cannot sit for any reason, use the supine position, which will not influence testing of the head and eye muscles. When the muscles of the oral cavity and throat are tested, however, the head should be elevated. The examiner stands or sits in front of the patient but slightly to one side. A stool on casters is preferred so the therapist can move about the patient quickly and efficiently.

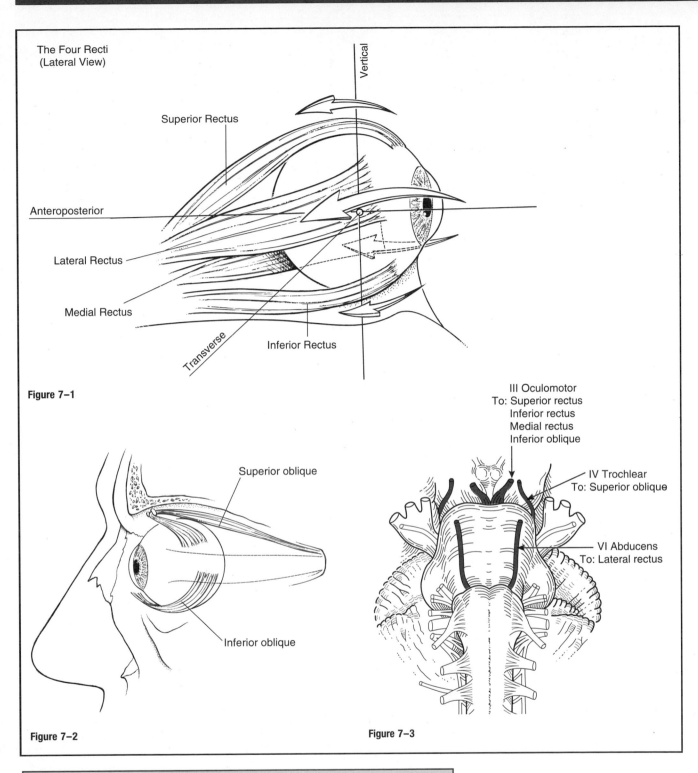

The Four Recti
(Lateral View)

Vertical

Superior Rectus

Anteroposterior

Lateral Rectus

Medial Rectus

Transverse

Inferior Rectus

Figure 7-1

Superior oblique

Inferior oblique

Figure 7-2

III Oculomotor
To: Superior rectus
Inferior rectus
Medial rectus
Inferior oblique

IV Trochlear
To: Superior oblique

VI Abducens
To: Lateral rectus

Figure 7-3

Table 7-1	EXTRAOCULAR MUSCLES	
Muscle	**Origin**	**Insertion**
6. Rectus superior	Sphenoid	Superior sclera
7. Rectus inferior	Sphenoid	Inferior sclera
8. Rectus medialis	Sphenoid	Medial sclera
9. Rectus lateralis	Sphenoid	Lateral sclera
10. Superior oblique	Sphenoid	Via a frontal bone pulley to the superolateral sclera
11. Inferior oblique	Maxilla (orbital surface)	Lateral sclera behind the Superior oblique

The six extraocular muscles of the eye move the eyeball in directions that depend on their attachments and on the influence of the movements themselves. It is probable that no muscle of the eye acts independently, and because these muscles cannot be observed, palpated, or tested individually, much of the knowledge of their function is derived from some variety of dysfunction.

The Axes of Eye Motion

The eyeball rotates in the orbital socket around one or more of three primary axes (Fig. 7–4), which intersect in the center of the eyeball.[1]

Vertical axis: Around this axis the lateral motions (abduction and adduction) take place in a horizontal plane.

Transverse axis: This is the axis of rotation for upward and downward motions.

Anteroposterior axis: Motions of rotation in the frontal plane occur around this axis.

The neutral position of the eyeball occurs when the gaze is straight ahead and far away. In this neutral position the axes of the two eyes are parallel. Normally, the motions of the two eyes are conjugate, that is, coordinated, and the two eyes move together.

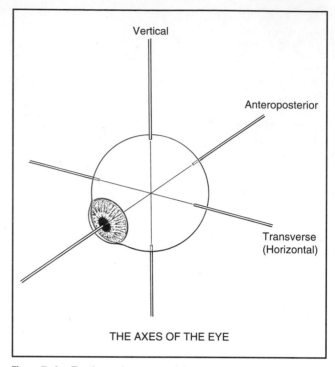

THE AXES OF THE EYE

Figure 7–4. The three primary axes of the eye.

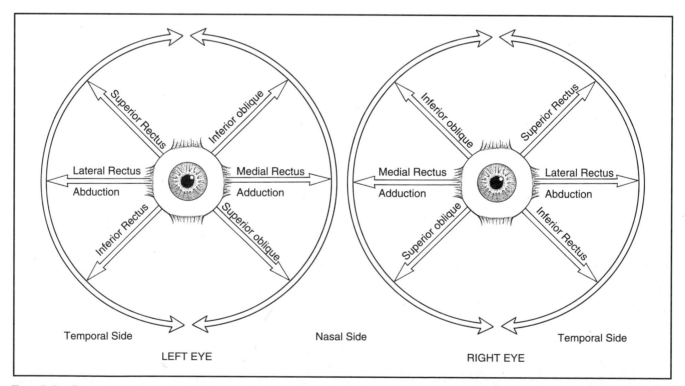

Figure 7–5. *Extraocular muscles and their actions.* The six extraocular muscles enable each eye to move in a circular arc, usually accompanied by head movements, though head position is static during testing. The traditional pairing of extraocular muscles is an oversimplification of their movement patterns. In any ocular rotation all six muscles change length. The reference point for description of the motions of the extraocular muscles is the center of the cornea.

Eye Motions

The extraocular muscles seem to work as a continuum; as the length of one changes, the length and tension of the others are altered, giving rise to a wide repertoire of movement.[2,3] Despite this continuous commonality of activity, the function of the individual muscles can be simplified and understood in a manner that does not detract from accuracy but simplifies the test procedure.

Conventional clinical testing assigns the following motions to the various extraocular muscles[1-3] (Fig. 7–5):

6. Rectus superior (III, Oculomotor)

Primary Movement: Elevation of the eyeball; movement is upward and outward

Secondary movements

1. Rotation of the adducted eyeball so the upper end of the vertical axis is inward (see Fig. 7–4)
2. Adduction of the eyeball to a limited extent

7. Rectus inferior (III, Oculomotor)

Primary Movement: Depression of the eyeball; movement is downward and outward

Secondary Movements

1. Adduction of the eye
2. Rotation of the adducted eyeball so the upper end of the vertical axis is outward

8. Rectus medialis (III, Oculomotor)

Primary Movement: Adduction of the eyeball

Secondary Movements: None

9. Rectus lateralis (VI, Abducent)

Primary Movement: Abduction of the eyeball

Secondary Movements: None. VI nerve lesions limit lateral movement. In paralysis the eyeball is turned medially and cannot be abducted.

10. Obliquus superior (IV, Trochlear)

Primary Movement: Depression of the eye

Secondary Movements

1. Abduction of the eyeball
2. IV nerve lesions limit depression, but abduction may be intact because abduction is the VI nerve.

11. Obliquus Inferior (III, Oculomotor)

Primary Movement: Elevation of the eye, particularly from adduction; movement is upward and inward.

Secondary Movements

1. Abduction of the eyeball
2. Rotation of the eyeball so the vertical axis is outward
3. In paralysis the eyeball is deviated downward and somewhat laterally; it cannot move upward when in abduction.
4. Note: In a III nerve lesion, the eye is outward and cannot be brought in. (This is often referred to irreverently as the "bum's eye," that is, down and out.) Such a lesion also results in ptosis, or drooping of the upper eyelid.[2,3]

Eye Tracking

Eye movements are tested by having the patient look in the *cardinal directions* (numbers in parentheses refer to tracks shown in Figure 7–6).[2] All pairs in tracking are antagonists.

Laterally (1)	Upward and laterally (5)
Medially (2)	Upward and medially (6)
Upward (3)	Downward and medially (7)
Downward (4)	Downward and laterally (8)

Ask the patient to follow the examiner's slowly moving finger (or a pointer or flashlight) in each of the following tests. The object the patient is to follow should be at a comfortable reading distance. First one eye is tested and then the other, covering the nontest eye. After single testing, both eyes are tested together for conjugate movements. Each test is started in the neutral position of the eye.

The range, speed, and smoothness of the motion should be observed as well as the ability to sustain lateral and vertical gaze.[2–4] The physical therapist will not be able to use these observational methods to distinguish movement deviations accurately because accuracy requires the sophisticated instrumentation used in ophthalmology. The tracking movements will appear normal or abnormal, but little else will be possible.

Position of Patient: Head and eyeball in neutral alignment, looking straight ahead at examiner's finger to start. Head must remain static. If the patient turns the head while tracking the examiner's finger, the head will have to be held still with the examiner's other hand or by an assistant.

Instructions to Patient: "Look at my finger. Follow it with your eyes" (Fig. 7–7).

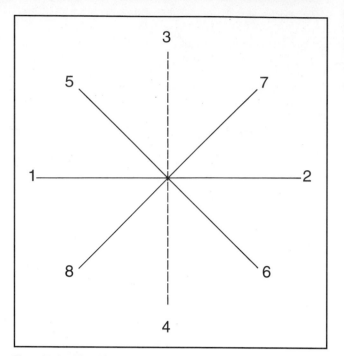

Figure 7–6   The eight cardinal directions of eye motion.

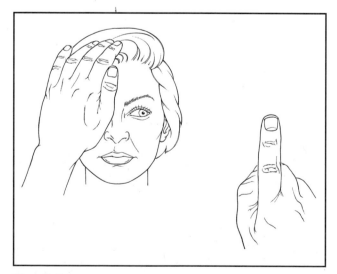

Figure 7–7.

Test: Test each eye separately by covering first one eye and then the other. Then test both eyes together.

Examples of two bilateral tests show conjugate motion in the two eyes when tracking upward and to the right (Fig. 7–8) and when tracking downward and to the left (Fig. 7–9).

Criteria for Grading

F: Immediate tracking in a smooth motion over the full range. Completes full excursion of the test movement.

WF and NF: Not possible to distinguish accurately from Grade F or Grade 0 without detailed diplopia testing.

0: Tracking motion in a given test absent.

The Muscles of the Face

The face should be observed for mobility of expression, and any asymmetry or inadequacy of muscles should be documented. A one-sided appearance when talking or smiling, a lack of tone (with or without atrophy), the presence of fasciculations, asymmetrical or frequent blinking, smoothness of the face or excessive wrinkling are all clues to VII nerve involvement.

The facial muscles (except for motions of the jaw) convey all emotions via voluntary and involuntary movements.

Figure 7–8. Patient tracks upward and to the right. The patient's right eye shows motion principally with the Superior rectus; the left eye shows motion principally with the inferior oblique.

Figure 7–9. Patient tracks downward and to the left. The right eye movement reflects principally the Superior oblique; the left eye shows motion principally with the Inferior rectus.

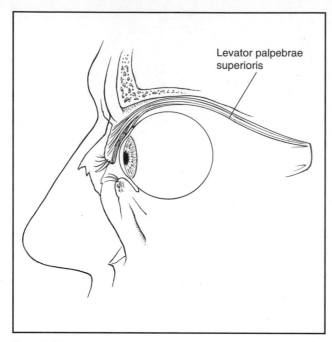

Figure 7–10

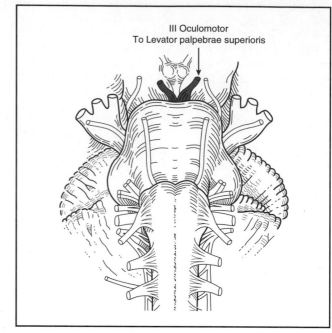

Figure 7–11

Table 7–2	MUSCLES OF THE EYELIDS AND EYEBROWS	
Muscle	**Origin**	**Insertion**
3. Levator palpebrae superioris	Sphenoid bone	Orbital septum aponeurosis Superior tarsus Rectus superior sheath
4. Orbicularis oculi	Frontal bone Maxilla (frontal process)	Lateral palpebral raphe Blends with Occipitofrontalis and Corrugator muscles
5. Corrugator supercilii	Frontal bone	Deep skin of eyebrow

Eye Opening (3. Levator palpebrae superioris)

Opening the eye by raising the upper eyelid is a function of the Levator palpebrae superioris. The muscle should be evaluated by having the patient open and close the eye with and without resistance. The function of this muscle is assessed by its strength in maintaining a fully opened eye against resistance.

The patient with an oculomotor (III) nerve lesion will lose the function of the levator muscle, and the eyelid will droop in a partial or complete ptosis. (A patient with cervical sympathetic pathology may have a ptosis but will be able to raise the eyelid voluntarily.) Ptosis is evaluated by observing the amount of the iris that is covered by the eyelid.

In the presence of a facial (VII) nerve lesion the levator sign may be present.[2] In this case, the patient is asked to look downward and then slowly close the eyes. A positive levator sign is noted when the upper eyelid on the weak side moves upward because the action of the Levator palpebrae superioris is unopposed by the Orbicularis oculi.

Test: Patient attempts to keep the eyelids open against manual resistance (Fig. 7–12). Both eyes are tested at the same time. **NEVER PRESS ON THE EYEBALL FOR ANY REASON!**

Manual Resistance: The thumb or index finger is placed lightly over the opened eyelid above the lashes, and resistance is given in a downward direction (to close the eye). The examiner is cautioned to avoid depressing the eyeball into the orbit while giving resistance.

Instructions to Patient: "Open your eyes wide. Hold it. Don't let me close them."

Criteria for Grading

F: Completes normal range of movement and holds against examiner's light manual resistance. Iris will be fully visible.

WF: Can open eye but only partially uncovers the iris and takes no resistance. Patient may alternately open and close the lids, but excursion is small. The Frontalis muscle also may contract as the patient attempts to open the eye.

NF: Unable to open the eye, and the iris is almost completely covered.

0: No eyelid opening.

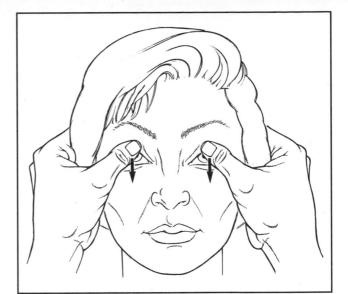

Figure 7–12

PERIPHERAL VS. CENTRAL LESIONS OF THE VII (FACIAL) NERVE

Involvement of the facial nerve may result from a lesion that affects the nerve or the nucleus, i.e., *a peripheral lesion.* Motor functions of the face also may be impaired after a *central* or supranuclear lesion. These two sites of interruption of the VII nerve lead to dissimilar clinical problems.[5]

The peripheral lesion results in a flaccid paralysis of all the muscles of the face *on the side of the lesion* (Occipitofrontalis, Corrugator, Orbicularis oculi, nose, and mouth muscles). The affected side of the face becomes smooth, the eye remains open, the lower lid sags, and blinking does not completely close the eye; the nose is depressed and may deviate to the opposite side. The cheek muscles are flaccid, so the cheek appears hollow and the mouth is drawn to one side. Eating and drinking are difficult because chewing and retention of fluids and saliva are impaired. Speech sounds, especially vowels or sounds that require pursing of the lips, are slurred.

When the VII nerve is affected central to the nucleus, there is paresis of the muscles of the lower face but sparing of the muscles of the upper face. This occurs because the nuclear center that controls the upper face has both contralateral and ipsilateral supranuclear connections, whereas that which controls the lower face has only contralateral supranuclear innervation. For this reason, a lesion in one cerebral hemisphere causes paresis of the lower part of the face *on the contralateral side* and there is sparing of the upper facial muscles. This may be called a "central VII syndrome."

One notable difference between peripheral and central disorders is that peripheral lesions often (but certainly not always) result in paralysis of all facial muscles; central lesions leave some function even of the involved muscles and are, therefore, a paretic and not a paralytic problem.

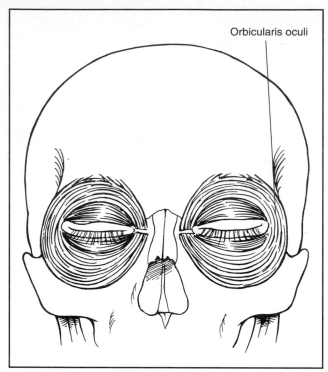

Orbicularis oculi

Figure 7–13

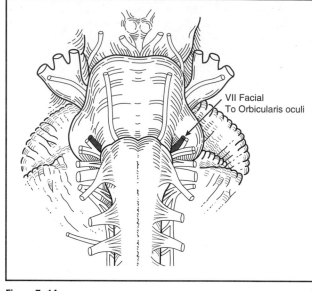

VII Facial
To Orbicularis oculi

Figure 7–14

Closing the Eye (4. Orbicularis oculi)

The Orbicularis oculi muscle is the sphincter of the eye.[1] Its palpebral portion closes the eyelids gently as in blinking and sleep. The orbital portion of the muscle closes the eyes with greater force as in winking. The lacrimal portion draws the eyelids laterally and compresses them against the sclera to receive tears. All portions act to close the eyes tightly. Observation of the patient without specific testing will detect weakness of the Orbicularis because the blink will be delayed on the involved side.

Test: Observe the patient opening and closing the eyes voluntarily, first singly and then together (Fig. 7–15). (Single eye closing is not a universal skill.) Patient closes eyes tightly, first singly, then together.

Rather than using resistance, the examiner may look at the depth to which the eyelashes are buried in the face when the eyes are closed tightly, noting whether the lashes are deeper on the uninvolved side.

Manual Resistance: Place the thumb and index finger below and above (respectively) each closed eye using a light touch (Fig. 7–16). The examiner attempts to open the eyelids by spreading the thumb and index finger apart. **NEVER PRESS ON THE EYEBALL FOR ANY REASON.**

Instructions to Patient: "Close your eyes as tightly as you can. Hold them closed. Don't let me open them" OR "Close your eye against my finger."

Criteria for Grading

F: Closes eyes tightly and holds against examiner's resistance. Iris may not be visible.

WF: Takes no resistance to eye closure; closure may be incomplete, but only a small amount of the sclera and no iris should be visible. There may be closure of the eye, but the eyelid on the weaker side may be delayed in contrast to the quick closure on the normal side.

NF: Unable to close eyes so that the iris is completely covered. (These patients may need adjunct tear drops to prevent drying of the eye.)

0: No evidence of Orbicularis oculi activity.

Figure 7–15

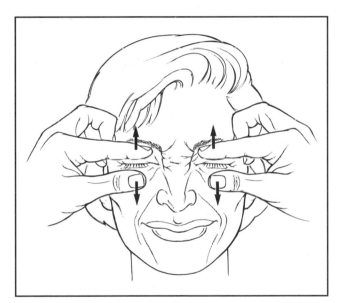

Figure 7–16

⊙ HELPFUL HINTS

If in closing the eyes tightly, the eyeball rotates upward, the patient is exerting effort to perform the test correctly. This upward rotation of the eyeball is called Bell's phenomenon. If the patient is not exerting effort, all protestations to the contrary, the eyeball will remain in the neutral position.

This observation may give the physical therapist a clue for other testing done with this kind of patient.

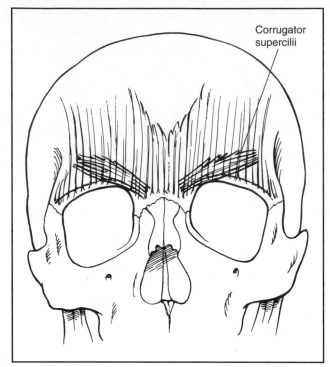

Corrugator
supercilii

Figure 7–17

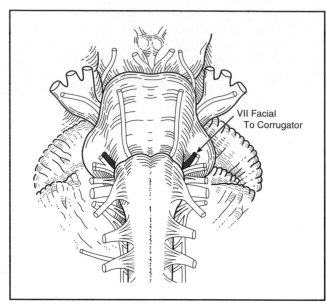

VII Facial
To Corrugator

Figure 7–18

Frowning (5. Corrugator supercilii)

To observe the action of the Corrugator muscle, the patient is asked to frown. Frowning draws the eyebrow down and medially, producing vertical wrinkling of the forehead.

Test: Patient is asked to frown; the eyebrows are drawn down and together (Fig. 7–19).

Manual Resistance: The examiner uses the thumb (or index fingers) of each hand placed gently at the nasal end of each eyebrow and attempts to move the eyebrows apart (smoothes away the frown) (Fig. 7–20).

Instructions to Patient: "Frown. Don't let me erase it."

Criteria for Grading

F: Completes normal range (wrinkles are prominent) and holds against slight resistance.

WF: Frowns, but wrinkles are shallow and not too obvious; is unable to take resistance.

NF: Slight motion detected.

0: No frown.

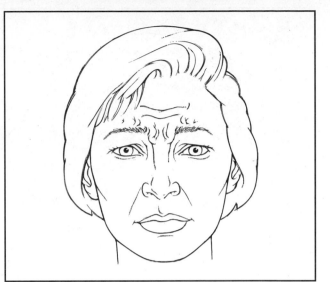

Figure 7–19

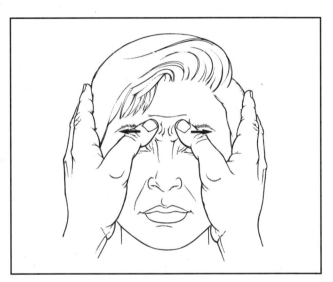

Figure 7–20

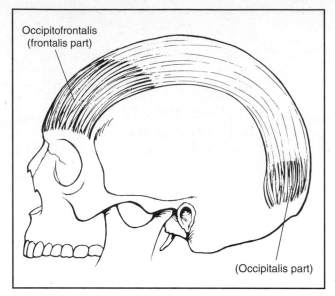

Occipitofrontalis
(frontalis part)

(Occipitalis part)

Figure 7–21

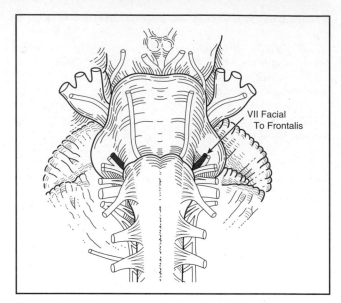

VII Facial
To Frontalis

Figure 7–22

Raising the Eyebrows (1. Occipitofrontalis, frontalis part)

To examine the frontal belly of the Occipito-frontalis muscle the patient is asked to create an expression of surprise where the forehead skin wrinkles horizontally. The occipital belly of the muscle is not tested usually, but it draws the scalp backward.

Test: Patient raises the eyebrows so that horizontal forehead lines appear (Fig. 7–23).

Manual Resistance: Examiner places the pad of a thumb above each eyebrow and applies resistance in a downward direction (smoothing the forehead) (Fig. 7–24).

Instructions to Patient: "Raise your eyebrows as high as you can. Don't let me pull them down."

Criteria for Grading

F: Completes movement; horizontal wrinkles are prominent. Tolerates considerable resistance.

WF: Wrinkles are shallow and easily erased by gentle resistance.

NF: Only slight motion detected.

0: No eyebrow raising.

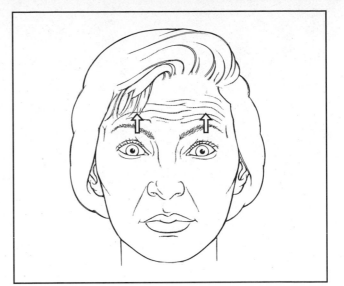

Figure 7–23

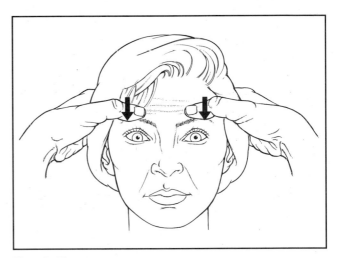

Figure 7–24

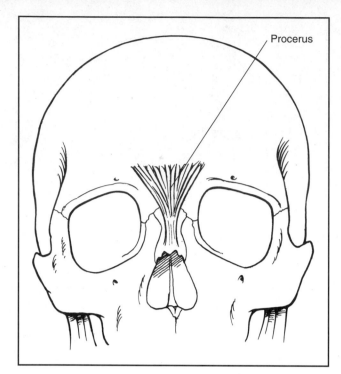

Figure 7–25

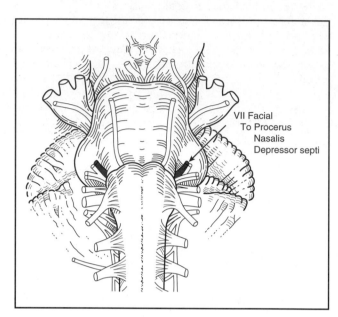

Figure 7–26

Table 7–3	MUSCLES OF THE NOSE	
Muscle	**Origin**	**Insertion**
12. Procerus	Nasal bone and cartilage	Skin over lower forehead between eyebrows Joins Occipitofrontalis
13. Nasalis		
Transverse part (compressor nares)	Maxilla (lateral to incisive fossa)	Aponeurosis over bridge of nose
Alar part (dilator nares)	Maxilla (above lateral incisor)	Ala nasi Skin at tip of nose
14. Depressor septi	Maxilla (above incisive fossa)	Nasal septum Alar cartilage

The three muscles of the nose are all innervated by the facial (VII) nerve. The Procerus draws the medial angle of the eyebrows downward, causing transverse wrinkles across the bridge of the nose. The Nasalis (compressor nares) depresses the cartilaginous portion of the nose and draws the ala down toward the septum. The Nasalis (dilator nares) dilates the nostrils. The Depressor septi draws the alae downward, constricting the nostrils.

Of the three nose muscles only the Procerus is tested clinically. The others are observed with respect to nostril flaring and narrowing in patients who have such talent.

Wrinkling the Bridge of the Nose (12. Procerus)

Test: Patient wrinkles nose as if expressing distaste (Fig. 7–27).

Manual Resistance: The pads of the thumbs are placed beside the bridge of the nose, and resistance is given laterally (smoothing the creases) (Fig. 7–28).

Instructions to Patient: "Wrinkle your nose as if to say 'yuk'."

Criteria for Grading

F: Prominent creases; patient tolerates some resistance.

WF: Shallow creases; patient yields to any resistance.

NF: Motion barely discernible.

0: No change of expression.

⊙ HELPFUL HINTS

Isolated wrinkling of the nose is rare, and most patients bring in other facial muscles to perform this expressive movement.

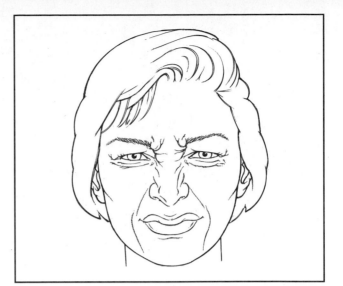

Figure 7–27

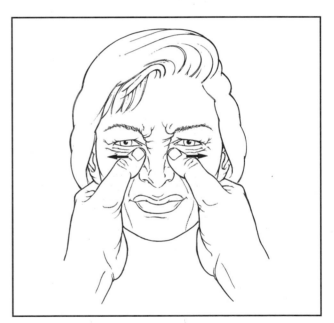

Figure 7–28

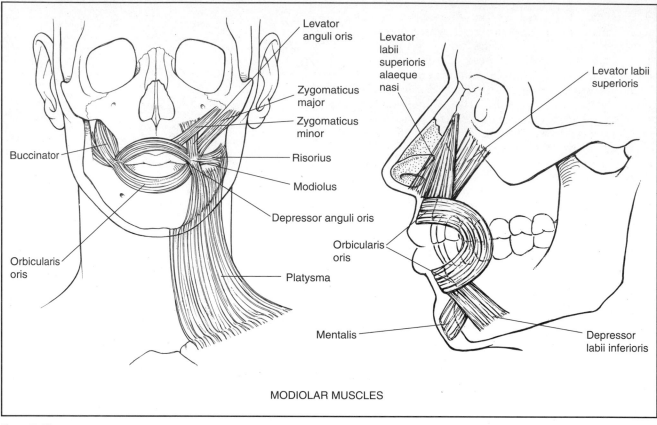

Figure 7–29

Figure 7–30

Table 7–4	MUSCLES OF THE MOUTH	
Muscle	**Origin**	**Insertion**
25. Orbicularis oris	Modiolus No bony attachment	Modiolus
15. Levator labii superioris	Orbit of eye (inferior) Maxilla Zygomatic bone	Upper lip
17. Levator anguli oris	Maxilla (canine fossa)	Modiolus Skin at angle of mouth
18. Zygomaticus major	Zygomatic bone	Modiolus
24. Depressor labii inferioris	Mandible	Modiolus
26. Buccinator	Maxilla and mandible (alveolar processes)	Modiolus
21. Mentalis	Mandible (incisive fossa)	Skin over chin
23. Depressor anguli oris	Mandible (oblique line)	Modiolus

Others
19. Zygomaticus minor
20. Risorius
22. Transverse menti
88. Platysma

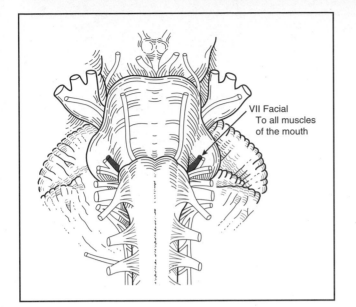

VII Facial
To all muscles
of the mouth

Figure 7–31

There are many muscles associated with the mouth, and all have some distinctive function, except perhaps the Risorius. Rather than detail a test for each, only definitive tests will be presented for the Buccinator and the Orbicularis oris (the sphincter of the mouth). The function of the remaining muscles is illustrated, and individual testing is left to the examiner. All muscles of the mouth are innervated by the Facial (VII) nerve.

THE MODIOLUS

The arrangement of the facial musculature often causes confusion and misunderstanding. This is not surprising since there are 14 small bundles of muscles running in various directions, with long names and unsupported functional claims. Of all the muscles of the face, those about the mouth may be the most important because they have responsibility for both ingestion of food and speech.

One major source of confusion is the relationship between the muscles around the mouth. The common description until recently was of uninterrupted circumoral muscles. In fact, the Orbicularis oris muscle is not a complete ellipse but rather contains fibers from the major extrinsic muscles that converge on the buccal angle as well as intrinsic fibers.[1,6,7] The authors and others do not describe complete ellipses, but most drawings illustrate such.[6]

The area on the face that has a large concentration of converging and diverging fibers from multiple directions lies immediately lateral and slightly above the corner of the mouth. Using the thumb and index finger on the outer skin and inside the mouth and compressing the tissue between them will quickly identify the knotlike structure known as the *modiolus*.[8-10]

The modiolus (from the Latin meaning nave of a wheel) is described as a muscular or tendinous node, a rather concentrated attachment of many muscles.[8-9] Its basic shape is conical (though this is oversimplified); it is about 1 cm thick and is found in most people about 1 cm lateral to the buccal angle. Its shape and size vary considerably with gender, race, and age. The muscular fibers enter and exit on different planes, superficial and deep, with some spiraling, but essentially they comprise a three-dimensional complexity.

Different classifications of modiolar muscles exist, but basically nine or ten facial muscles are associated with the structure:[9]

Radiating out from:	*Retractors of the upper lip:*
Levator anguli oris	Levator labii superioris
Orbicularis orls	Levator labii superioris alaeque nasi
Depressor anguli oris	Zygomaticus minor
Zygomaticus major	*Retractors and elevators of the upper*
Buccinator	*lip:*
	Mentalis
	Depressor labii inferioris

Frequently associated are the special fibers of the Orbicularis oris (Incisive superior, Incisive inferior), Platysma, and Risorius (the latter is not a constant feature in the facial musculature).

The Orbicularis oris and the Buccinator form an almost continuous muscular sheet, which can be fixed in a number of positions by the Zygomaticus major, Levator anguli oris, and Depressor anguli oris (the latter three being the "stays" used to immobilize the modiolus in any position).

When the modiolus is firmly fixed, the Buccinator can contract to apply force to the cheek teeth; the Orbicularis oris can contract against the arch of the anterior teeth, thus sealing the lips together and closing the mouth tightly.[9] Similarly, control of the modiolar active and stay muscles enables accurate and fine control of lip movements and pressures in speech.

Lip Closing (25. Orbicularis oris)

This circumoral muscle serves many functions for the mouth. It closes the lips, protrudes the lips, and holds the lips tight against the teeth. Furthermore, it shapes the lips for such functional uses as kissing, whistling, sucking, drinking, and the infinite shaping for articulation in speech.

Test: Patient compresses and protrudes the lips (Fig. 7–32).

Resistance: A tongue blade rather than a finger is used to provide resistance in deference to hygiene. The blade is placed diagonally across both upper and lower lips, and resistance is applied inward toward the oral cavity (Fig. 7–33).

Instructions to Patient: "Purse your lips. Hold it. Push against the tongue blade."

Criteria for Grading

F: Completely seals lips and holds against relatively strong resistance.

WF: Closes lips but is unable to take resistance.

NF: Has some lip movement but is unable to bring lips together.

0: No closure of the lips.

Figure 7–32

Figure 7–33

Cheek Compression (26. Buccinator)

The Buccinator is a prime muscle used for positioning food for chewing and for controlling the passage of the bolus. It also compresses the cheek against the teeth and acts to expel air when the cheeks are distended (blowing).

Test: Patient compresses the cheeks (bilaterally) by drawing them into the mouth (Fig. 7–34).

Resistance: A tongue blade is used for resistance. The blade is placed inside the mouth, its flat side lying against the cheek (Fig. 7–35). Resistance is given by levering the blade inward against the cheek (at the angle of the mouth), which will cause the flat blade to push the test cheek outward.

Alternatively, the gloved index fingers of the examiner may be used to offer resistance. In this case, the index fingers are placed in the mouth (the left finger to the inside of the patient's left cheek and vice versa). The fingers are used simultaneously to try to push the cheeks outward. Use caution in this form of the test for patients with cognitive impairment (lest they bite!) or with those who have a bite reflex.

Instructions to Patient: "Suck in your cheeks. Hold. Don't let me push them out."

Criteria for Grading

F: Performs movement correctly and holds against strong resistance.

WF: Performs movement but is unable to hold against any resistance.

NF: Movement is detectable but not complete.

0: No motion of cheeks occurs.

Figure 7–34

Figure 7–35

Other Oral Muscles

17. *Levator anguli oris*

This muscle elevates the angles of the mouth and reveals the teeth in smiling. When used unilaterally, it conveys the expression of sneering (Fig. 7–36). The muscle creates the nasolabial furrow, which deepens in expressions of sadness and with aging.

15. *Levator labii superioris*

16. *Levator labii superior alaeque nasi*

These two muscles elevate the upper lip (Fig. 7–37). The labii superioris also protracts the upper lip, and the alaeque nasi dilates the nostrils.

18. *Zygomaticus major*

The major Zygomaticus muscles draw the angles of the mouth upward and laterally as in laughing (Fig. 7–38).

Figure 7–36

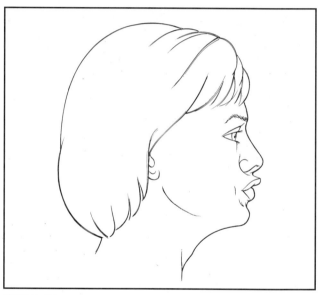

Figure 7–37

Figure 7–38

21. *Mentalis*

The Mentalis protrudes the lower lip as in pouting or sulking (Fig. 7–39).

23. *Depressor anguli oris*

88. *Platysma*

These muscles depress the lower lip and the buccal angle of the mouth to give an expression of grief or sadness (Fig. 7–40). The Platysma draws the lower lip backward, producing an expression of horror, and it pulls up the skin of the neck from the clavicle (evoking the expression of "Egad"). This muscle may be tested by asking the patient to open the mouth against resistance or bite the teeth together tightly.

24. *Depressor labii inferioris*

This muscle draws the lower lip down and laterally, producing an expression of melancholy or irony (Fig. 7–41).

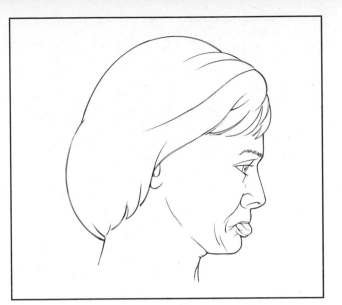

Figure 7–39

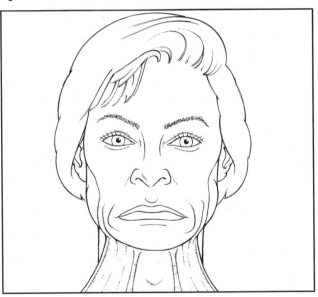

Figure 7–40

Figure 7–41

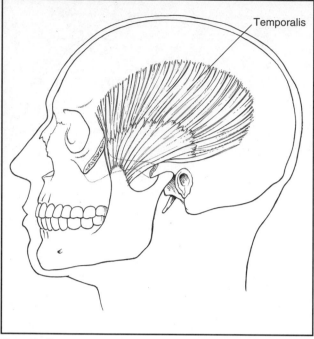

Figure 7–42

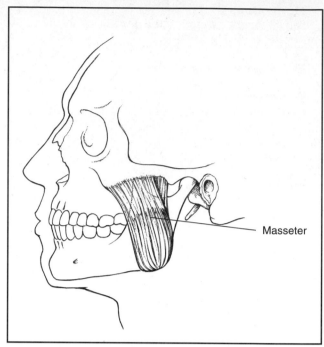

Figure 7–43

Table 7–5	MUSCLES OF MASTICATION	
Muscle	**Origin**	**Insertion**
28. Masseter		
Superficial	Maxilla (zygomatic process)	Mandible (ramus)
Intermediate	Zygomatic arch	Mandible (ramus)
Deep	Zygomatic arch	Mandible (ramus)
29. Temporalis	Temporal bone (fossa)	Mandible (coronoid process, ramus)
30. Lateral pterygoid		
Superior	Sphenoid bone (great wing)	Mandible (condylar neck)
Inferior	Mandible (lat. pterygoid plate)	Temporomandibular joint
31. Medial pterygoid	Sphenoid bone	Mandible (ramus)
	Palatine bone	
	Maxilla (tuberosity)	
75–78 Suprahyoids	Mandible	Hyoid bone

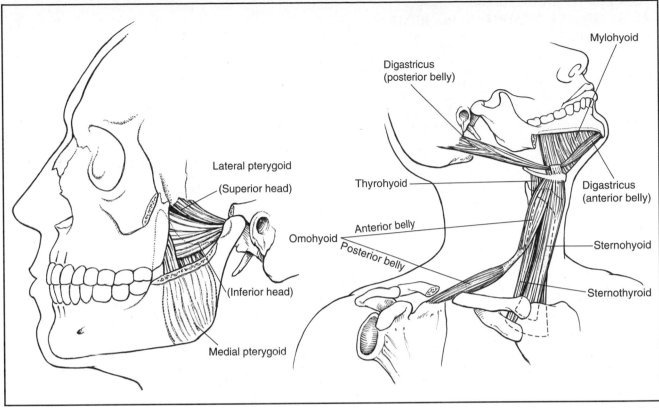

Figure 7–44

Figure 7–45

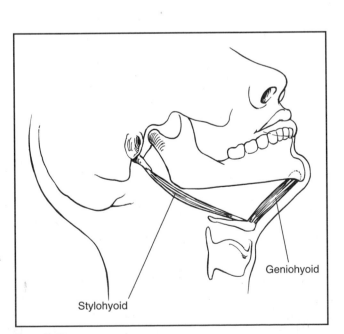

Figure 7–46

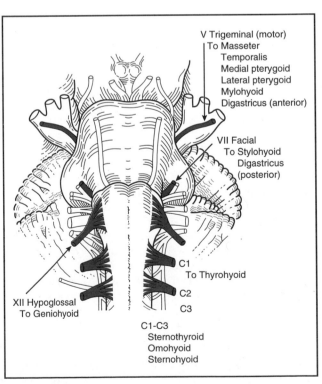

Figure 7–47

MUSCLES OF MASTICATION (28. MASSETER, 29. TEMPORALIS, 30, 31. PTERYGOIDS)

The mandible is the only moving bone in the skull, and mandibular motion is largely related to chewing and speech. The muscles that control the jaw are all near the rear of the mandible (on the various surfaces and processes of the ramus), where they contribute considerable force for chewing and biting.[1] The muscles of mastication move the mandible forward (protraction) and backward (retraction) as well as shifting it laterally. Excursion of the mandible is customarily limited somewhat except in trained singers, who learn to open the mouth very wide to add to their vocal repertoire. The velocity of motions used for chewing is relatively slow, but for speech motions it is very rapid.

The muscles of mastication are all innervated by the motor division of the V (trigeminal) nerve. The Masseter elevates and protrudes the mandible. The Temporalis elevates and retracts the mandible. The Lateral pterygoids, acting in concert, protrude and depress the mandible; when one acts alone, it causes lateral movement to the opposite side. The Medial pterygoids acting together elevate and protrude the mandible along with the Lateral pterygoids, but acting alone they draw the mandible forward with deviation to the opposite side (as in chewing). The suprahyoid muscles, acting via the hyoid bone, aid in jaw depression when the hyoid is fixed. The infrahyoids are weak accessories to jaw depression.

Lesions of the motor division result in weakness or paralysis of the motions of elevating, depressing, protruding, and rotating the mandible. In a unilateral lesion the jaw deviates to the weak side; in a bilateral lesion the jaw sags and is "paralyzed." The jaws should be examined for muscle tone, atrophy (jaw contour), and fasciculations.

Jaw Opening (Mandibular Depression) (30. Lateral pterygoid and 75–78. Suprahyoid Muscles)

Note: Prior to testing the jaw muscles, the temporomandibular joint should be checked for tenderness and crepitus. If either is present, all manual testing is avoided, and jaw opening and closing are simply observed.

Test: Patient opens the mouth as far as possible and holds against manual resistance.

Manual Resistance: One hand of the examiner is cupped under the chin; the other hand is placed on the crown of the head for stabilization (Fig. 7–48). Resistance is given in a vertical upward direction in an attempt to close the jaw.

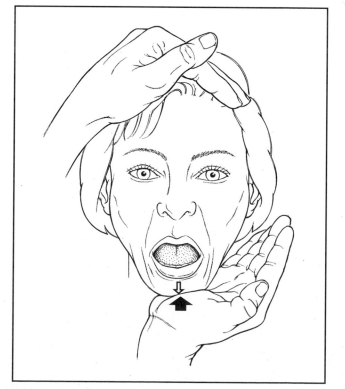

Figure 7–48

Instructions to Patient: "Open your mouth as wide as you can. Hold it. Don't let me close it."

Criteria for Grading

F: Completes available range and holds against strong resistance. Indeed, this muscle is so powerful that in the normal person it can rarely be overcome with manual resistance. The mouth opening should accommodate three (sometimes four) stacked fingers (on an average-sized person), or 35 to 40 mm. There should be no deviation except downward.

WF: Can open mouth to accommodate two or fewer stacked fingers and can take some resistance.

NF: Minimal motion occurs. The Lateral pterygoid can be palpated with a gloved finger inside the mouth, with the tip directed posteriorly past the last upper molar to the condyloid process of the ramus of the mandible. No resistance is tolerated.

0: No voluntary mandibular depression occurs.

Jaw Closure (Mandibular Elevation) (28. Masseter, 29. Temporalis, 31. Medial pterygoid)

Test: Patient clenches jaws tightly.

Manual Resistance: The chin of the patient is grasped between the thumb and index finger of the examiner and held firmly in the thumb web. The other hand is placed on top of the head for stability. Resistance is given vertically downward in an attempt to open the closed jaw (Fig. 7–49).

Instructions to Patient: "Clench (or hold) your teeth together as tightly as you can, keeping your lips relaxed. Hold it. Don't let me open your mouth."

Criteria for Grading

F: Patient closes mouth (jaw) tightly. Examiner should not be able to open the mouth. This is a very strong muscle group. Consider circus performers who hang by their teeth!

WF: Patient closes jaw, but examiner can open the mouth with less than maximal resistance.

NF: Patient closes mouth but tolerates no resistance. Masseter and Temporalis muscles are palpated on both sides. The Masseter is palpated under the zygomatic process on the lateral cheek above the angle of the jaw. The Temporalis muscle is palpated over the temple at the hairline, anterior to the ear and superior to the zygomatic bone.

Figure 7–49

0: Patient cannot completely close the mouth. This is more of a cosmetic problem (drooling, for example) than a significant clinical one.

In unilateral involvement, the jaw deviates to the strong side during attempts to close the mouth.

Alternate Test Procedure

The patient is asked to bite hard on a tongue blade with the molar teeth. Comparison of the depth of the bite marks from each side of the jaw is an indication of strength. If the examiner can pull out the tongue blade while the patient is biting, there is weakness of the Masseter, Temporalis, and Lateral pterygoid muscles. *Note:* This method of testing should never be used with a patient who has a bite reflex or the patient may break the blade and be injured by the splinters.

Lateral Jaw Deviation (30. Lateral and 31. Medial pterygoids)

When the patient deviates the jaw to the right, the acting muscles are the right Lateral pterygoid and the left Medial pterygoid. Deviation to the left is supported by the left Lateral pterygoid and the right Medial pterygoid.

With weakness of the pterygoids, when the patient opens his mouth there will be deviation to the side of the weakness.

The patient moves the jaw side to side against resistance. In V nerve involvement the patient can move the jaw to the paralyzed side but not to the unaffected side.

Test: Patient deviates jaw to the right and then to the left.

Manual Resistance: One hand of the examiner is used for resistance and is placed with the palmar side of the fingers against the jaw (Fig. 7–50). The other hand is placed with the fingers and palm against the opposite temple to stabilize the head. Resistance is given in a lateral direction to move the jaw toward the midline.

Criteria for Grading

F: The range of motion for jaw lateral deviation is variable. Deviation is assessed by comparing the relationship between the upper and lower incisor teeth when the jaw is moved laterally from the midline. Do not assess deviation by the position of the lips. A pencil or ruler lined up vertically with the center of the nose may indicate mandibular deviation.

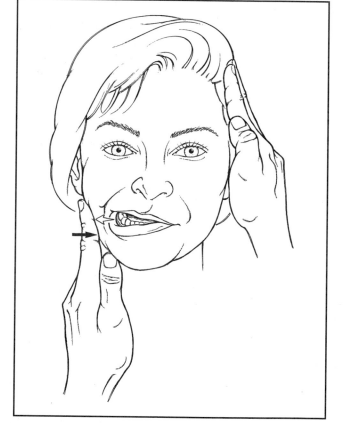

Figure 7–50

Most people can move the center point of the lower incisors laterally over three upper teeth (approximately 10 mm).[5] The patient tolerates strong resistance.

WF: Motion is decreased to lateral movement across one upper tooth, and resistance is minimal.

NF: Minimal deviation occurs, and no resistance is taken.

0: No motion occurs.

Jaw Protrusion (30., 31., Medial and Lateral pterygoids)

The Medial and Lateral pterygoids act to protrude the jaw, which gives the face a pugnacious expression. The protrusion causes a malocclusion of the teeth, the lower teeth projecting beyond the upper teeth. With a unilateral lesion, the protruding jaw deviates to the weak side.

Test: Patient protrudes jaw so the lower teeth project beyond the upper teeth.

Manual Resistance: This is a powerful motion. The examiner stabilizes the head with one hand placed behind the head (Fig. 7–51). The hand for resistance cups the chin in the thumb web with the thumb and index finger grasping the mandible. Resistance is given horizontally backward.

Instructions to Patient: "Push your jaw forward. Hold it. Don't let me push it back."

Criteria for Grading

F: Completes a range that moves the lower teeth in front of the upper teeth and can hold against strong resistance. There is sufficient space between the teeth in most people to see a gap between the upper and lower teeth.

WF: Moves jaw slightly forward but there is no discernible gap between the upper and lower teeth, and the patient tolerates only slight resistance.

NF: Minimal motion is detected, and patient takes no resistance.

0: No motion and no resistance occur.

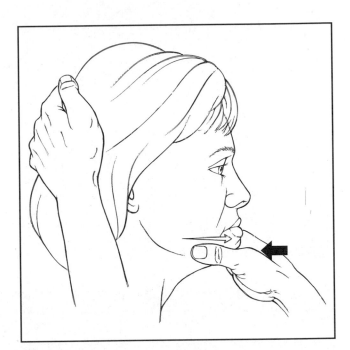

Figure 7–51

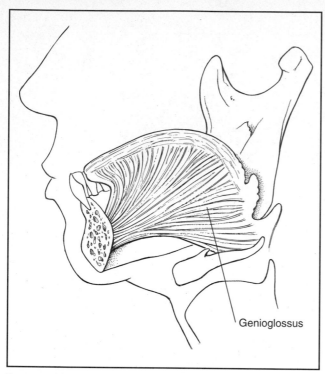

Figure 7-52

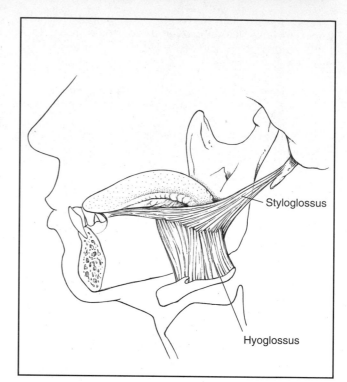

Figure 7-53

Table 7-6	MUSCLES OF THE TONGUE	
Muscle	**Origin**	**Insertion**
Extrinsic Muscles		
32. Genioglossus	Mandible (symphysis menti)	Hyoid bone Undersurface of tongue
33. Hyoglossus	Hyoid bone (greater horn)	Side of tongue (posterior)
34. Chondroglossus	Hyoid bone (lesser horn)	Blends with intrinsic tongue muscles
35. Styloglossus	Styloid process Stylomandibular ligament	Side of tongue
36. Palatoglossus	Soft palate (anterior)	Side of tongue
Others		
Suprahyoid muscles		
Intrinsic muscles		
37. Superior longitudinal	Root of tongue (superior surface)	Tip of tongue
38. Inferior longitudinal	Root of tongue (inferior surface)	Tip of tongue
39. Transverse	Median lingual septum	Dorsum and lateral margins
40. Vertical	Dorsum of tongue	Ventral surface

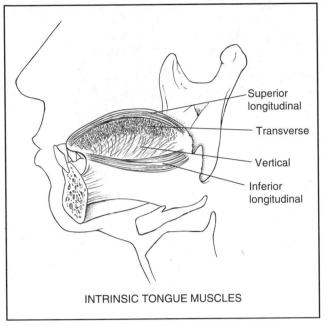

Superior
longitudinal

Transverse

Vertical

Inferior
longitudinal

INTRINSIC TONGUE MUSCLES

Figure 7–54

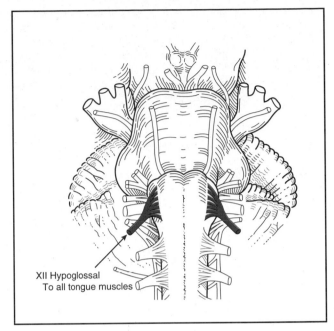

XII Hypoglossal
To all tongue muscles

Figure 7–55

The extrinsic and intrinsic muscles of the tongue all are innervated by the hypoglossal (XII) cranial nerve, a pure motor nerve. One XII nerve innervates half of the tongue (unilaterally). The hypoglossal nucleus, however, receives both crossed (mostly) and uncrossed (to a lesser extent) upper motor neuron fibers from the lowest part of the precentral gyrus via the internal capsule. Lesions of the XII nerve or its central connections may cause tongue paresis or paralysis.

Description of Tongue Muscles

The paired extrinsic muscles pass from the skull or the hyoid bone to the tongue. The intrinsic muscles rise and end within the tongue. The bulk of the tongue structure is muscle.

The principal muscle of the tongue is the Genioglossus. It is a triangular muscle whose apex arises from the apex of the mandible, which is hard and immobile; its base inserts into the base of the tongue, which is soft and mobile. The Genioglossus is the principal tongue protractor, and it has crossed supranuclear innervation. The posterior fibers of the paired Genioglossi draw the root of the tongue forward; a single Genioglossus pushes the tongue toward the opposite side. The anterior fibers of the paired muscles draw the tongue back into the mouth after protrusion and depress it. The Genioglossi acting together also depress the central part of the tongue, making it a tube.

The Hyoglossi (paired) and the Chondroglossi retract and depress the sides of the tongue, making the superior surface convex. The two Styloglossi draw the tongue upward and backward and elevate the sides, causing a dorsal transverse concavity.

The suprahyoid muscles influence the movements of the tongue via their action on the hyoid bone.

The intrinsic tongue muscles are similarly innervated by the XII nerve. The Superior longitudinal muscle shortens the tongue and curls its tip upward; the Inferior longitudinal shortens the tongue and curls its tip downward. Their combined function is to alter the shape of the tongue in almost infinite variations to provide the tongue with the versatility required for speech and swallowing.

One test of a tongue motion used by therapists is called "channeling" in which the tongue is curled longitudinally; this motion may be considered to assist in sucking and directing the bolus of food into the pharynx. The difficulty presented with this motion, however, is that it is not a constant motion but rather a dominantly inherited trait that only 50 percent of the population can perform. Testing it is all right as long as the inability to perform the motion is not considered a neurologic deficiency.

Examination of the Tongue

The tongue is a restless muscle, and when testing it, minor deviations are best ignored.[4] The test should start with observation of the tongue at rest on the floor of the mouth and then with the tongue protruded. The tongue is observed as it is curled up and down over the lip and then when the margins are elevated; both motions should be performed both slowly and rapidly. In all tests the ability to change the shape of the tongue is observed, but especially in tipping and channeling. One listens for difficulty in enunciation, especially of consonants.

The examiner must become familiar with the contour and mass of the normal tongue. The tongue should be examined for atrophy, which is evidenced by decreased mass, corrugations on the sides, and longitudinal furrowing. Unilateral atrophy is easy to detect and is usually accompanied by deviation to that side. When there is bilateral atrophy, the tongue will protrude weakly if at all, and deviation also will be weak.

Fasciculations are easily visible in the tongue at rest (the surface of the tongue appears to be in constant motion) and can be separated from the normal tremulous motions that occur in the protruded tongue. The "tremors" that are a part of supranuclear lesions disappear when the tongue is at rest in the mouth, whereas the fasciculations of motor neuron disease continue. The hyperkinesias of Parkinsonism are exaggerated when the tongue is protruded or during talking.

The therapist proceeds to examine protrusion and deviation of the tongue at slow and fast speeds. The normal tongue can move in and out (in the midline) with vigor and usually protrudes quite far beyond the lips.[11] The tongue deviates to the side of a weakness whether the cause of that weakness is a disturbance of the upper motor neuron (supranuclear disturbance) or the lower motor neuron (infranuclear disturbance).

Unilateral Weakness of the Tongue: At rest in the mouth, the tongue with a unilateral weakness may deviate slightly to the uninvolved side because of the unopposed action of the Styloglossus.[11] The protruded tongue will deviate to the weak side and show weakness or inability to deviate to the normal side. Tipping may be normal because the intrinsic muscles are preserved. These functions may be impossible to evaluate if the clinical picture includes facial and jaw muscle weakness.

Early in the course of the disorder, before the onset of atrophy, the weak side of the tongue may appear enlarged and may ride higher in the mouth. After the onset of atrophy the weak side becomes smaller, furrowed, and corrugated on the lateral edge. A unilateral weakness of the tongue may result in few functional problems, and speech and swallowing may be minimally disturbed if at all.

Bilateral Paresis: In persons with bilateral lesions, the tongue cannot be protruded or moved laterally. There will be indistinct speech, and swallowing may be difficult. Some patients experience interference with breathing when swallowing is impaired because the tongue may fall back into the throat. Total paralysis of the tongue muscles is rare (except in brain stem lesions or advanced motor neuron disease).

Supranuclear vs. Infranuclear Lesions: In the presence of a supranuclear XII nerve lesion (central), the protruded tongue will deviate toward the side of the weakness but to the side opposite the cerebral lesion. There is no atrophy of the tongue muscles. The tongue muscles also may evidence spasticity.[11]

In dyskinetic states (athetosis, chorea, seizures, and so on) the tongue may protrude involuntarily as well as deviate to the opposite side. This is accompanied by other generally slow involuntary tongue movements that make speech thick and slow and difficult to understand.

Patients with hemiparesis following a vascular lesion (a unilateral corticobulbar lesion) may have a variety of bulbar symptoms including tongue muscle dysfunction. In common with other bulbar manifestations, these symptoms generally are moderate and subside with time or are well compensated, so that little functional disability persists.[5] Only in patients with a second stroke or a bilateral stroke (because these muscles have bilateral cortical innervation) will the bulbar signs persist.

Inability to flick the tongue in and out of the mouth quickly (after some practice) may indicate a bilateral supranuclear lesion. In an infranuclear (peripheral) nerve lesion, the tongue will deviate to the side of the weakness, which also is the side of the lesion. There will be atrophy of the tongue muscles. Bilateral atrophy most commonly is caused by motor neuron disease. The tongue also may be weak in myasthenia gravis (fatiguing after a series of protrusions), but there will be no atrophy.

The distinction between a lower motor neuron lesion and an upper motor neuron lesion of the XII nerve depends on the presence of supporting evidence of other upper motor neuron signs and on the presence of classic lower motor neuron signs such as hemiatrophy, unilateral fasciculations, and obvious deviation to the side of the paralysis when the tongue is protruded.[4]

Tongue Protrusion, Deviation, Retraction, Posterior Elevation, Channeling, and Curling

Test for Protrusion (32. Genioglossus, Posterior Fibers)

Patient protrudes tongue so that the tip extends out beyond lips.

Manual Resistance: Examiner uses a tongue blade against the tip of the tongue and provides resistance in a backward direction to the forward motion of the tongue (Fig. 7–56).

Instructions to Patient: "Stick out your tongue. Hold it. Don't let me push it in."

Test for Tongue Deviation (32. Genioglossus and Other Muscles)

Patient protrudes tongue and moves it to one side and then to the other.

Manual Resistance: Using a tongue blade, resist the lateral tongue motion along the side of the tongue near the tip (Fig. 7–57). Resistance is given in the direction opposite to the attempted deviation.

Instructions to Patient: "Stick out your tongue and move it to the right." (Repeat for left side.)

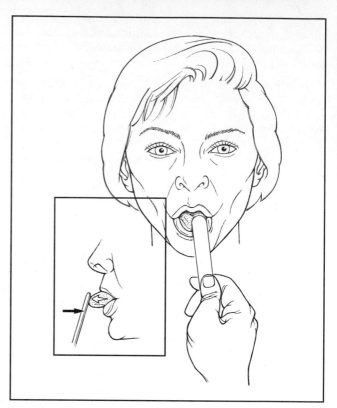

Figure 7–56

Figure 7–57

Test for Tongue Retraction (32. Genioglossus [anterior fibers] and 25. Styloglossus)

Patient retracts tongue from a protruded position.

Manual Resistance: Holding a 4 × 4 gauze pad, securely grasp the anterior tongue by its upper and under sides (Fig. 7–58). Resist retraction by holding the tongue firmly and gently pulling it forward. (The tongue is very slippery, but be careful not to pinch.)

Instructions to Patient: (Tell patient you are going to grasp the tongue.) "Stick out your tongue. Now pull your tongue back. Don't let me keep it out."

Test for Posterior Elevation of the Tongue (35. Palatoglossus and 36. Styloglossus)

Patient elevates (i.e., "humps") the dorsum of the posterior tongue.

Manual Resistance: Examiner places tongue blade on the superior surface of the tongue over the anterior one-third. Placing the blade too far back will initiate an unwanted gag reflex (Fig. 7–59).

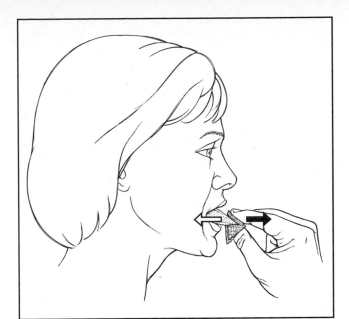

Figure 7–58

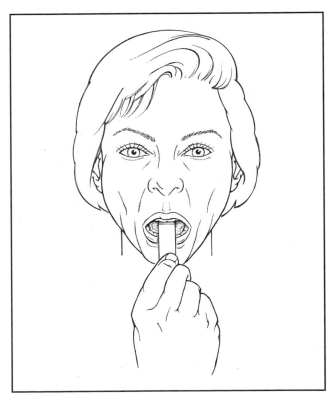

Figure 7–59

Resistance is applied in a down and backward direction, as in levering the tongue blade down, using the bottom teeth as a fulcrum (Fig. 7–60).

Instructions to Patient: This is a difficult motion for the patient to understand. After directions are given, time is allowed for practice.

Begin the test by rocking the tongue blade back and forth so the patient experiences pressure on the middle-back of the tongue.

"Push against this stick."

Test for Channeling the Tongue (32. Genioglossus and 37–40. Intrinsics)

Patient draws tongue downward and rolls sides up to make a longitudinal channel or tube, which is part of sucking and directing a bolus of food into the pharynx (Fig. 7–61). Inability to perform this motion should not be recorded as a deficit because the motion is a dominantly inherited trait, and its presence or absence should be treated as such.

Manual Resistance: None

Instructions to Patient: Demonstrate tongue motion to the patient.

"Make a tube with your tongue."

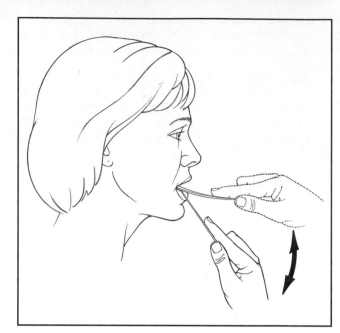

Figure 7–60

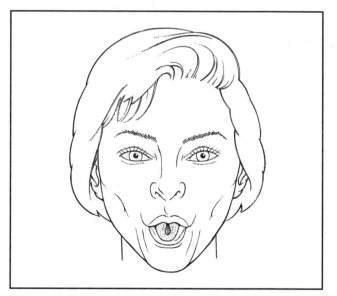

Figure 7–61

Test for "Tipping" or Curling the Tongue (37, 38. Superior and Inferior longitudinals)

Patient protrudes tongue and curls it upward to touch the philtrum and then downward to the chin (Fig. 7–62).

Manual Resistance: None

Instructions to Patient: "Touch above your upper lip with your tongue."
"Touch your tongue to your chin."

Criteria for Grading Tongue Motions

F: Patient completes available range and holds against resistance.

Protrusion: Tongue extends considerably beyond lips.
Deviation: Tongue reaches some part of the cheek or the lateral sulcus (pocket between teeth and cheek).
Retraction: Tongue returns to rest position in mouth against resistance.
Elevation: Tongue rises so that superior surface reaches the hard palate against considerable resistance; it blocks the oral cavity from the oropharynx.
Tipping: Tongue protrudes and touches area between upper lip and nasal septum (philtrum).

WF

Protrusion: Tongue reaches margin of lips.
Deviation: Tongue reaches corner(s) of mouth.
Retraction: Tongue returns to rest posture but with very very slight resistance.
Elevation: Tongue reaches hard palate with slight resistance, and oral cavity is blocked from oropharynx.
Tipping: Tongue protrudes and curls but does not reach philtrum.

NF

Protrusion: Minimal protrusion and tongue does not clear mouth.
Deviation: Tongue protrudes and deviates slightly to side.
Retraction: Tongue tolerates no resistance and retracts haltingly.
Elevation: Tongue moves toward hard palate but does not occlude oropharynx from oral cavity.

0

All motions: None

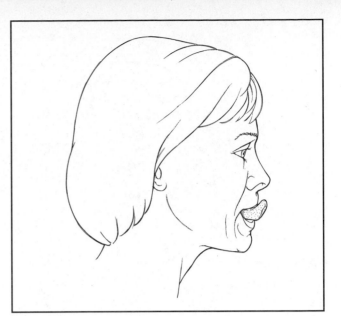

Figure 7–62

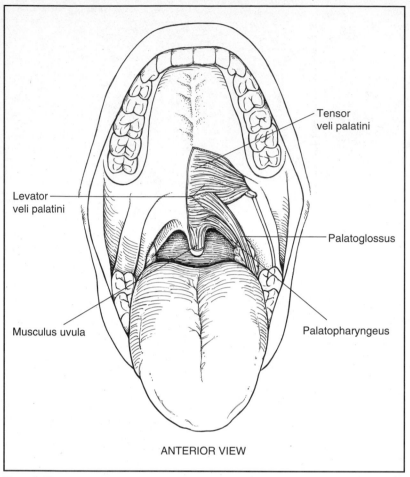

Tensor
veli palatini

Levator
veli palatini

Palatoglossus

Musculus uvula

Palatopharyngeus

ANTERIOR VIEW

Figure 7–63

Table 7–7	MUSCLES OF THE PALATE	
Muscle	**Origin**	**Insertion**
46. Levator veli palatini	Temporal bone Tympanic fascia Auditory tube	Palatine aponeurosis
47. Tensor veli palatini	Pterygoid process Auditory tube Sphenoid spine	Palatine aponeurosis Palatine bone
48. Musculus uvulae	Palatine bone Palatine aponeurosis	Uvula
49. Palatopharyngeus	Soft palate (pharyngeal aspect) Hard palate	Thyroid cartilage Side of pharynx

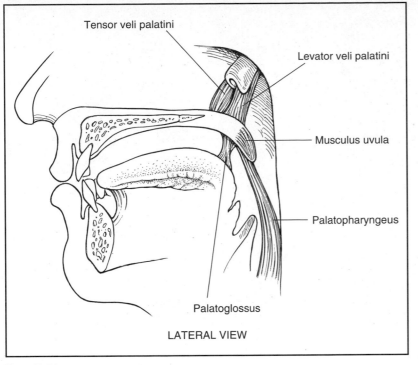

Figure 7–64

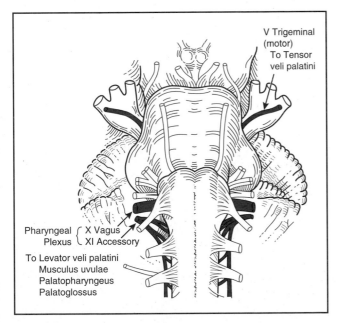

Figure 7–65

The muscles of the palate are innervated by the pharyngeal plexus (derived from the X [vagus] and XI [accessory] cranial nerves) with the single exception of the Tensor veli palatini, which derives its motor supply from the trigeminal (V) nerve.

The Tensor veli palatini elevates the soft palate, and paralysis of this muscle results in slight deviation of the uvula toward the unaffected side with its tip pointing toward the involved side. Weakness of the Tensor as an elevator of the palate may be masked if the pharyngeal muscles innervated by the pharyngeal plexus are intact.[1,11-12] In any event, the Levator veli palatini is a more important elevator of the palate than the Tensor.[12]

The Levator veli palatini also pulls the palate upward and backward to block off the nasal passages in swallowing. The Musculus uvulae shortens and bends the uvula to aid in blocking the nasal passages for swallowing. The Palatopharyngeus draws the pharynx upward and depresses the soft palate.

In the presence of a unilateral vagus (X) nerve lesion, the Levator veli palatini and the Musculus uvulae on the involved side are weak. There is a resultant lowering or flattening of the palatal arch, and the median raphe deviates toward the uninvolved side. With phonation, the uvula deviates to the uninvolved side.

With a bilateral vagus lesion, the palate cannot be elevated for phonation, but it does not sag because of the action of the Tensor veli palatini (V nerve).[12] The nasal cavity is not blocked off from the oral cavity with the bilateral lesion, which may lead to nasal regurgitation of liquids. Also, during speaking, air escapes into the nasal cavity, and the change in resonance gives a peculiar nasal quality to the voice. Dysphagia may be severe.

DESCRIPTION OF THE PALATE

The palate, or roof of the oral cavity, is viewed with the mouth fully open and the tongue protruding (Fig. 7–66). The palate has two parts: the hard palate is the vault over the front of the mouth, and the soft palate is the roof over the rear of the oral cavity.[1]

The hard palate is formed from the maxilla (palatine processes) and the horizontal plates of the palatine bones. Its boundaries are: anterolaterally, the alveolar arch and gums of the teeth, and posteriorly, the soft palate. The frontal mucosa is thick, pale, and corrugated; the posterior mucosa is darker, thinner, and not corrugated. The superior surface of the palate forms the nasal floor.

The soft palate is really a rather mobile soft tissue flap suspended from the hard palate, which slopes down and backward.[1] Its superior border is attached to (or continuous with) the posterior margin of the hard palate, and its sides blend with the pharyngeal wall. The inferior wall of the soft palate hangs free as a border between the mouth and the pharynx. The conical uvula drops from its posterior border.

The palatal arches are two curved folds of tissue containing muscles which descend laterally from the base of the uvula on either side. The anterior of these, the *palatoglossal arch*, holds the Palatoglossus and descends to end in the lateral sides of the tongue. The posterior fold, the *palatopharyngeal arch*, contains the Palatopharyngeus muscle and descends on the lateral wall of the oropharynx.[1,6] The palatine tonsils lie in a triangular notch between the diverging palatoglossal and palatopharyngeal arches.

The pharyngeal isthmus (or margin of the fauces) lies between the border of the soft palate and the posterior pharyngeal wall. The fauces forms the passageway between the mouth and the pharynx that includes the lumen as well as its boundary structures. The fauces closes during swallowing as a result of the elevation of the palate and contraction of the palatopharyngeal muscles (acting like a sphincter) and by elevation of the dorsum of the posterior tongue (Palatoglossus).

In examining the soft palate, observe the position of the palate and uvula at rest and during quiet breathing and then during phonation. If the palatine arches elevate symmetrically, minor deviations of the uvula are insignificant (for one thing, uvular changes often follow tonsillectomy).[11] Check for the presence of dysarthria and dysphagia (both liquids and solids).

Normally the uvula hangs in the midline and elevates in the midline during phonation.

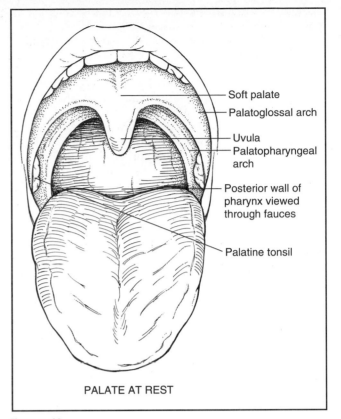

PALATE AT REST

Soft palate
Palatoglossal arch
Uvula
Palatopharyngeal arch
Posterior wall of pharynx viewed through fauces
Palatine tonsil

Figure 7–66

Elevation and Adduction of the Soft Palate
(46. Levator veli palatini, 47. Tensor veli palatini, 36. Palatoglossus, 48. Musculus uvulae)

Test: Patient produces a high-pitched "Ah-h-h" to cause the soft palate to elevate and adduct (the arches come closer together, narrowing the fauces) (Fig. 7–67).

To see the palate and fauces adequately, the examiner may need to place a tongue blade lightly on the tongue and use a flashlight to illuminate the interior of the mouth. Placing the tongue blade too far back or too heavily on the tongue may initiate a disagreeable gag reflex.

When this test does not give the desired information, the examiner may have to stimulate a gag reflex. Light-touch stimulation, done slowly and gradually with an applicator (preferably) or tongue blade placed on the posterior tongue or soft palate, will evoke a reflex and produce the desired motion when phonation fails to do so.

Remember that the gag reflex is not a constant finding. Some normal people do not have one, and many people have an exaggerated reflex.

Resistance: None.

Instructions to Patient: "Use a high-pitched (soprano) tone to say 'Ah-a-a-a'."

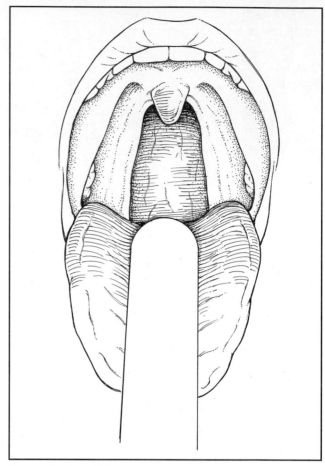

Figure 7–67. Say "Ah-h-h."

Criteria for Grading (Derived from Observation of Uvular and Arch Motion)

F: Uvula moves briskly and elevates while remaining in the midline. The palatoglossal and palatopharyngeal arches elevate and adduct to narrow the fauces.

WF: Uvula moves sluggishly and may deviate to one or the other side. Uvula deviation is toward the uninvolved side (Fig. 7–68). The arches may elevate slightly and asymmetrically.

NF: Almost imperceptible motion of both the uvula and the arches occurs.

0: No motion occurs, and the uvula is flaccid and pendulous.

Occlusion of the Nasopharynx (49. Palatopharyngeus)

Test: Aiming at the examiner's finger, the patient blows through the mouth with pursed lips to occlude the nasopharynx. Place a slim mirror above the upper lip (horizontally blocking off the mouth) to check for air escape from the nostrils (the mirror clouds). Alternatively, place a small feather fixed to a small plastic platform right under the nose, and the motion of the feather is used to detect air leakage.
Nasal speech is a sign of inability to close off the nasopharynx.

Resistance: None

Instructions to Patient: "Blow on my finger."

Criteria for Grading

F: No leakage of air through the nose.

WF: Minimal leakage of air. Slight mirror clouding or feather ruffling.

NF–0: Heavy mirror clouding or brisk feather ruffles.

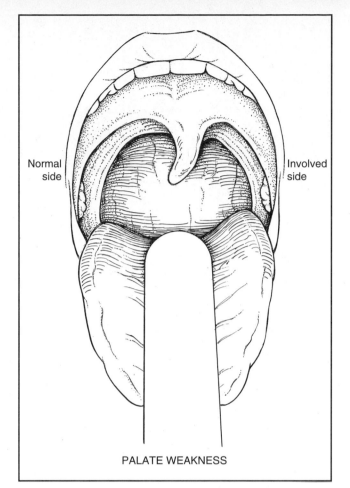

Normal side

Involved side

PALATE WEAKNESS

Figure 7–68

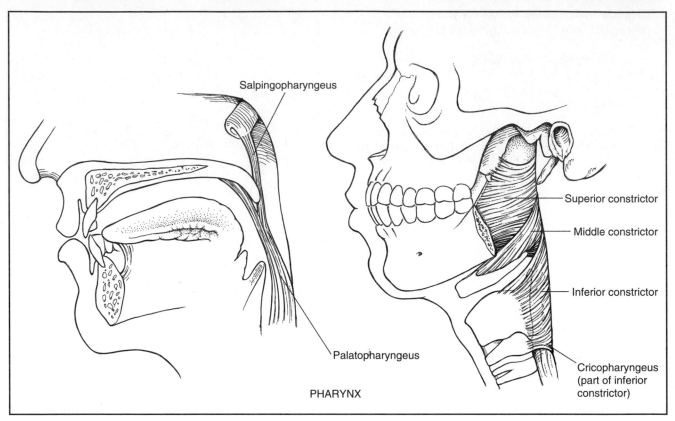

Figure 7–69

Figure 7–70

Table 7–8	MUSCLES OF THE PHARYNX	
Muscle	**Origin**	**Insertion**
41. Inferior constrictor	Cricoid cartilage Thyroid cartilage	Pharynx (posterior fibrous raphe)
42. Middle constrictor	Hyoid bone Stylohyoid ligament	Pharynx (median fibrous raphe)
43. Superior constrictor	Medial pterygoid plate Mandible Side of tongue	Pharynx (median fibrous raphe) Occipital bone
44. Stylopharyngeus	Styloid process	Thyroid cartilage
45. Salpingopharyngeus	Auditory tube	Blends with Palatopharyngeus

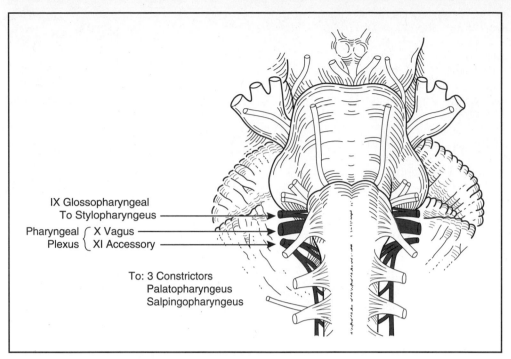

IX Glossopharyngeal
To Stylopharyngeus

Pharyngeal ⎧ X Vagus
Plexus   ⎩ XI Accessory

To: 3 Constrictors
Palatopharyngeus
Salpingopharyngeus

Figure 7–71

The function of the pharyngeal muscles is tested by observing their contraction during phonation and their elevation of the larynx during swallowing. The pharyngeal reflex also should be invoked and the nature of the muscle contraction noted. The manner in which the patient handles solid and liquid foods as well as the quality and character of speech should be described.

The motor parts of the glossopharyngeal (IX) cranial nerve go to the pharynx but probably innervate only the Stylopharyngeus muscle. The Stylopharyngeus elevates the upper lateral and posterior walls of the pharynx in swallowing.[16]

The remaining pharyngeal muscles (Inferior, Middle, and Superior constrictors, Palatopharyngeus, and Salpingopharyngeus) are innervated by the pharyngeal plexus composed of elements from the vagus (X) and accessory (XI) cranial nerves. The three constrictor muscles flatten and contract the pharynx in swallowing and are important participants in forcing the bolus of food into the esophagus, thereby initiating peristaltic activity in the gut. The Salpingopharyngeus blends with the Palatopharyngeus and elevates the upper portion of the pharynx.[1] Because the pharynx acts as a resonator box for sound, impairment of the pharyngeal muscles will alter the voice.

The Inferior constrictor has two parts, which often are referred to as if they were separate muscles.[1] One, the Cricopharyngeus, blends with the circular esophageal fibers to act as a distal pharyngeal sphincter in swallowing. These fibers prevent air from entering the esophagus during respiration and reflux of food from the esophagus back into the pharynx. It has been reported that when the system is at rest, the Cricopharyngeus is actively contracted to prevent air from entering the esophagus.[13] When a swallow is initiated, some form of neural inhibition causes the Cricopharyngeus to relax.[14,15] At the same time, the hyoid bone and the larynx elevate and move anteriorly, and the constrictor muscles act in a peristaltic manner, the sum of which permits passage of the bolus.[14]

The upper part of the Inferior constrictor is the Thyropharyngeus, which acts to propel the bolus of food downward.[1]

In unilateral lesions of the vagus (X) nerve, laryngeal elevation is decreased on one side, and in bilateral lesions it is decreased on both sides.

Constriction of the Posterior Pharyngeal Wall

Test: Patient opens mouth wide and says "Ah-h-h" with a high pitched tone.

This sound causes the posterior pharyngeal wall to contract (the soft palate adducts and elevates as well).

Because it is difficult to observe the posterior wall of the pharynx, use a flashlight to illuminate the interior of the mouth. A tongue blade will probably be needed to keep the tongue from obstructing the view, but care must be taken not to initiate a gag reflex.

Patients with weakness may have an accumulation of saliva in the mouth. Ask the patient to swallow, or, if this does not work, use mouth suctioning. If the patient has a nasogastric tube it will descend in front of the posterior wall and may partially obstruct a clear view.

If there is little or no motion of the pharyngeal wall, the examiner will have to stimulate the pharyngeal reflex to ascertain contractile integrity of the Superior constrictor and other muscles of the pharyngeal wall. Patients do not like this reflex test.

The Pharyngeal Reflex Test: The pharyngeal reflex is tested by applying a stimulus with an applicator to the posterior pharyngeal wall or adjacent structures (Fig. 7–72). The stimulus should be applied bilaterally. If positive, elevation and constriction of the pharyngeal muscles will occur along with retraction of the tongue.

Criteria for Grading

F: Brisk contraction of posterior pharyngeal wall.

WF: Decreased movement or sluggish motion of the pharyngeal wall.

NF: Trace of motion (easily missed).

0: No contractility of pharyngeal wall.

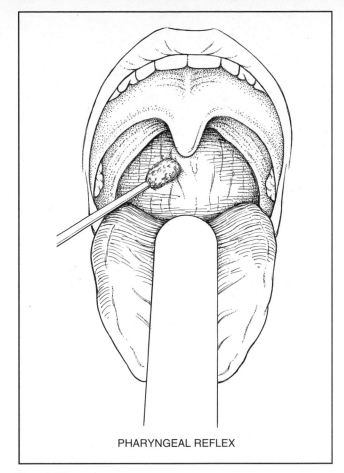

PHARYNGEAL REFLEX

Figure 7–72

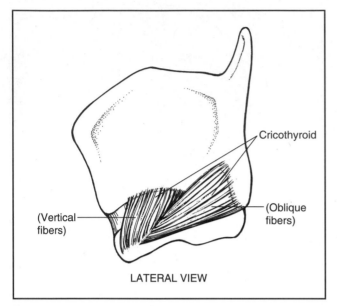

Cricothyroid

(Vertical fibers)

(Oblique fibers)

LATERAL VIEW

Figure 7–73

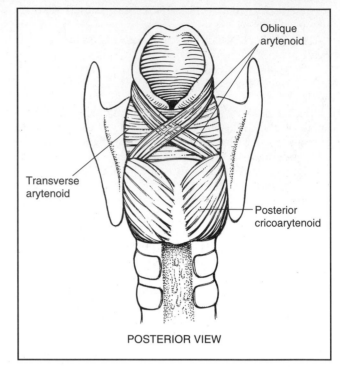

Oblique arytenoid

Transverse arytenoid

Posterior cricoarytenoid

POSTERIOR VIEW

Figure 7–74

Table 7–9	MUSCLES OF THE LARYNX	
Muscle	**Origin**	**Insertion**
50. Cricothyroid	Cricoid cartilage	Larynx (inferior cornu) Thyroid cartilage
51. Posterior cricoarytenoid	Cricoid cartilage	Arytenoid cartilage (posterior)
52. Lateral cricoarytenoid	Cricoid cartilage	Arytenoid cartilage (anterior)
53. Transverse arytenoid	Crosses transversely between the two arytenoid cartilages	
54. Oblique arytenoids	Arytenoid cartilage (posterior)	Arytenoid cartilage (apex) on other side
55. Thyroarytenoid	Thyroid cartilage Middle cricothyroid ligament	Arytenoid cartilage (anterior base)

Others
Infrahyoid muscles

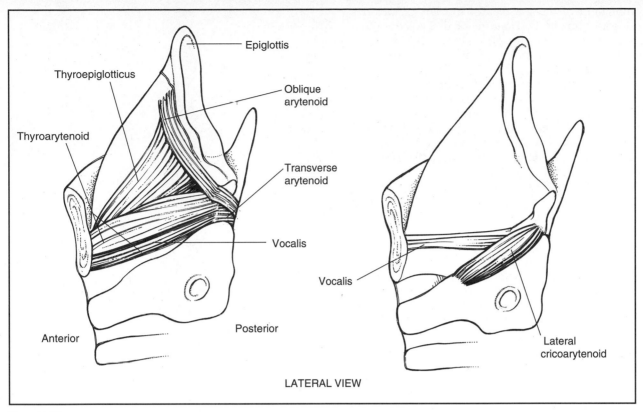

Epiglottis

Thyroepiglotticus

Oblique arytenoid

Thyroarytenoid

Transverse arytenoid

Vocalis

Vocalis

Anterior

Posterior

Lateral cricoarytenoid

LATERAL VIEW

Figure 7–75

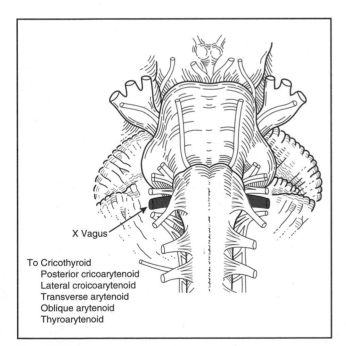

X Vagus

To Cricothyroid
Posterior cricoarytenoid
Lateral croicoarytenoid
Transverse arytenoid
Oblique arytenoid
Thyroarytenoid

Figure 7–76

Examination of the muscles of the larynx includes assessing the quality and nature of the voice, noting any abnormalities of phonation or articulation; impairment of coughing (see accompanying box); and any problems with respiration. Also important is the rate of opening and closing of the glottis.

Some general definitions are in order. *Phonation* is the production of vocal sounds without the formation of words; phonation is a function of the larynx.[5] *Articulation,* or the formation of words, is a joint function of the larynx along with the pharynx, palate, tongue, teeth, and lips.

The laryngeal muscles all are innervated by the recurrent branches of the X (vagus) cranial nerve with the exception of the Cricothyroid, which receives its motor innervation from the superior laryngeal nerve. The laryngeal muscles regulate the tension of the vocal cords and open and close the glottis by abducting and adducting the vocal cords. The vocal cords normally are open (abducted) during inspiration and adducted while speaking or during coughing.

The Cricothyroids (paired) are the principal tensors owing to their action in lengthening the vocal cords.[1,5,11] The Posterior cricoarytenoids (paired) are the main abductors and glottis openers; the Lateral cricoarytenoids (paired) are the main adductors and glottis closers. The Thyroarytenoids (paired) shorten and relax the vocal cords by drawing the arytenoid cartilages forward. The unpaired Arytenoid (transverse and oblique heads) draws the arytenoid cartilages together; the oblique head acts as the sphincter of the upper larynx (called the aryepiglottic folds), and the transverse head acts as the sphincter of the lower larynx.

Paralysis of the laryngeal muscles on one side does not cause an appreciable change in the voice in contrast to the difficulty resulting from bilateral weakness. Loss of the Cricothyroids leads to loss of the high tones and the voice sounds deep and hoarse and fatigues readily, but respiration is normal. Loss of the Thyroarytenoids bilaterally changes the shape of the glottis and results in a hoarse voice, but again, respiration is normal.

With bilateral paralysis of the Posterior cricoarytenoids, both vocal cords will lie close to the midline and cannot be abducted, leading to severe dyspnea and difficult inspiratory effort (inspiratory stridor).[5] Expiration is normal.

In bilateral adductor paralysis (lateral Cricoarytenoids), inspiration is normal because abduction is unimpaired. The voice, however, is lost or has a whisper quality.

With unilateral loss of both abduction and adduction the involved vocal cords are motionless, and the voice is low and hoarse. In bilateral loss all vocal cords are quiescent, and speech and coughing are lost. Marked inspiratory stress occurs, and the patient is dyspneic.

THE FUNCTIONAL ANATOMY OF COUGHING

Cough is an essential procedure to maintain airway patency and to clear the pharynx and bronchial tree when secretions accumulate. A cough may be a reflex or voluntary response to irritation anywhere along the airway downstream from the nose.

The cough reflex occurs as a result of stimulation of the mucous membranes of the pharynx, larynx, trachea, or bronchial tree. These tissues are so sensitive to light touch that any foreign matter or other irritation initiates the cough reflex. The sensory (afferent) limb of the reflex carries the impulses set up by the irritation via the glossopharyngeal and vagus cranial nerves to the fasciculus solitarius in the medulla, from which the motor impulses (efferent) then move out to the muscles of the pharynx, palate, tongue, and larynx as well as to those of the abdominal wall, chest, and Diaphragm. The reflex response is a deep inspiration (about 2.5 liters of air) followed quickly by a forced expiration during which the glottis closes momentarily, trapping air in the lungs.[13] The Diaphragm contracts spasmodically as do the abdominal muscles and intercostal muscles. This raises intrathoracic pressure (to above 200 mmHg) until the vocal cords are forced open, and the explosive outrush of air expels mucus and foreign matter. The expiratory airflow at this time may reach a velocity of 75 mph or higher.[13] Important to the reflex action is the fact that the bronchial tree and laryngeal walls collapse because strong compression of the lungs causes an invagination such that the high linear velocity of airflow moving past and through these tissues dislodges mucus or foreign particles, thus producing an effective cough.

The three phases of cough—inspiration, compression, and forced expiration—are mediated by the muscles of the thorax and abdomen as well as of the pharynx, larynx, and tongue. The deep inspiratory effort is supported by the Diaphragm, intercostals, and the arytenoid abductor muscles (the Posterior cricoarytenoids), permitting inhalation of upward of 1.5 liters of air.[16] The Palatoglossus and Styloglossus elevate the tongue and close off the oropharynx from the nasopharynx.

The compression phase requires the Lateral cricoarytenoid muscles to adduct and close the glottis.

The strong expiratory movement is augmented by strong contractions of the thorax muscles, particularly the Latissimus dorsi and the oblique and transverse abdominal muscles. The abdominal muscles raise intra-abdominal pressure, forcing the relaxing Diaphragm up and drawing the lower ribs down and medially. Elevation of the Diaphragm raises intrathoracic pressure to about 200 mmHg, and the explosive expulsion phase begins with forced abduction of the glottis.

MUSCLES OF THE LARYNX

Elevation of the Larynx in Swallowing

Test: The larynx elevates during swallowing. The examiner lightly grasps the larynx with the thumb and index finger on the anterior throat to determine the fact of elevation and its extent (Fig. 7–77). **DO NOT PRESS DIRECTLY ON THE FRONT OF THE LARYNX AND NEVER USE MUCH PRESSURE ON THE NECK.**

Resistance: None

Instructions to Patient: "Swallow."

Criteria for Grading

F: Larynx elevates at least 20 mm in most people.[17] The motion is quick and controlled.

WF: Laryngeal excursion may be normal or slightly limited. The motion is sluggish and may be irregular.

NF: Excursion is perceptible but less than normal. Aspiration may occur.

0: No laryngeal elevation occurs. (Aspiration will result in this event.)

Vocal Cord Abduction and Adduction (51. Posterior and 52. Lateral Cricoarytenoids)

In this test the examiner is looking for hoarseness, pitch and tone range, breathlessness, breathiness, nasal-quality speech, dysarthria, and articulation or phonation disturbances.

Test and Instructions to Patient: The patient is asked to respond to four different commands to determine the nature of airflow control during respiration, vocalization, and coughing.

1. "State your name."
Patient should be able to say his or her name completely without running out of breath.
2. "Sing several notes in the musical scale" (do, re, mi, etc.) first at a low pitch and then at a higher pitch.
Patient should be able to sustain a tone (even when "can't carry a tune") and vary the pitch.
3. "Repeat five times a hard staccato, interrupted sound: 'Akh, Akh, Akh.'
Examiner must demonstrate this sound to the patient. Patient should make and break sounds crisply with a definite halt between each sound in the series.
4. "Cough."

Figure 7–77

Evaluation of Cough in the Context of Laryngeal Function: Refer to accompanying box on cough. Examiner determines whether the patient has a voluntary and effective cough. A voluntary cough is initiated on command. A reflex cough, because it cannot be initiated on command, must be evaluated when it occurs, which may be outside of the test session. The reflex cough occurs in response to irritation of the membranes of the postnasal air passages.

An effective or functional cough, voluntary or reflex, clears secretions from the lungs or airways. A functional cough is dependent on the coordination of the respiratory and laryngeal muscles.

Control of inspiration must be sufficient to fill the lungs with the necessary volume of air to produce a cough. Effective expiration of air during a cough is dependent on forceful contraction of the abdominal muscles. The vocal cords must adduct tightly to prevent air loss. Adduction of the vocal cords must be maintained prior to the expulsion of air.

A nonfunctional cough resulting from laryngeal deficiency sounds like clearing the throat or a low guttural sound, or there may be no cough sound at all.

The kinesiology of swallowing is the subject of continued controversy. Many of the rapid actions described as sequential are close to simultaneous events. The means of studying swallowing are limited to a great extent by the limitations inherent in palpation, following ingested food, videofluoroscopy, manometry, and acoustic measures.

MUSCLE ACTIONS IN SWALLOWING

Ingestion of Food and Formation of Bolus (Oral Preparatory Phase)

- The food or liquid is placed in the oral cavity, and the Orbicularis oris contracts to maintain a labial seal and to prevent drooling. The Palatoglossus maintains a posterior seal by maintaining the tongue against the soft palate, which prevents leakage too early into the pharynx.[18]
- Foods are broken down mechanically by integrated action of the muscles of the tongue, jaw, and cheeks.
- Liquids: Intrinsic tongue muscles squirt fluids into the back of the mouth. The Mylohyoid raises and bulges the back of the tongue into the oropharynx. Lips must be closed to retain fluids.
- Solids: Muscles of the tongue and cheek (Buccinator) place the food between the teeth, which bite, crush, and grind it via action of the muscles of mastication (see Table 7–5). The food, when mixed with saliva (by the tongue intrinsics), forms a bolus behind the tip of the tongue.
- The tongue muscles (see Table 7–6) raise the anterior tongue and press it against the hard palate, which pushes the bolus back into the fauces.

The Oral Phase

- In this phase of swallowing the bolus is squeezed against the hard palate by the tongue, the lip seal is maintained, and the Buccinator continues to prevent pocketing or lodging of food in the lateral sulci.
- The tongue is drawn up and back by the Styloglossus.
- The palate muscles (see Table 7–7) depress the soft palate down onto the tongue to "grip" the bolus.
- The hyoid bone and the larynx are elevated and moved forward by the suprahyoid muscles.
- The palatal arches are adducted by the paired Palatoglossi.
- The bolus is driven back into the oropharynx.
- As a prelude to the act of swallowing, the hyoid bone is raised slightly, and this action is accompanied by a quiescence of all muscle action: chewing, talking, food movement in the mouth, capital and cervical motion, facial movements. Even respiration is momentarily arrested.[15,18]

- The soft palate is raised (Levator veli palatini) and tightened (Tensor veli palatini) to be firmly fixed against the posterior pharyngeal wall. This leads to a tight closure of the pharyngeal isthmus (Palatopharyngeus and Superior constrictor), which prevents the bolus from rising into the nasopharynx.

The Engulfing Actions Through the Pharynx (Pharyngeal Phase)

- The epiglottis moves upward and forward, coming to a halt at the root of the tongue, and literally bends backward (possibly because of the weight of the bolus) to cover the laryngeal inlet. The bolus of food slides over its anterior surface. (The epiglottis in man is not essential to swallowing, which is normal even in the absence of an epiglottis.[1])
- The fauces narrow (Palatoglossi).
- *Note:* The pharyngeal isthmus is at the border of the soft palate and the posterior pharyngeal wall and is the communication between the nasal and oral parts of the pharynx. Its closure is effected by the approximation of the two Palatopharyngeus muscles and the Superior constrictor, which form a palatopharyngeal sphincter.
- The larynx and pharynx are raised up behind the hyoid (Salpingopharyngeus, Stylopharyngeus, Thyrohyoid, and Palatopharyngeus).
- The arytenoid cartilages are drawn upward and forward (Oblique arytenoids and Thyroarytenoids), and the aryepiglottic folds approximate, which prevents movement of the bolus into the larynx.
- During swallowing the thyroid cartilage and hyoid bone are approximated, and there is a general elevation of the pharynx, larynx, and trachea. This causes the many laryngeal folds to bulge posteriorly into the laryngeal inlet, thus narrowing it during swallowing.[18,19]
- The bolus then slips further over the epiglottis, and partly by gravity and partly by the action of the constrictor muscles it passes into the lowest part of the pharynx. Passage is aided by contraction of the Palatopharyngei, which elevate and shorten the pharynx, thus angling the posterior pharyngeal wall to allow the bolus to slide easily downward.[20]
- The laryngeal passage is narrowed by the aryepiglottic folds (Posterior cricoarytenoids, Oblique arytenoids, and Transverse arytenoids), which close the laryngeal vestibule (glottis) and also form lateral channels to direct the bolus toward the esophagus.
- When the Posterior cricoarytenoids are weak or paralyzed, the laryngeal inlet is not closed off in swallowing, the aryepiglottic folds move medially, and fluid or food enters the larynx (aspiration).

The Esophageal Phase

- At the beginning of this phase the compressed bolus is in the distal pharynx. The Inferior constrictor pushes the bolus inferiorly (peristaltic action) to enter the esophagus. The distal fibers of the Inferior constrictor, called the Cricopharyngeus, are a distal sphincter and therefore must relax to allow the bolus to pass, but the mechanism of this action is in dispute.[20,21]

- After the passage of the bolus the intrinsic tongue muscles move saliva around the mouth to cleanse away debris.

TESTING SWALLOWING

Swallowing is tested only when there is good cause to suspect that the swallowing mechanisms are faulty. Do not make an *a priori* assumption that the presence of a nasogastric tube, a gastrostomy, or a liquid diet precludes swallowing. The examiner also should review information from the patient's history and current chart to identify the site of the lesion, the presence of upper respiratory tract infections, and similar facts, which will assist the direction of the evaluation.

When a patient has a tracheostomy, a suctioning machine is essential, and expertise with its use is required.

The examiner will have some prior information about patients from direct observation, such as how they handle saliva (swallowing it or drooling), whether and how they imbibe liquids and solids at mealtime, reports from nursing staff and family, and the nature of reported problems about swallowing. These all will suggest a starting point for testing.

In most swallowing tests use a bib around the patient's neck to prevent soiling. Remember to protect yourself from sudden aspirates! Damp washcloths or tissues should be available for clean-up.

Position of Patient: Sitting preferred, supine if necessary, but head and trunk should be elevated to at least 30°. Maintain head and neck in neutral position.

Position of Therapist: Sitting in front of and slightly to one side of patient.

Table 7–10 COMMON SWALLOWING PROBLEMS AND MUSCLE INVOLVEMENT	
Problem	**Possible Anatomic Cause**
Drooling	Weakness of Orbicularis oris
Pocketing in the lateral sulci	Weakness of the Buccinator and intrinsic and extrinsic tongue muscles
Decreased ability to break down food mechanically during the oral preparatory phase	Weakness of the muscles of mastication
Decreased ability to form bolus	Weakness of the intrinsic and extrinsic tongue muscles Weakness of the Buccinator
Decreased ability to maintain bolus in the oral cavity during the oral preparatory phase	Weakness of the Palatoglossus and/or Styloglossus
Nasal regurgitation	Weakness of the Palatopharyngeus, Levator veli palatini, and/or Tensor veli palatini
Posterior pharyngeal wall residual after the swallow	Weakness of the pharyngeal constrictor(s)
Coughing or choking prior to the swallow	Food may spill into an unprotected airway secondary to: 1. Weakness of the intrinsic and/or extrinsic tongue muscles resulting in decreased ability to form a bolus (lack of bolus formation may result in spillage of the oral contents without initiation of a swallow); or 2. Weakness of the Palatoglossus and Styloglossus muscles resulting in decreased ability to maintain the bolus in the oral cavity prior to initiation of a swallow
Coughing or choking during the swallow	Weakness of the muscles responsible for closing the true vocal folds, false vocal folds, and aryepiglottic folds
Coughing or choking after the swallow	Decreased strength of the Genioglossus resulting in decreased tongue retraction with vallecular residual, which spills after the swallow into an unprotected airway Pharyngeal constrictor weakness with residual spillage from the pharyngeal walls after the swallow into an unprotected airway Decreased cricopharyngeal opening with overflow from the pyriform sinus after the swallow into an unprotected airway

Preliminary Procedures to Determine Clinically the Safety of Ingestion of Food or Liquids

Test Sequence 1

Laryngeal Elevation: Examiner lightly grasps the larynx between the thumb and index finger on the anterior surface of the throat. Ask the patient to swallow. Ascertain if there is laryngeal elevation and its extent.

Criteria for Grading

F: Larynx elevates at least 20 mm. Motion is quick and controlled.

WF: Laryngeal excursion may be normal or slightly limited. The motion may be sluggish or appear irregular.

NF: Elevation is perceptible but significantly less than normal.

0: No laryngeal elevation occurs.

Implications of Grade: If the patient is graded F (functional) or WF (weak functional), proceed with the swallowing assessment. If the patient is graded NF (nonfunctional) or zero and does not have a tracheostomy, discontinue the swallowing assessment. For patients with a tracheostomy, add a blue vegetable dye to the bolus to facilitate identification of any aspirated bolus during suctioning.

Test Sequence 2

Initial Ingestion of Water

Prerequisites: The patient has a grade of F or WF on Test 1.

There also must be at least a grade of WF or higher on the tests for posterior elevation of the tongue (see page 295) and constriction of the posterior pharyngeal wall (see page 307).

Procedure: There are several ways to get water into the mouth to test swallowing. It does not matter which is used.

The first trial of swallowing begins with a small amount (1 to 3 ml) of water. The rationale is that should the patient not be able to swallow the water correctly and it is aspirated, the lungs can absorb this small quantity without penalty. There also is evidence accumulating that differences in pH of water can cause damage to the lungs, so the small amount of water is very important. Each procedure should be repeated at least three or four times.

1. If the patient is cognitively clear, offer a glass or cup containing a tiny amount of water and allow the patient to sip. The test is successful if the water can be swallowed with one attempt, the swallow is inaudible, and the water is swallowed without any choking or coughing.
   If successful, proceed to Test sequence 3.
2. If the patient cannot sip from a cup, offer a straw and ask the patient to suck a small amount. The shorter and wider the straw, the easier the task. If the swallowing attempt is successful as described in 1 above, proceed to Test sequence 3.
3. If the patient cannot sip or suck, trap water in a straw and place the straw in the side of the patient's mouth between the cheek and lower teeth. Tell the patient you are going to release the water and request a swallow. If successful, proceed to Test sequence 3.
4. If the patient is not cognitively clear, control the amount of water available. This is most readily done by trapping water in a straw to give to the patient.
5. For the patient who cannot handle fluid, try thickening the water with gelatin to a consistency of thin gruel or thick pea soup.

Outcomes: If any of these trials are successful proceed cautiously to a trial of pureed food. If none of these tests are successful and the patient does not have a tracheostomy, DO NOT give the patient food by mouth until further testing (e.g., fluoroscopy) can be conducted.

If the procedures with water are not successful and the patient has a tracheostomy (through which aspirated food can be suctioned), proceed cautiously to the use of pureed food, which usually is easier to swallow than water.

Test Sequence 3

Pureed Food

The most palatable commercial pureed foods are the pureed baby food fruits, possibly because they taste like what the label states they are! The pureed meats and vegetables are totally unseasoned, which is unfamiliar and usually unpalatable to adults. Avoid milk products initially because they thicken the saliva. Ask about patient food preferences and try to use something enjoyable.

A suctioning machine is essential if the patient has a tracheostomy. It is recommended that the food be colored with vegetable dye (blue is readily seen and is not confused with body secretions or fluids) so that any aspiration can be readily detected as the color appears in tracheostomy secretions.

Criteria for Initiating Trials with Pureed Foods

1. Laryngeal elevation is Functional (F) or Weak functional (WF).
2. Posterior pharyngeal wall constriction is at least WF.
3. Patient has been successful in handling water in Test sequence 2 or by observation.
4. Patient *must* have a functional cough (voluntary or reflex), or a tracheostomy. Some patients have a depressed gag reflex, but cough is the essential component in swallowing. The examiner cannot assume that a hyperactive gag reflex is synonymous with a functional cough.
5. The patient must have adequate cognition to attend to feeding.
6. There cannot be any respiratory problem present, such as aspiration pneumonia, that might be compromised by additional aspiration.

Procedures

1. Place a small amount (1/2 teaspoon) of food on the front of the tongue. Ask the patient to swallow, and observe ability to manipulate food in the mouth to position it for swallowing.

   Allow the patient to place the food in the mouth if possible because this will better coordinate feeding with the respiratory cycle.
2. If the patient cannot move the food in the mouth, push it back slightly with a tongue blade, being careful not to initiate a gag reflex. Ask the patient to swallow, while lightly palpating the larynx to check laryngeal elevation.
3. Ask the patient to open the mouth, and check to see that food has indeed been swallowed and that none of it has pooled in the pharyngeal isthmus or oral cavity.

4. To check for a clear airway ask the patient to repeat three sequential crisp sounds: "Agh, Agh, Agh." Any gurgling indicates that food is in the airway and ask the patient to swallow again.

   Repeat this procedure a number of times and check each response.

   After four or five trials with pureed food, pause for about 10 minutes to ascertain that the patient does not have delayed coughing because of food collecting in the pharynx, larynx, or trachea. A blue aspirate from the tracheostomy tube may occur sometime after the actual ingestion of food.

Outcomes: If the patient has no immediate or delayed coughing, choking, or positive aspirate after swallowing and the airway is clear, the test is successful.

If the patient repeatedly coughs, chokes, or has a positive aspirate, this is solid evidence that there is inadequacy of swallowing, and the test should be terminated and no other food administered.

For patients who have been on a nasogastric tube and have demonstrated the ability to swallow water and pureed food without aspiration, proceed with feeding the pureed food until at least three fourths of the jar has been consumed. For the next meal, order a tray of pureed food. Observe the patient during eating; look for any problems and assess fatigue.

Use of a Mechanical Soft Diet: A "Bulbar Mechanical Soft (BMS)" diet may be substituted for regular consistency food for patients with any of the following: lack of teeth or dentures, poor intraoral control for chewing, fatigue during mastication (e.g., postpolio or Guillain-Barré), limited jaw range of motion, limited attention span to complete the oral preparatory phase.

REFERENCES

1. Williams PL, Warwick R, Dyson M, Bannister LH. *Gray's Anatomy,* 37th British ed. Edinburgh, Churchill-Livingstone, 1989.
2. Walsh FB. *Clinical Neuroopthalmology,* 2nd ed. Baltimore, Williams & Wilkins, 1957.
3. Bender MB, Rudolph SH, Stacy CB. The neurology of the visual and oculomotor systems. *In* Joynt RJ (Ed). *Clinical Neurology.* Philadelphia: J.B. Lippincott, 1993.
4. Van Allen MW. *Pictorial Manual of Neurologic Tests.* Chicago: Year Book, 1969.
5. Haerer AF. *DeJong's The Neurologic Examination,* 5th ed. Philadelphia: J.B. Lippincott, 1992.
6. Clemente CD. *Gray's Anatomy* 30th American ed. Philadelphia: Lea & Febiger, 1985.
7. Hollingshead WH. *Functional Anatomy of the Limbs and Back.* Philadelphia: W.B. Saunders, 1969.
8. DuBrul EL. *Sicher and DuBrul's Oral Anatomy,* 8th ed. St. Louis: Ishiyaku EuroAmerica, 1988.
9. Nairn RI. The circumoral musculature: Structure and function. Br Dental J 138:49–56, 1975.
10. Lightoller GH. Facial muscles: The modiolus and muscles surrounding the rima oris with remarks about the panniculus adiposus. J Anat 60:1–85, 1925.
11. Brodal A. *Neurological Anatomy in Relation to Clinical Medicine.* London: Oxford University Press, 1981.
12. Misuria VK. Functional anatomy of the tensor palatini and levator palatini muscles. Ann Otolaryngol 102:265, 1975.
13. Guyton AC. *Textbook of Medical Physiology,* 8th ed. Philadelphia: W.B. Saunders, 1991.
14. Miller AJ. Neurophysiological basis of swallowing. Dysphagia 1:91–100, 1986.
15. Doty R. Neural organization of deglutition. *In Handbook of Physiology,* Section 6, Alimentary Canal. Washington DC: American Physiologic Society, 1968.
16. Starr JA. Manual techniques of chest physical therapy and airway clearance techniques. *In* Zadai CC. *Pulmonary Management in Physical Therapy. Clinics in Physical Therapy.* New York: Churchill-Livingstone, 1992.
17. Jacob P, Kahrilas PJ, Logemann JA, Shah V, Ha T. Upper esophageal sphincter opening and modulation during swallowing. Gastroenterology 97:1469–1478, 1989.
18. Logemann JA. *Evaluation and Treatment of Swallowing Disorders.* San Diego: College-Hill Press, 1983.
19. Bosma J. Deglutition: Pharyngeal stage. Physiol Rev 37:275–300, 1957.
20. Buthpitiya AG, Stroud D, Russell COH. Pharyngeal pump and esophageal transit. Dig Dis Sci 32:1244–1248, 1987.
21. Kilman WJ, Goyal RK. Disorders of pharyngeal and upper esophageal sphincter motor function. Arch Intern Med 136:592–601, 1976.

Upright Motor Control

The Test for Upright Control
Flexion Control Test
Extension Control Test

Chapter 8

The manual muscle tests described in Chapters 2 to 5 of this book are not germane to the evaluation of muscle activity when there is dysfunction of the central nervous system (CNS). In patients with CNS disorders, the muscles have normal innervation, but their control is disturbed because of damage to the CNS either in the brain or in the spinal cord. These persons have upper motor neuron disorders that are characterized by one or any combination of the following:

Abnormal limb movement patterns
Disturbed muscle tone (spasticity, rigidity)
Aberrations in the selection, amplitude, or timing of synergistic muscle activity, duration and rate (velocity) of activity in individual muscles
Impaired tactile sensation: paresthesias, anesthesias, or hypesthesias
Disturbed proprioception and kinesthesia
Impaired spatial discrimination
Impaired body image
Disturbed central balance mechanisms and abnormal postural reactions
Abnormal reflex activity

Analysis of a patient with some combination of these problems is a complex task. Manual muscle testing was not designed for such patients and should not be used to evaluate them.[1] *Manual muscle testing was (and is) designed to evaluate patients with a lower motor neuron disorder manifested by flaccid weakness or paralysis.* Its use in persons with CNS dysfunction yields spurious clinical results that have little or no relevance to function. Indeed, muscle testing scores in patients with lower motor neuron disorders do not necessarily relate to, or predict, function.

An obvious exclusion from this blanket assertion is patients with both CNS and lower motor neuron disorders. Two good examples are the patient with a spinal cord injury and the patient with amyotrophic lateral sclerosis.

Evaluation of muscle performance, however, is an important tool of the physical therapist in the patient with CNS derangement. One such tool was developed to test lower extremity control during standing.[2] It can be used in patients who have selective control, patterned motion, or a combination of the two.

Selective control is the ability to move a single joint without activating movement in an adjacent or neighboring joint of the same extremity. For example, the patient should be able to flex the elbow without incurring simultaneous motion at the shoulder or wrist.

Patterned motion is the inability to perform a fractionated motion (i.e., wrist extension without movement at the elbow or fingers). For example, following a stroke or brain injury a flexor pattern of movement is common in the upper extremity, as follows (the pattern is named after the prevailing motion at the elbow):

Shoulder abduction or extension
Elbow flexion
Forearm supination
Wrist and finger flexion

It is also common to see an extensor pattern of motion in the lower extremity:

Hip extension
Knee extension
Plantarflexion and inversion

These patterns are fairly stereotyped, but studies reveal multiple variations in the participating muscles and their amplitude in a "typical" flexion or extensor pattern.[3-5]

The Upright Motor Control Test was designed to incorporate the effects of upright posture and weight bearing.[2] It simulates the activity required for walking (i.e., flexion, which includes the factor of speed, and extension, which assesses joint stability.) Inter-tester reliability has been established at 96 percent agreement for the flexion portion of the test and 90 percent agreement for the extension portion of the test.[2] Validity with respect to prediction of gait performance from test data has not been established.

The Test for Upright Control

One examiner and an assistant are required to conduct this evaluation properly. The assistant should be a physical therapist or a person who has received extensive instruction in methods of positioning himself and the patient to provide appropriate (neither too little nor too much) stabilization and support. The patient must be able to understand all test instructions, as verified by appropriate responses to verbal commands or demonstration. The patient also must require no more than the assistance of one person for either single- or double-limb stance.

The test itself has two major sections: the flexion control test and the extension control test. Each of these sections has three parts, one each for the hip, the knee, and the ankle.

Flexion Control Test (in Parts 1, 2, 3)

The purpose of this portion of the upright motor control test is to ascertain flexion control of the nonweight-bearing extremity (i.e., for limb advancement in the swing phase of gait).

The test is conducted bilaterally unless there is unequivocal evidence that one side is without neurologic deficit. The assistant provides manual balance support by holding the patient's hand, positioning his arm so that the hand is at about the level of the greater trochanter. The support is given on the side contralateral to that being tested and should be sufficient for the patient to maintain standing balance during this segment of the test.

For the patient who has bilateral lower extremity involvement, external stabilization for contralateral hip and knee extension may be required during the single limb flexion test. This can be done manually by preventing knee flexion and holding the patient in hip extension; an external support such as a "knee immobilizer" may be used.

The examiner may stand in front of and facing the patient, or, if the patient has side confusion, he or she may stand slightly in front but facing in the same direction. The examiner demonstrates each test part as many times as necessary to ensure patient understanding. The patient then is allowed no more than two practice trials to avoid fatigue.

The actual data collection (graded trial) is limited to one trial per limb segment. Just prior to grading, the patient's test limb should be positioned in neutral at both hip and knee (0° at the hip and 0° at the knee). If the patient cannot reach neutral, a position of maximal extension range should be used.

PART 1: HIP FLEXION

Instructions to Patient: "Stand as straight as you can. Bring your knee up toward your chest, as high and as fast as you can."

Grading. The hip flexion motion must occur at the hip joint. Do not allow substitution or other contamination of the motion such as backward lean or pelvic tilt (see Table 8–1).

Table 8–1 HIP FLEXION

Score	Criteria
Weak (W)	No motion, or patient actively flexes less than 30°.
	Three repetitions through any range that requires as a group, more than 10 seconds to complete.
Moderate (M)	Actively completes an arc of hip flexion from 0° (or maximal extension angle) to between 30° and 60° three times within 10 seconds.
Strong (S)	Actively completes an arc of hip flexion from 0° (or maximal extension angle) to more than 60° three times within 10 seconds.

PART 2: KNEE FLEXION

Instructions to Patient: "Stand as straight as you can. Bring your knee up toward your chest three times, as high and as fast as you can."

Grading. See Table 8–2.

Table 8–2 KNEE FLEXION

Score	Criteria
Weak (W)	No motion, or knee flexes less than 30°.
	Completes three repetitions through any range but requires, as a group, more than 10 seconds.
Moderate (M)	Actively completes an arc of knee flexion from 0° to between 30° and 60° three times within 10 seconds.
Strong (S)	Knee flexes more than 60° three times within 10 seconds.

PART 3: ANKLE FLEXION (Dorsiflexion)

Instructions to Patient: "Stand as straight as you can. Bring your knee and foot up toward your chest as high and as fast as you can."

Grading. See Table 8–3.

Table 8–3 DORSIFLEXION

Score	Criteria
Weak (W)	No motion, or actively dorsiflexes to less than a right angle. (Examiner is cautioned not to confuse forefoot or toe extension with true ankle motion.)
	Completes three repetitions through any range but requires, as a group, more than 10 seconds.
Moderate (M)	This grade is not used because range of dorsiflexion is so limited and very little dorsiflexion is used in the swing phase of gait.
Strong (S)	Actively dorsiflexes to a right angle or greater three times in 10 seconds.

Extension Control Test (in Parts 4, 5, & 6)

The purpose of this portion of the upright motor control test is to ascertain extension control of a single weight-bearing extremity (i.e., for single limb stance in gait).

Instructions and procedures for the test are similar to those used in the flexion control test. The examiner demonstrates each segment sufficiently to ensure patient understanding but allows only two practice trials per segment to avoid fatigue. Only one graded trial per segment is permitted.

The starting position for this test is double-limb stance with both limbs in neutral alignment or the patient's maximal available extension range. The patient is required to bring the nontest limb off the floor; if this is not possible, help in flexing the nontest limb should be provided by the assistant.

The assistant helps to stabilize or provide hand support as described under each test part.

If the patient has a fixed equinus contracture that is greater than the neutral ankle position, the contracture must be accommodated by placing a hard wedge under the heel. The purpose of the wedge is to align the tibia into a vertical position.

If a stable plantigrade platform cannot be maintained (with manual support or with an ankle-foot orthosis), the examiner should give a score of UT (unable to test) at the hip and knee. The ankle score should be noted as E (excessive tone). That is to say, if excessive tone precludes the foot from assuming a position flat on the floor, the extension control test cannot be conducted.

PART 4: HIP EXTENSION

Positioning and Stabilization. The examiner is positioned beside the patient to offer hand support and to ensure that the patient begins from a position of neutral alignment or from the patient's maximal hip extension range (Fig. 8–1).

The assistant provides manual stabilization to maintain neutral knee extension and a stable ankle. Remember that plantigrade positioning of the foot is required.

Instructions to Patient: "Stand on both legs as straight as you can."

"Now stand as straight as you can on just your right/left leg." (*Note: This is the weaker limb if the test is to be unilateral.*)

"Lift *this* leg up (*point to or touch desired leg*) . . . keep standing as straight as you can."

Grading. When the patient is balanced on the test limb, the examiner gradually decreases the amount of hand support to determine the degree of hip control (Table 8–4).

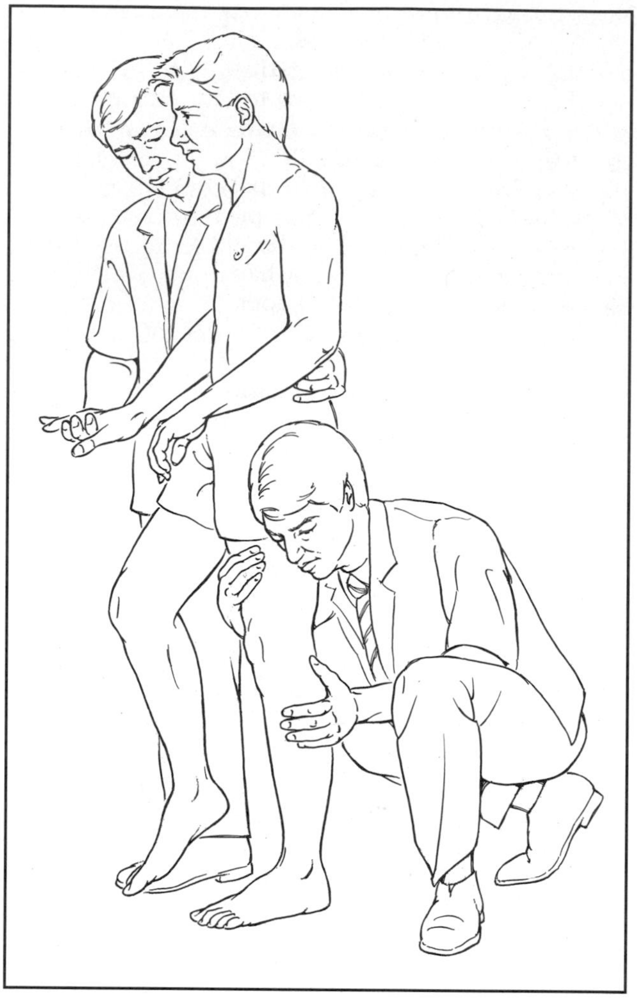

Figure 8–1. Hip extension test. The patient, aligned in neutral, raises the nontest limb. The examiner (on patient's right) maintains trunk and limb alignment in neutral, and if the knee or ankle (or both) are unstable, manual support is provided by the assistant, as illustrated.

Table 8–4   HIP EXTENSION

Score	Criteria
Weak (W)	Uncontrolled trunk flexion on hip occurs. (Examiner must prevent continued forward motion of the trunk by providing additional hand support.)
Moderate (M)	Patient is unable to maintain trunk completely erect or at the end of the available hip extension range. The patient is, however, able to stop the forward trunk momentum. Alternatively, the trunk wobbles back and forth *or* the patient hyperextends the trunk on the hip.
Strong (S)	Patient maintains trunk erect or at the end of the available hip extension range.

PART 5: KNEE EXTENSION

Positioning and Stabilization. The assistant is positioned behind the patient to provide hand support for balance and to maintain the trunk erect on the hip (Fig. 8–2).

The examiner positions the patient's knees in 30° of flexion bilaterally. If the patient is unable to maintain both feet flat on the floor with approximately 30° of knee flexion, a hard wedge should be placed under the heel to compensate for the limited dorsiflexion range of motion.

Instructions to Patient. "Stand on both feet with your knees bent. Keep your knees bent and lift your right/left leg." *(Note: the raised leg should be the stronger limb.)*

If the patient can support body weight on a flexed knee during single limb support without further collapse into flexion, proceed to the test for Strong (S) (Table 8–5).

Grading. See Table 8–5. When a knee flexion contracture is present, the grade awarded can never exceed Moderate (M).

Table 8–5   KNEE EXTENSION

Score	Criteria
Weak (W)	Patient is unable to maintain body weight on a flexed knee; therefore, the knee continues to collapse into flexion or the heel rises.
Moderate (M)	Patient supports body weight on a flexed knee without either further collapse into flexion or heel rise.
Strong (S)	Patient supports body weight on a flexed knee and on request straightens that knee to the end of available knee extension range. Hyperextension is allowed.
Excessive (E)	It is not possible to position the knee in flexion because of severe extensor thrust or extensor tone.
Unable to Test (UT)	Absence of plantigrade foot or other condition renders test invalid.

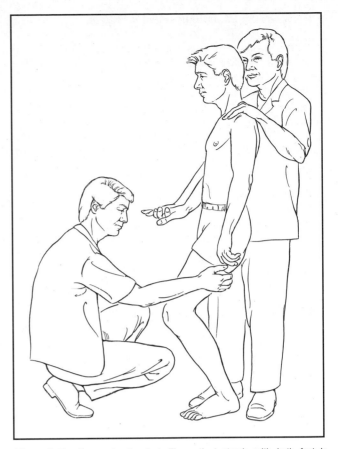

Figure 8–2. Knee extension test. The patient stands with both feet in plantigrade position. The examiner, kneeling in front, gives manual cues to the patient to flex both knees to 30 degrees. The assistant stands behind the patient to offer balance support to one of the patient's hands and uses the other hand to cue the patient to maintain erect posture.

PART 6: ANKLE EXTENSION (Plantar Flexion)

The purpose of this part of the extension test is to identify ankle control relative to maintaining a vertical tibial position.

If the patient has a knee flexion contracture in the test limb, the test cannot be conducted in a correct manner. With the knee flexed, the quadriceps muscle group can maintain single-limb stance despite the presence or absence of activity at the ankle.

Positioning and Stabilization. The assistant is positioned behind the patient to maintain the trunk in an erect posture over the hip (Fig. 8–3). The examiner is positioned to prevent knee hyperextension (i.e., ankle plantarflexion).

The passive range of ankle motion must be measured with the knee extended. If necessary, accommodate lack of dorsiflexion range (as occurs with a plantarflexion contracture) by placing a hard wedge under the heel. This will place the ankle in more plantarflexion, thus providing some relative dorsiflexion range for the purpose of this test.

Instructions to Patient. "Stand on both legs as straight as you can. Lift and hold up your right/left leg." *(Note: The raised leg should be the stronger limb.)*

If the patient can control the tibia with the knee in neutral, proceed to ask for a heel rise while the knee is kept at 0°:

"Keep your knee straight and go up on your toes as high as you can."

Grading. See Table 8–6.

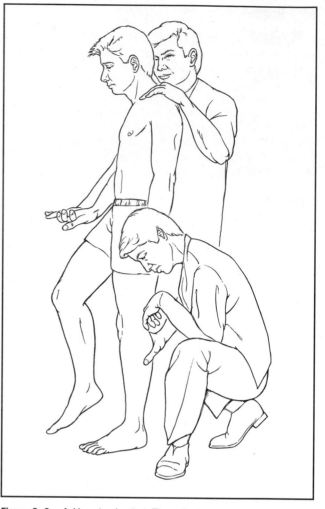

Figure 8–3. Ankle extension test. The patient stands erect in plantigrade position, then raises the nontest limb. The examiner kneels alongside or slightly behind to keep the knee from hyperextending. The assistant stands behind the patient to offer balance support and to cue the patient to maintain erect posture.

Table 8–6 PLANTAR FLEXION

Score	Criteria
Weak (W)	Patient is unable to maintain knee in neutral position; knee collapses into flexion and the ankle into dorsiflexion so that the tibia is displaced forward. Alternatively, the knee or ankle segment wobbles back and forth between flexion and extension or hyperextension. The presence of an extensor thrust that cannot be controlled by examiner also may indicate lack of adequate ankle control.
Moderate (M)	Patient can control the knee in a neutral (0°) position and the ankle in a neutral (90°) position so that the tibia is vertical.
Strong (S)	Patient maintains knee at neutral and lifts heel off floor on command. (Any degree of heel rise while maintaining the knee at neutral is acceptable.
Excessive (E)	Severity of equinus or varus is so great that patient cannot maintain a stable plantigrade ankle.
Unable to Test (UT)	Patient has a knee flexion contracture.

REFERENCES

1. Lovett RW, Martin EG. Certain aspects of infantile paralysis and a description of a method of muscle testing. JAMA 66:729–733, 1916.
2. Montgomery J. Assessment and treatment of locomotor deficits in stroke. *In* Duncan P, Radke M (eds): *Stroke Rehabilitation*. Chicago: Year Book, 1987.
3. Perry J, Giovan P, Harris LJ, et al. The determinants of muscle action in the hemiparetic lower extremity. Clin Orthop 131:71–89, 1978.
4. Sawner K, LaVigne JM. *Brunnstrom's Movement Therapy in Hemiplegia*. Philadelphia: JB Lippincott, 1992.
5. Knutsson E, Richards C. Different types of disturbed motor control in gait of hemiparetic patients. Brain 102:405–430, 1979.

Ready
Reference
Anatomy

Chapter 9

Using This Ready Reference Section

This section of the book is intended as a quick source of information about muscles, their anatomic description, participation in motions, and innervation. This information is not intended to be comprehensive, and for depth of subject matter, the reader is referred to any of the major texts of human anatomy. The authors of this book relied on both the American[1] and British[2] versions of *Gray's Anatomy* as principal references but also used Sobotta,[3] Clemente,[4] Netter,[5] Hollingshead,[6] Grant,[7] and Moore,[8] among others. The final arbiter in all cases is the 37th edition of *Gray's Anatomy* (British) by Williams et al.[2]

The variations in text descriptions of individual muscles remain exceedingly diverse so the authors at times consolidated information to provide abstracted descriptions.

Origins, insertions, descriptions, and functions of individual muscles are abbreviated but should allow the reader to place the muscle correctly and visualize its most common actions; this in turn may help the reader to recall more detailed anatomy.

(Nomina Anatomica nomenclature for the muscles appears in parentheses when a more common usage is listed.)

MUSCLE REFERENCE NUMBERS

Each skeletal muscle in the body has been given a number that is used with that muscle throughout the book. The order of numbering is derived from the regional sequence of muscles used in the first part of this reference. The numbering should, however, permit the reader to refer quickly to any one of the summaries or to cross-check information between summaries. In the first part of the Ready Reference Section the muscles are listed in alphabetical order, and this is followed by a list of muscles by region. In each muscle test the participating muscles also are preceded by their assigned reference number.

PART I. ALPHABETICAL LIST OF MUSCLES

A

159. Abductor digiti minimi (hand)
215. Abductor digiti minimi (foot)
224. Abductor hallucis
171. Abductor pollicis brevis
166. Abductor pollicis longus

180. Adductor brevis
225. Adductor hallucis
179. Adductor longus
181. Adductor magnus
173. Adductor pollicis

144. Anconeus
 27. Auriculares
201. Articularis genus

B

140. Biceps brachii
192. Biceps femoris

141. Brachialis
143. Brachioradialis

 26. Buccinator
120. Bulbospongiosus

C

 34. Chondroglossus
116. Coccygeus
139. Coracobrachialis
  5. Corrugator supercilii
 50. Cricothyroid (Cricothyroideus)
117. Cremaster

D

133. Deltoid (Deltoideus)
 23. Depressor anguli oris
 24. Depressor labii inferioris
 14. Depressor septi

101. Diaphragm
 78. Digastricus

E

1. and 2. Epicranius
149. Extensor carpi radialis brevis
148. Extensor carpi radialis longus
150. Extensor carpi ulnaris
147. Extensor digiti minimi
154. Extensor digitorum
212. Extensor digitorum brevis
211. Extensor digitorum longus
221. Extensor hallucis longus
155. Extensor indicis
168. Extensor pollicis brevis
167. Extensor pollicis longus

F

151. Flexor carpi radialis
153. Flexor carpi ulnaris

160. Flexor digiti minimi brevis (hand)
216. Flexor digiti minimi brevis (foot)
214. Flexor digitorum brevis
213. Flexor digitorum longus
157. Flexor digitorum profundus
156. Flexor digitorum superficialis
223. Flexor hallucis brevis
222. Flexor hallucis longus
170. Flexor pollicis brevis
169. Flexor pollicis longus

G

205. Gastrocnemius

190. Gemellus inferior
189. Gemellus superior

32. Genioglossus
77. Geniohyoid (Geniohyoideus)

182. Gluteus maximus
183. Gluteus medius
184. Gluteus minimus
178. Gracilis

H

33. Hyoglossus

I

176. Iliacus

66. Iliocostalis cervicis
89. Iliocostalis thoracis
90. Iliocostalis lumborum

41. Inferior pharyngeal constrictor (Constrictor pharyngis inferior)
38. Inferior longitudinal [tongue] (Longitudinalis inferior)
84–87. Infrahyoids (see *Sternothyroid, Thyrohyoid, Sternohyoid, Omohyoid*)
136. Infraspinatus

102. Intercostales externi
103. Intercostales interni
104. Intercostales intimi

164. Interossei, dorsal [hand] (Interossei dorsales)
219. Interossei, dorsal [foot] (Interossei dorsales)
165. Interossei, palmar or volar (Interossei palmares)
220. Intcrossei, plantar (Interossei plantares)

69. Interspinales cervicis
97. Interspinales thoracis
98. Interspinales lumborum

70. Intertransversarii cervicis
99. Intertransversarii thoracis
99. Intertransversarii lumborum

121. Ischiocavernosus

L

52. Lateral cricoarytenoid (Cricoarytenoideus lateralis)
30. Lateral pterygoid (Pterygoideus lateralis)
130. Latissimus dorsi

115. Levator ani
17. Levator anguli oris
15. Levator labii superioris
16. Levator labii superioris alaeque nasi
3. Levator palpebrae superioris
127. Levator scapulae
46. Levator veli palatini
107. Levatores costarum

60. Longissimus capitis
64. Longissimus cervicis
91. Longissimus thoracis

74. Longus capitis
79. Longus colli

163. Lumbricales (hand)
218. Lumbricales (foot)

M

28. Masseter
31. Medial pterygoid (Pterygoideus medialis)
21. Mentalis
42. Middle pharyngeal constrictor (Constrictor pharyngis medius)
94. Multifidi
48. Musculus uvulae
75. Mylohyoid (Mylohyoideus)

N

13. Nasalis

O

54. Oblique arytenoid (Arytenoideus obliquus)
59. Obliquus capitis inferior
58. Obliquus capitis superior
110. Obliquus externus abdominis
11. Obliquus inferior
111. Obliquus internus abdominis
10. Obliquus superior

188. Obturator externus (Obturatorius externus)
187. Obturator internus (Obturatorius internus)

1. Occipitofrontalis
87. Omohyoid (Omohyoideus)

161. Opponens digiti minimi
172. Opponens pollicis

4. Orbicularis oculi
25. Orbicularis oris

P

36. Palatoglossus
49. Palatopharyngeus

162. Palmaris brevis
152. Palmaris longus

177. Pectineus
131. Pectoralis major
129. Pectoralis minor

209. Peroneus brevis
208. Peroneus longus
210. Peroneus tertius

186. Piriformis
207. Plantaris
88. Platysma

202. Popliteus
51. Posterior cricoarytenoid (Cricoarytenoideus posterior)
12. Procerus

147. Pronator quadratus
146. Pronator teres

174. Psoas major
175. Psoas minor
114. Pyramidalis

Q

191. Quadratus femoris
100. Quadratus lumborum
217. Quadratus plantae

196–200. Quadriceps femoris (see *Rectus femoris, Vastus intermedius, Vastus medialis longus, Vastus medialis oblique,* and *Vastus lateralis*)

R

113. Rectus abdominis
72. Rectus capitis anterior

73. Rectus capitis lateralis
56. Rectus capitis posterior major
57. Rectus capitis posterior minor

196. Rectus femoris

7. Rectus inferior
9. Rectus lateralis
8. Rectus medialis
6. Rectus superior

125. Rhomboid major (Rhomboideus major)
126. Rhomboid minor (Rhomboideus minor)
20. Risorius

71. Rotatores cervicis
96. Rotatores lumborum
95. Rotatores thoracis

S

45. Salpingopharyngeus
195. Sartorius

80. Scalenus anterior
81. Scalenus medius
82. Scalenus posterior

194. Semimembranosus

62. Semispinalis capitis
65. Semispinalis cervicis
93. Semispinalis thoracis

193. Semitendinosus

128. Serratus anterior
109. Serratus posterior inferior
108. Serratus posterior superior

206. Soleus

123. Sphincter ani externus
122. Sphincter urethrae

63. Spinalis capitis
68. Spinalis cervicis
92. Spinalis thoracis

61. Splenius capitis
67. Splenius cervicis

83. Sternocleidomastoid (Sternocleidomastoideus)
86. Sternohyoid (Sternohyoideus)
84. Sternothyroid (Sternothyroideus)

35. Styloglossus
76. Stylohyoid (Stylohyoideus)
44. Stylopharyngeus
84. Stylothyroid (Stylothyroideus)

132. Subclavius

105. Subcostales
134. Subscapularis

43. Superior pharyngeal constrictor (Constrictor pharyngis superior)
37. Superior longitudinal [tongue] (Longitudinalis superior)
145. Supinator

75–78. Suprahyoids (see *Mylohyoid, Stylohyoid, Geniohyoid, Digastricus*)
135. Supraspinatus

T

29. Temporalis
2. Temporoparietalis

185. Tensor fasciae latae
47. Tensor veli palatini

138. Teres major
137. Teres minor

85. Thyrohyoid (Thyrohyoideus)
55. Thyroarytenoid (Thyroarytenoideus)

203. Tibialis anterior
204. Tibialis posterior

39. Transverse lingual (Transversus linguae)
112. Transversus abdominis
53. Transverse arytenoid (Arytenoideus transversus)
22. Transversus menti
119. Transversus perinei profundus
118. Transversus perinei superficialis

106. Transversus thoracis

124. Trapezius
142. Triceps brachii

U

48. Uvula *(see Musculus uvulae)*

V

198. Vastus intermedius
199. Vastus medialis longus
200. Vastus medialis oblique
197. Vastus lateralis

40. Vertical [tongue] (Verticalis linguae)

Z

18. Zygomaticus major
19. Zygomaticus minor

PART II. LIST OF MUSCLES BY REGION

HEAD AND FOREHEAD

1. Occipitofrontalis
2. Temporoparietalis

EYELIDS

3. Levator palpebrae superioris
4. Orbicularis oculi
5. Corrugator supercilii

OCULAR MUSCLES

6. Rectus superior
7. Rectus inferior
8. Rectus medialis
9. Rectus lateralis
10. Obliquus superior
11. Obliquus inferior

NOSE

12. Procerus
13. Nasalis
14. Depressor septi

MOUTH

15. Levator labii superioris
16. Levator labii superioris alaeque nasi
17. Levator anguli oris
18. Zygomaticus major
19. Zygomaticus minor
20. Risorius
21. Mentalis
22. Transversus menti
23. Depressor anguli oris
24. Depressor labii inferioris
25. Orbicularis oris
26. Buccinator

EAR

27. Auriculares

JAW/MASTICATION

28. Masseter
29. Temporalis
30. Lateral pterygoid
31. Medial pterygoid

TONGUE

32. Genioglossus
33. Hyoglossus
34. Chondroglossus
35. Styloglossus
36. Palatoglossus

37. Superior longitudinal
38. Inferior longitudinal
39. Transverse lingual
40. Vertical lingual

PHARYNX

41. Inferior pharyngeal constrictor
42. Middle pharyngeal constrictor
43. Superior pharyngeal constrictor
44. Stylopharyngeus
45. Salpingopharyngeus
49. Palatopharyngeus (see under *Palate*)

PALATE

46. Levator veli palatini
47. Tensor veli palatini
48. Musculus uvulae
36. Palatoglossus (see under *Tongue*)
49. Palatopharyngeus

LARYNX

50. Cricothyroid
51. Posterior cricoarytenoid
52. Lateral cricoarytenoid
53. Transverse arytenoid
54. Oblique arytenoid
55. Thyroarytenoid
    55a. Vocalis
    55b. Thyroepiglotticus

NECK

56. Rectus capitis posterior major
57. Rectus capitis posterior minor
58. Obliquus capitis superior
59. Obliquus capitis inferior
60. Longissimus capitis
61. Splenius capitis
62. Semispinalis capitis
63. Spinalis capitis

64. Longissimus cervicis
65. Semispinalis cervicis
66. Iliocostalis cervicis
67. Splenius cervicis

68. Spinalis cervicis
69. Interspinales cervicis
70. Intertransversarii cervicis
71. Rotatores cervicis

72. Rectus capitis anterior
73. Rectus capitis lateralis
74. Longus capitis

75. Mylohyoid
76. Stylohyoid
77. Geniohyoid
78. Digastricus

79. Longus colli
80. Scalenus anterior
81. Scalenus medius
82. Scalenus posterior

83. Sternocleidomastoid

84. Sternothyroid
85. Thyrohyoid
86. Sternohyoid
87. Omohyoid

88. Platysma

BACK

61. Splenius capitis (see under *Neck*)
67. Splenius cervicis (see under *Neck*)

66. Iliocostalis cervicis (see under *Neck*)
89. Iliocostalis thoracis
90. Iliocostalis lumborum

60. Longissimus capitis (see under *Neck*)
64. Longissimus cervicis (see under *Neck*)
91. Longissimus thoracis

63. Spinalis capitis
68. Spinalis cervicis
92. Spinalis thoracis

62. Semispinalis capitis (see under *Neck*)
65. Semispinalis cervicis (see under *Neck*)
93. Semispinalis thoracis

94. Multifidi

71. Rotatores cervicis
95. Rotatores thoracis
96. Rotatores lumborum

69. Interspinalis cervicis
97. Interspinalis thoracis
98. Interspinalis lumborum

70. Intertransversarii cervicis

99. Intertransversarii thoracis
99. Intertransversarii lumborum

100. Quadratus lumborum

THORAX (RESPIRATION)

101. Diaphragm
102. Intercostales externi
103. Intercostales interni
104. Intercostales intimi
105. Subcostales
106. Transversus thoracis
107. Levatores costarum
108. Serratus posterior superior
109. Serratus posterior inferior

ABDOMEN

110. Obliquus externus abdominis
111. Obliquus internus abdominis
112. Transversus abdominis
113. Rectus abdominis
114. Pyramidalis

PERINEUM

115. Levator ani
116. Coccygeus
117. Cremaster
118. Transversus perinei superficialis
119. Transversus perinei profundus
120. Bulbospongiosus
121. Ischiocavernosus

122. Sphincter urethrae
123. Sphincter ani externus

UPPER EXTREMITY

Shoulder Girdle

124. Trapezius
125. Rhomboid major
126. Rhomboid minor
127. Levator scapulae

128. Serratus anterior
129. Pectoralis minor

Vertebrohumeral

130. Latissimus dorsi
131. Pectoralis major

Shoulder

132. Subclavius
133. Deltoid
134. Subscapularis
135. Supraspinatus
136. Infraspinatus

137. Teres minor
138. Teres major

Elbow

139. Coracobrachialis
140. Biceps brachii
141. Brachialis
142. Triceps brachii
143. Brachioradialis
144. Anconeus

Forearm

145. Supinator
146. Pronator teres
147. Pronator quadratus
140. Biceps brachii (see under *Elbow*)

Wrist

148. Extensor carpi radialis longus
149. Extensor carpi radialis brevis
150. Extensor carpi ulnaris

151. Flexor carpi radialis
152. Palmaris longus
153. Flexor carpi ulnaris

Fingers

154. Extensor digitorum
155. Extensor indicis
156. Flexor digitorum superficialis
157. Flexor digitorum profundus

163. Lumbricales
164. Interossei (dorsal)
165. Interossei (palmar)

Little Finger and Hypothenar Muscles

158. Extensor digiti minimi
159. Abductor digiti minimi
160. Flexor digiti minimi brevis
161. Opponens digiti minimi
162. Palmaris brevis

Thumb and Thenar Muscles

166. Abductor pollicis longus
167. Extensor pollicis longus
168. Extensor pollicis brevis
169. Flexor pollicis longus
170. Flexor pollicis brevis
171. Abductor pollicis brevis
172. Opponens pollicis
173. Adductor pollicis

LOWER EXTREMITY

Hip/Thigh

174. Psoas major
175. Psoas minor
176. Iliacus
177. Pectineus

178. Gracilis
179. Adductor longus
180. Adductor brevis
181. Adductor magnus

182. Gluteus maximus
183. Gluteus medius
184. Gluteus minimus
185. Tensor fasciae latae
186. Piriformis
187. Obturator internus
188. Obturator externus
189. Gemellus superior
190. Gemellus inferior
191. Quadratus femoris

192. Biceps femoris
193. Semitendinosus
194. Semimembranosus
195. Sartorius

Knee

196–200. Quadriceps femoris
196. Rectus femoris
197. Vastus lateralis
198. Vastus intermedius
199. Vastus medialis longus
200. Vastus medialis oblique

201. Articularis genus

192. Biceps femoris
193. Semitendinosus
194. Semimembranosus

202. Popliteus

Ankle

203. Tibialis anterior
204. Tibialis posterior

205. Gastrocnemius
206. Soleus
207. Plantaris

208. Peroneus longus
209. Peroneus brevis
210. Peroneus tertius

Lesser Toes

211. Extensor digitorum longus
212. Extensor digitorum brevis
213. Flexor digitorum longus
214. Flexor digitorum brevis
215. Abductor digiti minimi
216. Flexor digiti minimi brevis
217. Quadratus plantae

218. Lumbricales

219. Interossei (dorsal)
220. Interossei (plantar)

Great Toe (Hallux)

221. Extensor hallucis longus
222. Flexor hallucis longus
223. Flexor hallucis brevis
224. Abductor hallucis
225. Adductor hallucis

PART III. SKELETAL MUSCLES OF THE HUMAN BODY

HEAD 335
  Scalp (Forehead) 335
  Eyelids 335
  Ocular 336
  Nose 338
  Mouth 339
  Jaw (Mastication) 344
  Ear 344
  Tongue 346
  Pharynx 347
  Palate 349
  Larynx 350

NECK 351

TRUNK 360
  Back 360
  Respiration 363

Abdomen 366
Perineum 368

UPPER EXTREMITY 371
  Scapula 371
  Vertebrohumeral 373
  Scapulohumeral 374
  Elbow 376
  Forearm 378
  Wrist 378
  Fingers 380
  Thumb 385

LOWER EXTREMITY 387
  Hip/thigh 387
  Knee 393
  Ankle 395
  Lesser toes 397
  Great toe 400

Muscles of the Forehead

THE EPICRANIUS (TWO MUSCLES)

1. Occipitofrontalis
2. Temporoparietalis

1. OCCIPITOFRONTALIS

Occipital belly (Occipitalis):

Origin:

Occiput (superior nuchal line, lateral 2/3)
Temporal bone (mastoid)

Insertion:

Galea aponeurotica

Frontal belly (Frontalis):

Origin:

No bony attachments
Median fibers continuous with Procerus
Intermediate fibers join Corrugator supercilii and Orbicularis oculi
Lateral fibers also join Orbicularis oculi

Insertion:

Galea aponeurotica

Description:

The Epicranius consists of the Occipitofrontalis with its 4 thin branches on either side of the head; the broad aponeurosis called the Galea aponeurotica; and the Temporoparietalis with its 2 slim branches. The medial margins of the two bellies join above the nose and run together upward and over the forehead.

The Galea aponeurotica covers the cranium between the frontal belly and the occipital belly of the Epicranius and between the two occipital bellies over the occiput. It is adhered closely to the dermal layers (scalp), which allows the scalp to be moved freely over the cranium.

Function:

Contracting together, both bellies draw the scalp up and back, thus raising the eyebrows (surprise!) and assisting with wrinkling the forehead

Working alone, the frontal belly raises the eyebrow on the same side

Innervation:

Facial (VII) nerve
    Temporal branches: to Frontalis
    Posterior auricular branch: to Occipitalis

2. TEMPOROPARIETALIS

Origin:

Temporal fascia (superior and anterior to external ear, then fanning out and up over temporal fascia)

Insertion:

Galea aponeurotica (lateral border)
Into skin and temporal fascia somewhere high on lateral side of head

Description:

A thin broad sheet of muscle in 2 bellies lying on either side of head. Highly variable. See also description of Occipitofrontalis.

Function:

Tightens scalp
Draws back skin over temples
Raises auricula of the ear
In concert with Occipitofrontalis, raises the eyebrows, widens the eyes, and wrinkles the skin of the forehead (in expressions of surprise and fright)

Innervation:

Facial (VII) nerve (temporal branches)

Muscles of the Eyelids

3. Levator palpebrae superioris
4. Orbicularis oculi
5. Corrugator supercilii

3. LEVATOR PALPEBRAE SUPERIORIS

Origin:

Sphenoid (inferior surface of small wing)

Insertion:

Into three lamellae:
    Aponeurosis of the orbital septum
    Superior tarsus
    Sheath of the Rectus superior (and with it, blends with the superior fornix of the conjunctiva)

Description:

Thin and flat muscle lying posterior and superior to the orbit. At its origin it is tendinous, broadening out to end in a wide aponeurosis that splits into three lamellae.

Function:

Raises upper eyelid

Innervation:

Oculomotor (III) nerve

4. ORBICULARIS OCULI

Origin:

Frontal bone (nasal part)
Maxilla (frontal process in front of lacrimal groove)
Medial palpebral ligament

Insertion:

Palpebral part: lateral palpebral raphe
Orbital part: after forming a complete ellipse, blends into the Occipitofrontalis and Corrugator muscles. Some fibers (Depressor supercilii) also insert into skin of eyebrow and participate in drawing eyebrow down.
Lacrimal part: Superior and Inferior tarsi

Description:

Forms a broad thin layer that fills the eyelids and surrounds the circumference of the orbit but also spreads over the temple and cheek. Orbital portion acts like a sphincter.

Function:

Palpebral part: closes lids as in blinking and sleep
Orbital part: closes lids but with greater force, as in winking
Lacrimal part: draws the eyelids and lacrimal canals medially, compressing them against the globe of the eye to receive tears

Also compresses lacrimal sac during blinking
Entire muscle draws skin of forehead, temple, and cheek toward medial angle of orbit, tightly closing eye
The orbital muscles around the eye are important because they cause blinking, which keeps the eye lubricated and prevents dehydration of the conjunctiva

Innervation:

Facial (VII) nerve (temporal and zygomatic branches)

5. CORRUGATOR SUPERCILII

Origin:

Frontal bone (superciliary arch, medial end)

Insertion:

Skin (deep surface) of eyebrow over middle of orbital arch

Description:

Fibers of this small pyramidal muscle pass upward and laterally from nasal aspect of orbit to a location under the lateral surface of Occipitofrontalis

Function:

Draws eyebrow down and medially, producing vertical wrinkles of forehead (frowning)

Innervation:

Facial (VII) nerve (temporal branch)

Ocular Muscles

    6. Rectus superior
    7. Rectus inferior
    8. Rectus medialis
    9. Rectus lateralis
   10. Obliquus superior
   11. Obliquus inferior

6–9 THE FOUR RECTI (Fig. 9–1)

Rectus superior, inferior, medialis, and lateralis

Origin:

Sphenoid (from a tubercle on greater wing via insertion of common tendon) via common annular tendon ringing the superior, medial, and inferior margins of optic foramen (canal)

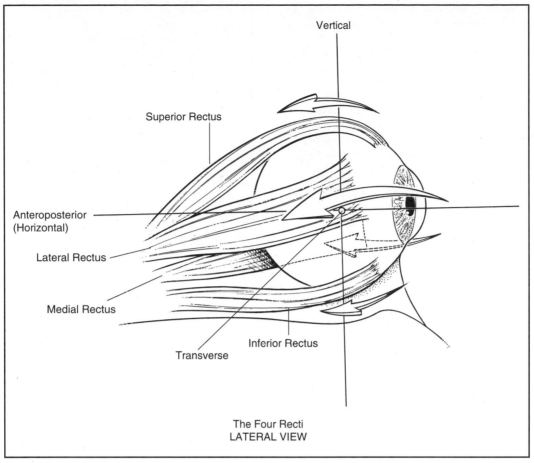

Vertical

Superior Rectus

Anteroposterior
(Horizontal)

Lateral Rectus

Medial Rectus

Transverse

Inferior Rectus

The Four Recti
LATERAL VIEW

Figure 9–1. The four recti.

The ring is completed by a lower fibrous extension (tendon of Zinn), which gives origin to the Rectus inferior, part of the Rectus medialis, and the lower head of origin of the Rectus lateralis. An upper fibrous expansion gives rise to the Rectus superior, part of the Rectus medialis, and the upper head of the Rectus lateralis.

Insertion:

Each of the Recti passes anteriorly in the position indicated by its name and inserts via a tendinous expansion into the sclera a short distance behind the cornea

Description:

The four Recti share a common origin and insert at different points on the sclera (Fig. 9–1). The Rectus superior is the smallest and thinnest and inserts on the superoanterior sclera under the orbital roof. The inferior muscle inserts on the inferoanterior sclera just above the orbital floor. The Rectus medialis is the broadest of the recti and inserts on the medial scleral wall well in front of the equator. The Rectus lateralis, the

longest of the recti, courses around the lateral side of the eyeball to insert well forward of the equator.

Function:

The ocular muscles rotate the eyeball in directions that depend on the geometry of their relationships and that can be altered by the eye movements themselves. Eye movements also are accompanied by head motions, which assist with the incredibly complex varieties of stereoscopic vision.

The ocular muscles are not subject to direct inspection or routine assessment. It is essential to know that a change in the tension of one of the muscles alters the length-tension relationships of all six ocular muscles. It is likely that all six muscles are continuously involved, and consideration of each in isolation is not a functional exercise. It is, however, interesting to consider the functional relationship between the four Recti and the two Obliquui as two differing synergies.

The Rectus superior, inferior, and medialis act together as adductors or convergence muscles.

The lateral Rectus, together with the two Obliquui, act as muscles of abduction or divergence.

Convergence generally is associated with elevation of the visual axis, and divergence with lowering of the visual axis.

These descriptions are necessarily brief, and brevity may assign falsely simple functions to the extraocular muscles. Readers are referred for more detail to the British edition of *Gray's Anatomy*.[2] For specific functions refer also to Chapter 7 (Extraocular Muscles).

Innervation:

Oculomotor (III) nerve: Rectus superior, inferior, and medialis, and Obliquus inferior
Abducens (VI) nerve: Rectus lateralis
Trochlear (IV) nerve: Obliquus superior

10. OBLIQUUS SUPERIOR OCULI

Origin:

Sphenoid bone (superior and medial to canal and Rectus superior)

Insertion:

Frontal bone (via a round tendon that inserts through a pulley [a cartilaginous ring called the trochlea] that inserts in the trochlear fovea)
Sclera (behind the equator on the superolateral surface)

Description:

The superior oblique lies superomedially in the orbit (Fig. 9–2). It passes forward, ending in the round tendon that loops through the trochlear pulley, which is attached to the trochlear fovea. It then turns abruptly posterolaterally and passes thence to the sclera to end between the Rectus superior and the Rectus lateralis.

Function:

The superior oblique acts on the eye from above, whereas the inferior oblique acts on the eye directly below; the superior oblique elevates the posterior aspect of the eyeball, and the inferior oblique depresses it. The superior oblique, therefore, rotates the visual axis downward, whereas the inferior oblique rotates it upward, both motions occurring around the transverse axis.

Innervation:

Trochlear (IV) nerve

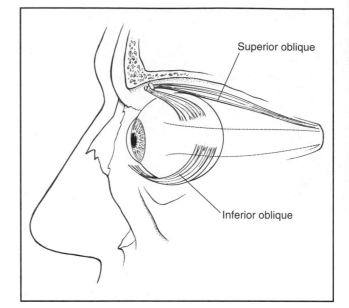

Figure 9–2. The oblique extraocular muscles.

11. OBLIQUUS INFERIOR OCULI

Origin:

Maxilla (orbital surface, lateral to the lacrimal groove)

Insertion:

Sclera (lateral part) between the insertions of the Rectus superior and lateralis and near, but behind, the insertion of the Superior oblique

Description:

Located near the anterior margin of the floor of the orbit, it passes laterally under the eyeball between the inferior rectus and the bony orbit. It then bends upward on the lateral side of the eyeball, passing under the lateral rectus to insert on the sclera beneath that muscle (Fig. 9–2).

Function:

See under *Obliquus superior.*

Innervation:

Oculomotor (III) nerve

Muscles of the Nose

12. Procerus
13. Nasalis
14. Depressor septi

12. PROCERUS

Origin:

Nasal bone (dorsum of nose, lower part)
Nasal cartilage (lateral, upper part)

Insertion:

Skin over lower part of forehead between eyebrows
Joins Occipitofrontalis

Description:

From its origin over bridge of nose it courses straight upward to blend with Frontalis.

Function:

Produces transverse wrinkles over bridge of nose
Draws eyebrows downward

Innervation:

Facial (VII) nerve (buccal branch)

13. NASALIS

Transverse Part (Compressor nares)

Origin:

Maxilla (above and lateral to incisive fossa)

Insertion:

Aponeurosis over bridge of nose joining with muscle on opposite side

Alar Part (Dilator nares)

Origin:

Maxilla (above lateral incisor tooth)
Alar cartilage

Insertion:

Ala nasi
Skin at tip of nose

Description:

Muscle has two parts that cover the distal and medial surfaces of the nose. Fibers from each side rise upward and medially, meeting in a narrow aponeurosis near the bridge of the nose.

Function:

Transverse part: depresses cartilaginous portion of nose and draws alae toward septum

Alar part: dilates nostrils (during breathing it resists tendency of nares to close from atmospheric pressure)
Noticeable in anger or labored breathing

Innervation:

Facial (VII) nerve (buccal branches)

14. DEPRESSOR SEPTI

Origin:

Maxilla (above and lateral to incisive fossa, i.e., central incisor)

Insertion:

Nasal septum (mobile part) and alar cartilage

Description:

Fibers ascend vertically from central maxillary origin. Muscle lies deep to the superior labial mucous membrane. It often is considered part of the dilator nares (of the Nasalis).

Function:

Draws alae of nose downward (constricting nares)

Innervation:

Facial (VII) nerve (buccal branches)

Muscles of the Mouth

15. Levator labii superioris
16. Levator labii superioris alaeque nasi
17. Levator anguli oris
18. Zygomaticus major
19. Zygomaticus minor
20. Risorius
21. Mentalis
22. Transversus menti
23. Depressor anguli oris
24. Depressor labii inferioris
25. Orbicularis oris
26. Buccinator

COMMENTARY

The muscles of the face are different from most skeletal muscles in the body because they are cutaneous muscles located in the deep layers of the skin and frequently have no bony attachments. All of them (scalp, eyelids, nose, lips, cheeks, mouth, and pinna) give rise to "expressions" and convey "thought," the most visible of the body language systems.

The orbital muscles of the mouth are important for speech, drinking, and ingestion of solid foods. Although the Buccinator is described in this section, it is not a muscle of expression but does serve an important role in regulating the position of, and action on, food in the mouth.

These muscles are continuously tonic to provide the facial skin with tension, and the skin becomes baggy or flabby (resulting in, for example, crow's feet or wattles) when it is denervated or in the presence of the atrophic processes associated with aging. There are wide differences in these muscles among individuals and among racial groups, and to deal with such variations craniofacial and plastic surgeons often classify the facial muscles differently (e.g., in single vs. multiple heads) from the system presented here.

Continuous skin tension also gives rise to the gaping wounds that occur with facial lacerations, and surgeons take great care to understand the planes of the muscles to minimize scarring in the repair of such wounds.

The facial muscles all arise from the mesoderm of the second branchial (hyoid) arch. The muscles lie in all parts of the face and head but retain their innervation by the VII cranial (facial) nerve.

15. LEVATOR LABII SUPERIORIS

Origin:

Orbit of eye (inferior margin)
Maxilla
Zygomatic bone

Insertion:

Upper lip

Description:

Converging from a rather broad place of origin on the inferior orbit, the fibers converge and descend into the upper lip between the other levator muscles and the Zygomaticus minor.

Function:

Elevates and protracts upper lip.

THE MODIOLUS

The arrangement of the facial musculature often causes confusion and misunderstanding. This is not surprising because there are 14 small bundles of muscles running in various directions, with long names and unsupported functional claims. Of all the muscles of the face, those about the mouth may be the most important because they are responsible both for ingestion of food and speech.

One major source of confusion is the relationship that exists among the muscles around the mouth. The common description until recently was of uninterrupted circumoral muscles. In fact, the Orbicularis oris muscle is not a complete ellipse but contains fibers from the major extrinsic muscles that converge on the buccal angle as well as intrinsic fibers.[1,6] These authors and others do not describe complete ellipses, but most drawings illustrate such.[1]

The area on the face that has a large concentration of converging and diverging fibers from multiple directions lies immediately lateral and slightly above the corner of the mouth. Using the thumb and index finger on the outer skin and inside the mouth and compressing the tissue between them quickly will identify a knotlike structure known as the modiolus.[9-11]

The modiolus (from the Latin meaning nave of a wheel) has been described as a muscular or tendinous node, a rather concentrated attachment of many muscles[9] (Fig. 9-3). Its basic shape is conical (although this description is oversimplified); it is about 1 cm thick and is found in most people about 1 cm lateral to the buccal angle. Its shape and size vary considerably with gender, race, and age. The muscular fibers enter and exit on different planes, superficial and deep, with some spiraling but essentially comprising three-dimensional complexity.

Different classifications of modiolar muscles exist, but basically nine or ten facial muscles are associated with the structure:[10]

RADIATING FROM THE MODIOLUS

Levator anguli oris
Orbicularis oris
Depressor anguli oris
Zygomaticus major
Buccinator

RETRACTORS OF UPPER LIP

Levator labii superioris
Levator labii superioris alaeque nasi
Zygomaticus minor

RETRACTORS AND DEPRESSORS OF LOWER LIP

Mentalis
Depressor labii inferioris

Frequently associated are the special fibers of the Orbicularis oris (Incisive superior, Incisive inferior), Platysma, and Risorius (the latter is not a constant feature in the facial musculature).

The Orbicularis oris and the Buccinator form an almost continuous muscular sheet that can be fixed in a number of positions by the Zygomaticus major, Levator anguli oris, and Depressor anguli oris (the latter three muscles being the "stays" used to immobilize the modiolus in any position).

When the modiolus is firmly fixed, the Buccinator can contract to apply force to the cheek teeth; the Orbicularis can contract against the arch of the anterior teeth, thus sealing the lips together and closing the mouth tightly.[10] Similarly, the function of the modiolar active and stay muscles allows accurate and fine control of lip movements and pressures in speech.

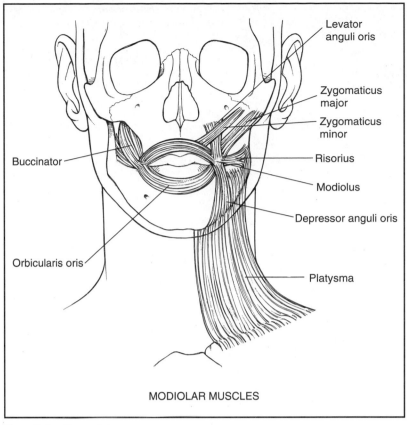

Figure 9–3. The modiolus.

Innervation:

Facial (VII) nerve (buccal branch)

16. LEVATOR LABII SUPERIORIS ALAEQUE NASI

Origin:

Maxilla (frontal process)

Insertion:

Ala of nose and upper lip

Description:

Muscle fibers descend obliquely lateral and divide into two slips: one to the greater alar cartilage of the nose and one to blend with the Levator labii superioris and Orbicularis oris (thence to the modiolus)

Function:

Dilates nostrils
Elevates upper lip

Innervation:

Facial (VII) nerve (buccal branch)

17. LEVATOR ANGULI ORIS

Origin:

Maxilla (canine fossa)

Insertion:

Modiolus
Dermal attachment at angle of mouth

Description:

Muscle descends from maxilla, inferolateral to orbit, down to modiolus. It lies partially under the Zygomaticus minor.

Function:

Raises angle of mouth and by so doing displays teeth in smiling
Contributes to nasolabial furrow (from side of nose to corner of upper lip). Deepens in sadness (and age).

Innervation:

Facial (VII) nerve (buccal branch)

18. ZYGOMATICUS MAJOR

Origin:

Zygomatic bone (in front of zygomaticotemporal suture)

Insertion:

Modiolus

Description:

Descends obliquely lateral to blend with other modiolar muscles. A small and variable group of superficial fascicles called the Malaris are considered part of this muscle.

Function:

Draws angle of mouth lateral and upward (as in laughing)

Innervation:

Facial (VII) nerve (buccal branch)

19. ZYGOMATICUS MINOR

Origin:

Zygomatic bone (malar surface) medial to origin of Zygomaticus major

Insertion:

Upper lip; blends with Levator labii superioris Modiolus

Description:

Descends initially with Zygomaticus major, then moves medially on top of Levator labii superioris, with which it blends.

Function:

Elevates and curls upper lip, exposing the maxillary teeth (as in sneering, expressions of contempt, and smiling)
Deepens nasolabial furrow

Innervation:

Facial (VII) nerve (buccal branch)

20. RISORIUS

Origin:

Masseteric fascia

Insertion:

Modiolus

Description:

This muscle is so highly variable, it hardly deserves to be called a muscle. When present, it passes forward almost horizontally. It may vary from a few fibers to a wide, thin, superficial, fan-shaped sheet. It is usually called the muscle of laughing, but that is no more true for the Risorius than for any other modiolar muscle.

Function:

When present, retracts angle of mouth

Innervation:

Facial (VII) nerve (buccal branch)

21. MENTALIS

Origin:

Mandible (incisive fossa)

Insertion:

Skin over chin

Description:

Descends medially from its origin just lateral to labial frenulum to center of integument of chin.

Function:

Wrinkles skin over chin
Protrudes lower lip (as in sulking)

Innervation:

Facial (VII) nerve (mandibular branch)

22. TRANSVERSUS MENTI

Origin:

Skin of the chin, laterally

Insertion:

Skin of the chin
Blends with contralateral muscle

Description:

As frequently absent as it is present. Very small muscle traverses chin inferiorly and therefore is called the mental sling. Often continuous with Depressor anguli oris.

Function:

Depresses angle of mouth; supports skin of chin

Innervation:

Facial (VII) nerve (mandibular branch)

23. DEPRESSOR ANGULI ORIS

Origin:

Mandible (oblique line)

Insertion:

Modiolus

Description:

Ascends in a curve from its broad origin below tubercle of mandible to a narrow fasciculus into modiolus. Often continuous below with Platysma.

Function:

Depresses lower lip
Depresses buccal angle of mouth (as in sadness)

Innervation:

Facial (VII) nerve (mandibular branch)

24. DEPRESSOR LABII INFERIORIS

Origin:

Mandible (between symphysis and mental foramen)

Insertion:

Lower lip; angle of mouth
Modiolus

Description:

Passes upward and medially from a broad origin; then narrows and blends with Orbicularis oris and Depressor labii inferioris of opposite side.

Function:

Draws lower lip down and laterally. Contributes to expressions of irony, sorrow, and melancholy.

Innervation:

Facial (VII) nerve (mandibular branch)

25. ORBICULARIS ORIS

Origin:

No fascial attachments except the modiolus. This is a composite muscle with contributions from other muscles of the mouth, which form a complex sphincterlike structure, but it is not a true sphincter. Via its incisive components, the muscle attaches to the maxilla (Incisivus labii superioris) and mandible (Incisivus labii inferioris).

Insertion:

Modiolus

Description:

This muscle is not a complete ellipse of muscle surrounding the mouth. The fibers actually form four separate functional quadrants on each side that provide great diversity of oral movements. There is overlapping function among the quadrants (upper, lower, left, and right). The muscle is connected with the maxillae and septum of the nose by lateral and medial accessory muscles.

The *Incisivus labii superioris* is a lateral accessory muscle of the upper lip within the Orbicularis oris, and there is a similar accessory muscle, the *Incisivus labii inferioris,* for the lower lip. These muscles have bony attachments to the floor of the maxillary incisive (superior) fossa and the mandibular incisive (inferior) fossa. They arch laterally between the Orbicularis fibers on the respective lip and, after passing the buccal angle, insert into the modiolus. The modiolus acts as a force-transmission system to the lips from muscles attached to it.

The Orbicularis oris has another accessory muscle, the *Nasolabialis,* which lies medially and connects the upper lip to the nasal septum. (The interval between the contralateral Nasolabialis corresponds to the philtrum, the depression on the upper lip beneath the nasal septum.)

Function:

Closes lips
Protrudes lips
Holds lips tight against teeth
Shapes lips for whistling, kissing, sucking, drinking, etc.
Alters shape of lips for articulation

Innervation:

Facial (VII) nerve (buccal branches, bilaterally)
This innervation is of interest because when one facial nerve is injured distal to the stylomastoid foramen, only half of the Orbicularis oris muscle is paralyzed. When this occurs, as in Bell's palsy, the mouth droops and may be drawn medially on the side of injury.

26. BUCCINATOR

Origin:

Maxilla and mandible (external surfaces of alveolar processes)

Insertion:

Modiolus

Description:

The principal muscle of the cheek is classified as a facial muscle (because of its innervation) despite its role in mastication. The Buccinator forms the lateral wall of the oral cavity, lying deep to the other facial muscles and filling the gap between the maxilla and the mandible.

Function:

Compresses cheek against the teeth
Expels air when cheeks are distended (in blowing)
Acts in mastication to control passage of food

Innervation:

Facial (VII) nerve (buccal branch)

Extrinsic Muscles of the Ear

27. THE AURICULARES

Auricularis anterior

Origin:

Anterior fascia in temporal area (lateral edge of epicranial aponeurosis)

Insertion:

Spine of helix

Auricularis superior

Origin:

Temporal fascia

Insertion:

Auricle (cranial surface)

Auricularis posterior

Origin:

Mastoid process of temporal bone via a short aponeurosis

Insertion:

Auricle (cranial surface, concha)

Function (all):

Limited function in humans except at parties! The anterior muscle elevates the auricle and moves it forward; the superior muscle elevates the au-

ricle slightly, and the posterior draws it back. Auditory stimuli may evoke minor responses from these muscles.

Innervation:

Facial (VII) nerve (posterior auricular branch)

Muscles of Jaw and Mastication

28. Masseter
29. Temporalis
30. Lateral pterygoid
31. Medial pterygoid

28. MASSETER

Superficial part:

Origin:

Maxilla (zygomatic process via an aponeurosis)
Zygomatic bone (maxillary process and inferior border of arch)

Insertion:

Mandible (angle and lower half of lateral surface of ramus)

Intermediate part:

Origin:

Zygomatic arch (inner surface of anterior 2/3)

Insertion:

Mandible (ramus, central part)

Deep part:

Origin:

Zygomatic arch (posterior 1/3 continuous with intermediate part)

Insertion:

Mandible (ramus [superior half] and coronoid process)

Description:

A thick muscle connecting the upper and lower jaws and consisting of three layers that blend anteriorly. The superficial layer descends backward to the angle of the mandible and the lower mandibular ramus. (The middle and deep layers compose the deep part cited in *Nomina Anatomica*.) The muscle is easily palpable and lies under the parotid gland posteriorly; the anterior margin overlies the Buccinator.

Function:

Elevates the mandible (occlusion of the teeth in mastication)
Up and down biting motion

Innervation:

Trigeminal (V) nerve (mandibular division, masseteric branches)

29. TEMPORALIS

Origin:

Temporal bone (all of temporal fossa)
Temporal fascia (deep surface)

Insertion:

Mandible (coronoid process, medial surface, apex, and anterior border; anterior border of ramus almost to 3rd molar)

Description:

A broad muscle that radiates like a fan on the side of the head from most of the temporal fossa, converging downward to coronoid process of the mandible. The descending fibers converge into a tendon that passes between the zygomatic arch and the cranial wall. The more anterior fibers descend vertically, but the more posterior the fibers the more oblique their course until the most posterior fibers are almost horizontal.

Function:

Elevates mandible to close mouth and approximate teeth (biting motion)
Retracts mandible (posterior fibers)
Participates in lateral grinding motion

Innervation:

Trigeminal (V) nerve (mandibular division)

30. LATERAL PTERYGOID

Origin:

Superior head: Sphenoid bone (greater wing, infratemporal crest and surface)
Inferior head: Mandible (lateral pterygoid plate, lateral surface)

Insertion:

Mandible (condylar neck, depression in anterior part)
Temporomandibular joint (anterior margin of articular capsule and disc)

Description:

A short, thick muscle with two heads that runs posterolaterally to the mandibular condyle, neck, and disc of the TMJ. The fibers of the upper head are directed downward and laterally, while those of the lower head course horizontally. The muscle lies under the mandibular ramus.

Function:

Protracts mandibular condyle and disc of TMJ joint forward while the mandibular head rotates on disc (aids opening of mouth)
The Lateral pterygoid, acting with the elevators of the mandible, protrudes the jaw, causing malocclusion of the teeth (i.e., the lower teeth project in front of the upper teeth).
When the Lateral and Medial pterygoids on the same side act jointly, the mandible and the jaw (chin) rotate to the opposite side (chewing motion).

Innervation:

Trigeminal (V) nerve (mandibular division)

31. MEDIAL PTERYGOID

Origin:

Sphenoid bone (lateral pterygoid plate)
Palatine bone (grooved surface of pyramidal process)
Maxilla (tuberosity) joins in slip from lateral surfaces of pyramidal processes of palatine bone)

Insertion:

Mandible medial surface of ramus via a strong tendon, reaching as high as mandibular foramen)

Description:

This short, thick muscle occupies the position on the inner side of the mandibular ramus, whereas the Masseter occupies the outer position. The Medial pterygoid is separated by the Lateral pterygoid from the mandibular ramus. The deep fibers rise from the palatine bone; the more superficial fibers arise from the maxilla and lie superficial to the Lateral pterygoid. The fibers descend posterolaterally to the mandibular ramus.

Function:

Elevates mandible to close jaws (biting)
Protrudes mandible (with Lateral pterygoid)
Unilaterally the Medial and Lateral pterygoids together rotate the mandible forward and to the

opposite side. This alternating motion is chewing.

The Medial pterygoid and Masseter are situated to form a sling that suspends the mandible. This sling is a functional articulation in which the temporomandibular joint acts as a guide. As the mouth opens and closes, the mandible moves on a center of rotation established by the sling and the sphenomandibular ligament.

Innervation:

Trigeminal (V) nerve (mandibular division, medial pterygoid branch)

Muscles of the Tongue

EXTRINSIC TONGUE MUSCLES

32. Genioglossus
33. Hyoglossus
34. Chondroglossus
35. Styloglossus
36. Palatoglossus

32. GENIOGLOSSUS

Origin:

Mandible (symphysis menti on inner surface of superior mental spine)

Insertion:

Hyoid bone via a thin aponeurosis (inferior fibers)
Middle pharyngeal constrictor muscle (intermediate fibers)
Undersurface of tongue, whole length mingling with the intrinsic musculature (superior fibers)

Description:

The tongue is separated into lateral halves by the lingual septum, which extends along its full length and inserts inferiorly into the hyoid bone. The extrinsics extend outside the tongue.

The Genioglossus is a thin, flat muscle that fans out backward from its mandibular origin, running parallel with and close to the midline. The lower fibers run downward to the hyoid; the median fibers run posteriorly and join the middle constrictor of the pharynx; and the superior fibers run upward to insert on the whole length of the underside of the tongue. The muscles of the two sides are blended anteriorly and separated posteriorly by the medial lingual septum.

Function:

Forward traction of tongue (tip protrudes beyond mouth)

Depression of central part of tongue (with bilateral action)

Innervation:

Hypoglossal (XII) nerve

33. HYOGLOSSUS

Origin:

Hyoid bone (side of body and whole length of greater horn)

Insertion:

Side of tongue

Description:

Thin, quadrilateral muscle whose fibers run almost vertically

Function:

Depression of tongue

Innervation:

Hypoglossal (XII) nerve

34. CHONDROGLOSSUS

Origin:

Hyoid bone (lesser horn, medial side and base)

Insertion:

Blends with intrinsic muscles between Hyoglossus and Genioglossus

Description:

A very small muscle (about 2 cm long) that is sometimes considered part of the Hyoglossus

Function:

Assists in tongue depression

Innervation:

Hypoglossal (XII) nerve

35. STYLOGLOSSUS

Origin:

Temporal bone (styloid process, apex)
Stylomandibular ligament (styloid end)

Insertion:

Side of tongue near dorsal surface to blend with intrinsics (longitudinal portion)

Overlaps Hyoglossus and blends with it (oblique portion)

Description:

Shortest and smallest of extrinsic tongue muscles. The muscle curves down anteriorly and divides into longitudinal and oblique portions. It lies between the internal and external carotid arteries.

Function:

Draws tongue up and backward

Innervation:

Hypoglossal (XII) nerve

36. PALATOGLOSSUS

Origin:

Soft palate (anterior surface)

Insertion:

Side of tongue intermingling with intrinsic muscles

Description:

Technically an extrinsic muscle of the tongue, this muscle is functionally closer to the palate muscles. It is a small fasciculus, narrower in the middle than at its ends. It passes anteroinferiorly and laterally in front of the tonsil to reach the side of the tongue. Along with the mucous membrane covering it, the Palatoglossus forms the palatoglossal arch or fold.

Function:

Elevates root of tongue
Approximates palatoglossal arch to close the oral cavity from the oropharynx

Innervation:

Hypoglossal (XII) nerve

INTRINSIC TONGUE MUSCLES

37. Superior longitudinal
38. Inferior longitudinal
39. Transverse lingual
40. Vertical lingual

37. SUPERIOR LONGITUDINAL

Attachments and description:

Oblique and longitudinal fibers run immediately under the mucous membrane on dorsum of tongue.

Arises from submucous fibrous layer near epiglottis and from the median lingual septum. Fibers run anteriorly to the edges of the tongue.
Function of intrinsics: See *40. Vertical lingual.*

38. INFERIOR LONGITUDINAL

Attachments and description:

Narrow band of fibers close to the inferior lingual surface. Extends from the root to the apex of the tongue. Some fibers connect to hyoid body. Blends with Styloglossus anteriorly.

39. TRANSVERSE LINGUAL

Attachments and description:

Passes laterally across tongue from the median lingual septum to the edges of the tongue. Blends with Palatopharyngeus.

40. VERTICAL LINGUAL

Attachments and description:

Located only at the anterolateral regions and extends from the dorsal to the ventral surfaces of the tongue.

Function of Intrinsics:

These muscles change the shape and contour of the tongue. The longitudinal muscles tend to shorten it. The Superior longitudinal also turns the apex and sides upward, making the dorsum concave. The Inferior longitudinal pulls the apex and sides downward to make the dorsum convex. The Transverse muscle narrows and elongates the tongue. The Vertical muscle flattens and widens it.
These almost limitless alterations give the tongue the incredible versatility and precision necessary for speech and swallowing functions.

Innervation of Intrinsics:

Hypoglossal (XII) nerve

Muscles of the Pharynx

41. Inferior pharyngeal constrictor
42. Middle pharyngeal constrictor
43. Superior pharyngeal constrictor
44. Stylopharyngeus
45. Salpingopharyngeus

49. Palatopharyngeus (see under *Muscles of the Palate*)

41. INFERIOR PHARYNGEAL CONSTRICTOR

Origin:

Cricoid cartilage (sides)
Thyroid cartilage (oblique line on the side as well as from inferior cornu)

Insertion:

Pharynx (posterior median fibrous raphe, along with opposite number)

Description:

The thickest and largest of the pharyngeal constrictors, the muscle has two parts: the Cricopharyngeus and the Thyropharyngeus. Both parts spread posteromedially and join the muscle of the opposite side at the fibrous median raphe. The lowest fibers run horizontally and circle the narrowest part of the pharynx. The other fibers course obliquely upward to overlap the Middle constrictor.
During swallowing the Cricopharyngeus acts like a sphincter; the Thyropharyngeus uses peristaltic action to propel food downward.

Function:

During swallowing all constrictors act as general sphincters and in peristaltic action.

Innervation:

Pharyngeal plexus formed by components of vagus [X], accessory [XI], glossopharyngeal [IX], and external laryngeal nerves

42. MIDDLE PHARYNGEAL CONSTRICTOR

Origin:

Hyoid bone (whole length of superior border of greater cornu)
Stylohyoid ligament

Insertion:

Pharynx (posterior median fibrous raphe)

Description:

From their origin the fibers fan out in three directions: the lower ones descend to lie under the Inferior constrictor; the medial ones pass transversely, and the superior fibers ascend to overlap the Superior constrictor. At its insertion it joins with the muscle from the opposite side.

Function:

Serves as a sphincter and acts during peristaltic functions in deglutition

Innervation:

Pharyngeal plexus (see under *Inferior constrictor*)

43. SUPERIOR PHARYNGEAL CONSTRICTOR

Origin:

Splenoid bone (medial pterygoid plate and its hamulus)
Pterygomandibular raphe
Mandible (mylohyoid line)
Side of tongue

Insertion:

Median pharyngeal fibrous raphe
Occipital bone (pharyngeal tubercle on basilar part)

Description:

The smallest of the constrictors, the fibers of this muscle curve posteriorly and are elongated by an aponeurosis to reach the occiput. The attachments of this muscle are differentiated as pterygopharyngeal, buccopharyngeal, mylopharyngeal, and glossopharyngeal.
The interval between the superior border of this muscle and the base of the skull is closed by the pharyngobasilar fascia known as the *sinus of Morgagni.*
A small band of muscle blends with the Superior constrictor from the upper surface of the palatine aponeurosis and is called the palatopharyngeal sphincter. This band is visible when the soft palate is elevated; often it is hypertrophied in individuals with cleft palate.

Function:

Acts as a sphincter and has peristaltic functions in swallowing

Innervation:

Pharyngeal plexus

44. STYLOPHARYNGEUS

Origin:

Temporal bone (styloid process, medial side of base)

Insertion:

Thyroid cartilage (posterior border)
Blends with pharyngeal constrictors and Palatopharyngeus.

Description:

A long thin muscle that passes downward along the side of the pharynx and between the

Superior and Middle constrictors to spread out beneath the mucous membrane.

Function:

Elevation of upper lateral pharyngeal wall in swallowing

Innervation:

Glossopharyngeal (IX) nerve

45. SALPINGOPHARYNGEUS

Origin:

Auditory tube (inferior aspect of cartilage near orifice)

Insertion:

Blends with Palatopharyngeus.

Description:

Small muscle whose fibers pass downward, lateral to the uvula, to blend with fibers of the Palato-pharyngeus

Function:

Elevates pharynx to move a bolus of food

Innervation:

Pharyngeal plexus

Muscles of the Palate

46. Levator veli palatini
47. Tensor veli palatini
48. Musculus uvulae
49. Palatopharyngeus
36. Palatoglossus (see under *Muscles of the Tongue*)

46. LEVATOR VELI PALATINI
   ### (Levator palati)

Origin:

Temporal bone (inferior surface of petrous bone)
Tympanic fascia
Auditory tube cartilage

Insertion:

Palatine aponeurosis (upper surface, where it blends with opposite muscle at the midline)

Description:

Fibers of this small muscle run downward and medially from the petrous temporal bone to

pass above the margin of the Superior pharyngeal constrictor and anterior to the Salpingopharyngeus.

Function:

Elevates soft palate

Innervation:

Pharyngeal plexus (accessory [XI] nerve, cranial part, and vagus [X] nerve)

47. TENSOR VELI PALATINI
   ### (Tensor palati)

Origin:

Sphenoid bone (pterygoid process, scaphoid fossa)
Auditory tube cartilage (lateral lamina)
Sphenoid spine (medial part)

Insertion:

Palatine aponeurosis
Palatine bone (horizontal plate)

Description:

This small thin muscle lies lateral to the Levator veli palatini and the auditory tube. It descends vertically between the medial pterygoid plate and the Medial pterygoid muscle, converging into a delicate tendon, which turns medially around the pterygoid hamulus.

Function:

Draws soft palate to one side (unilateral)
Tightens palate, depressing it and flattening its arch (with its contralateral counterpart)

Innervation:

Trigeminal (V) nerve (mandibular branch)

48. MUSCULUS UVULAE
   ### (Azygos uvulae)

Origin:

Palatine bones (posterior nasal spine)
Palatine aponeurosis

Insertion:

Uvula

Description:

A bilateral muscle, its fibers descend into the uvular mucosa.

Function:

Elevates and retracts uvula

Innervation:

Pharyngeal plexus

49. PALATOPHARYNGEUS (Pharyngopalatinus)

Origin:

Anterior fasciculus:

Soft palate (palatine aponeurosis)
Hard palate (posterior border)

Posterior fasciculus:

Pharyngeal aspect of soft palate (palatine aponeurosis)

Insertion:

Thyroid cartilage (posterior border)
Side of pharynx on an aponeurosis

Description:

Along with its overlying mucosa, it forms the palatopharyngeal arch. It arises by two fasciculi separated by the Levator veli palatini, all of which join in the midline with their opposite muscles. The two muscles unite and are joined by the Salpingopharyngeus to descend behind the tonsils. The muscle forms an incomplete longitudinal wall on the internal surface of the pharynx.

Function:

Elevates pharynx and pulls it forward, thus shortening it during swallowing. The muscles also approximate the palatopharyngeal arches.

Innervation:

Pharyngeal plexus (accessory [XI] and vagal [X] nerves)

36. PALATOGLOSSUS

See under *Muscles of the Tongue*

Muscles of the Larynx (Intrinsics)

50. Cricothyroid
51. Posterior cricoarytenoid
52. Lateral cricoarytenoid
53. Transverse arytenoid
54. Oblique arytenoids

55. Thyroarytenoid
    Vocalis
    Thyroepiglotticus

50. CRICOTHYROID

Origin:

Cricoid cartilage (front and lateral)

Insertion:

Inferior cornu of larynx (anterior border)
Thyroid cartilage (lamina lower border)

Description:

The fibers of this paired muscle are arranged in two groups: a lower oblique group (pars obliqua), which slants posterolaterally to the inferior cornu, and a superior group (pars recta or vertical fibers), which ascends backward to the lamina.

Function:

Regulates tension of vocal folds
Stretches vocal ligaments by raising the cricoid arch, thus increasing tension in the vocal folds

Innervation:

Vagus (X) nerve (superior laryngeal nerve, external laryngeal branch)

51. POSTERIOR CRICOARYTENOID

Origin:

Cricoid cartilage (broad depression on corresponding half of posterior surface)

Insertion:

Arytenoid cartilage (back of muscular process)

Description:

The fibers pass cranially and laterally to converge on the back of the arytenoid cartilage on the same side. The lowest fibers are nearly vertical and become oblique and finally almost transverse at the superior border.

Function:

Regulates tension of vocal folds
Opens glottis by rotating arytenoid cartilages laterally and separating (abducting) the vocal folds
Retracts arytenoid cartilages, thereby helping to tense the vocal folds

Innervation:

Vagus (X) nerve (recurrent laryngeal nerve)

52. LATERAL CRICOARYTENOID

Origin:

Cricoid cartilage (cranial border of arch)

Insertion:

Arytenoid cartilage (front of muscular process)

Description:

Fibers run obliquely upward and backward.

Function:

Closes glottis by rotating arytenoid cartilages medially, approximating (adducting) the vocal folds

Innervation:

Vagus (X) nerve (recurrent laryngeal nerve)

53. TRANSVERSE ARYTENOID

Attachments and description:

A single muscle that crosses transversely between the two arytenoid cartilages. Often considered a branch of an Arytenoid muscle. It attaches to the back of the muscular process and the adjacent lateral borders of both arytenoid cartilages.

Function:

Approximates (adducts) the arytenoid cartilages, closing the glottis

Innervation:

Vagus (X) nerve (recurrent laryngeal branch)

54. OBLIQUE ARYTENOIDS

Origin:

Arytenoid cartilage (back of muscular process)

Insertion:

Arytenoid cartilage (apex on opposite side)

Description:

A pair of muscles lying superficial to the Transverse arytenoid. Arrayed as two fasciculi that cross on the posterior midline. Often considered part of an Arytenoid muscle. Fibers that continue laterally around the apex of the Arytenoid are sometimes termed the Aryepiglottic muscle.

Function:

Acts as a sphincter for the laryngeal inlet (by adducting the aryepiglottic folds and approximating the arytenoid cartilages)

Innervation:

Vagus (X) nerve (recurrent laryngeal branch)

55. THYROARYTENOID

Origin:

Thyroid cartilage (caudal half of angle)
Middle cricothyroid ligament

Insertion:

Arytenoid cartilage (base and anterior surface)

Description:

Lies lateral to the vocal fold, ascending posterolaterally. Many fibers are carried to the aryepiglottic fold.
The lower and deeper fibers, which lie medially, appear to be differentiated as a band inserted into the vocal process of the arytenoid cartilage. This band frequently is called the *Vocalis* muscle. It is adherent to the vocal ligament, to which it is lateral and parallel.
Other fibers of this muscle continue as the *Thyroepiglotticus* and insert into the epiglottic margin; other fibers that swing along the wall of the sinus to the side of the epiglottis are termed the *Superior thyroarytenoid*.

Function:

Regulates tension of vocal folds
Draws arytenoid cartilages toward thyroid cartilage, thus shortening and relaxing vocal ligaments
Rotates the arytenoid cartilages medially to approximate vocal folds
The *Vocalis* relaxes the posterior vocal folds while the anterior folds remain tense, thus raising the pitch of the voice.
The *Thyroepiglotticus* widens the laryngeal inlet via action on the aryepiglottic folds.

Innervation:

Vagus (X) nerve (recurrent laryngeal nerve)

Muscles of the Neck

CAPITAL EXTENSOR MUSCLES

This group of eight muscles consists of suboccipital muscles extending between the atlas, axis, and skull and large overlapping muscles from the 6th thoracic to the 3rd cervical vertebrae and rising to the skull.

56. Rectus capitis posterior major
57. Rectus capitis posterior minor
58. Obliquus capitis superior
59. Obliquus capitis inferior
60. Longissimus capitis
61. Splenius capitis
62. Semispinalis capitis
63. Spinalis capitis

The capital extensor muscles control the head as an entity separate from the cervical spine.[12]

56. RECTUS CAPITIS POSTERIOR MAJOR

Origin:

Axis (spinous process)

Insertion:

Occiput (lateral part of inferior nuchal line; surface just inferior to nuchal line)

Description:

Starts as a small tendon and broadens as it rises upward and laterally (review suboccipital triangle in any anatomy text).

Function:

Capital extension of head
Rotation of head to same side
Lateral bending of head to same side

Innervation:

C1 spinal nerve (suboccipital nerve, dorsal rami)

57. RECTUS CAPITIS POSTERIOR MINOR

Origin:

Atlas (tubercle on posterior arch)

Insertion:

Occiput (medial portion of inferior nuchal line; surface between inferior nuchal line and foramen magnum)

Description:

Begins as a narrow tendon, which broadens into a wide band of muscle as it ascends.

Function:

Extension of head
Lateral bending of head to same side

Innervation:

C1 spinal nerve (suboccipital nerve, dorsal rami)

58. OBLIQUUS CAPITIS SUPERIOR

Origin:

Atlas (transverse process, superior surface), where it joins insertion of Obliquus capitis inferior.

Insertion:

Occiput (between superior and inferior nuchal lines; lies lateral to Semispinalis capitis)

Description:

Starts as a narrow muscle and then widens as it rises upward and medially.

Function:

Extension of head on atlas (muscle on both sides)
Lateral bending to same side (muscle on that side)

Innervation:

C1 spinal nerve (suboccipital nerve, dorsal rami)

59. OBLIQUUS CAPITIS INFERIOR

Origin:

Axis (apex of spinous process)

Insertion:

Atlas (transverse process, inferior and dorsal part)

Description:

Passes laterally and slightly upward. This is the larger of the two obliquui.

Function:

Capital extension
Lateral bending (muscle on that side)
Rotation of head to same side

Innervation:

C1 spinal nerve (suboccipital nerve, dorsal rami)

60. LONGISSIMUS CAPITIS

Origin:

T1–T5 vertebrae (transverse processes)
C4–C7 vertebrae (articular processes)

Insertion:

Temporal bone (mastoid process [posterior margin])

Description:

A muscle with several tendons lying under the Splenius cervicis. Sweeps upward and laterally and is considered a continuation of the Sacrospinalis.

Function:

Extension of head
Lateral bending and rotation of head to same side

Innervation:

C3–C8 cervical nerves with variations (dorsal rami)

61. SPLENIUS CAPITIS

Origin:

Ligamentum nuchae at C3–C7
C7–T4 vertebrae (spinous processes) with variations

Insertion:

Temporal bone (mastoid process)
Occiput (surface below lateral 1/3 of superior nuchal line)

Description:

Fibers directed upward and laterally as a broad band deep to Sternocleidomastoid.

Function:

Extension of head
Rotation of head to same side (debated)
Lateral bending of head to same side

Innervation:

C3–C6 cervical nerves with variations (dorsal rami)

62. SEMISPINALIS CAPITIS

Origin:

C7 and T1–T6 vertebrae (variable) as series of tendons from tips of transverse processes
C4–C6 vertebrae (articular processes)

Insertion:

Occiput (between superior and inferior nuchal lines)

Description:

Tendons unite to form a broad muscle in the upper posterior neck, which passes vertically upward.

Function:

Capital extension (muscles on both sides)
Rotation of head to opposite side (debated)
Lateral bending of head to same side

Innervation:

C2–T1 spinal nerves (dorsal rami)

63. SPINALIS CAPITIS

Origin:

C5–C7 and T1–T3 vertebrae (variable) (spinous processes)

Insertion:

Occiput (between superior and inferior nuchal lines)

Description:

The smallest and thinnest of the Erector spinae, these muscles lie closest to the vertebral column. The Spinales are inconstant and are difficult to separate.

Function:

Extension of head

Innervation:

C3–T1 spinal nerves (dorsal rami)

CERVICAL EXTENSOR MUSCLES

This group of eight overlapping cervical muscles arises from the thoracic vertebrae or ribs and inserts into the cervical vertebrae.

64. Longissimus cervicis
65. Semispinalis cervicis
66. Iliocostalis cervicis
67. Splenius cervicis
68. Spinalis cervicis
69. Interspinales cervicis
70. Intertransversarii cervicis
71. Rotatores cervicis

124. Trapezius (see under *Shoulder Girdle*)

This group of muscles is responsible for cervical spine extension as opposed to head (capital) extension.

64. LONGISSIMUS CERVICIS

Origin:

T1–T5 vertebrae (variable) (tips of transverse processes)

Insertion:

C2–C6 vertebrae (posterior tubercles of transverse processes)

Description:

A continuation of the Sacrospinalis group, the tendons are long and thin, and the muscle courses upward and slightly medially.

Function:

Extension of the cervical spine
Lateral bending of cervical spine to same side
Accessory muscle for depression of ribs

Innervation:

C3–T6 spinal nerves (variable) (dorsal rami)

65. SEMISPINALIS CERVICIS

Origin:

T1–T5 vertebrae (variable) (transverse processes)

Insertion:

Axis (C2) to C5 vertebrae (spinous processes)

Description:

A narrow, thick muscle arising from a series of tendons and ascending vertically

Function:

Extension of the cervical spine
Rotation of cervical spine to opposite side
Lateral bending to same side

Innervation:

C2–T5 spinal nerves (dorsal rami)

66. ILIOCOSTALIS CERVICIS

Origin:

Ribs 3 to 6 (angles); sometimes ribs 1 and 2 also.

Insertion:

C4–C6 vertebrae (transverse processes, posterior tubercles)

Description:

Flattened tendons arise from ribs on dorsum of back and become muscular as they rise and turn medially to insert on cervical vertebrae. The muscle lies lateral to the Longissimus cervicis. The Iliocostales form the lateral column of the Sacrospinalis group.

Function:

Extension of the cervical spine
Lateral bending to same side
Depression of ribs (accessory)

Innervation:

C4–T6 spinal nerves (variable) (dorsal rami)

67. SPLENIUS CERVICIS

Origin:

T3–T6 vertebrae (spinous processes)

Insertion:

C1–C3 vertebrae (variable) (transverse processes, posterior tubercles)

Description:

Narrow tendinous band arises from bone and intraspinous ligaments and forms a broad sheet along with the Splenius capitis. This muscle rises upward and laterally under the Trapezius and Rhomboids and medially to the Levator scapulae.

Function:

Extension of the cervical spine
Rotation of cervical spine to same side
Lateral bending to same side
Synergistic with opposite Sternocleidomastoid

Innervation:

C4–C8 spinal nerves (variable) (dorsal rami)

68. SPINALIS CERVICIS

Origin:

C6–C7 and T1–T2 vertebrae (spinous processes)

Insertion:

C1–C3 vertebrae (spinous processes)

Description:

The smallest and thinnest of the Erector spinae, they lie closest to the vertebral column. They are inconstant and difficult to separate.

Function:

Extension of the cervical spine

Innervation:

C3–C8 spinal nerves (dorsal rami)

69. INTERSPINALES CERVICIS

Origin and Insertion:

Spinous processes of contiguous cervical vertebrae

Six pairs occur between the axis and the first thoracic vertebrae.

Function:

Extension of the cervical spine

Innervation:

C3–C8 spinal nerves (dorsal rami)

70. INTERTRANSVERSARII CERVICIS

Origin and Insertion:

Both anterior and posterior pairs occur at each segment. The anterior muscles interconnect the anterior tubercles of contiguous transverse processes and are innervated by the ventral primary rami. The posterior muscles interconnect the posterior tubercles of contiguous transverse processes and are innervated by the dorsal primary rami.

Description:

Small fasciculi lie between the transverse processes of the contiguous vertebrae. The cervicis is the most developed of this group.

Function:

Extension of spine (muscles on both sides)
Lateral bending to same side (muscle on one side)

Innervation:

Anterior cervicis: C3–C8 spinal nerves (ventral rami)
Posterior cervicis: C3–C8 spinal nerves (dorsal rami)

THE ROTATORES

These are the deepest muscles of the Transversospinalis group, lying as 11 pairs of very short muscles beneath the Multifidus. The fibers run obliquely upward and medial or almost horizontal. They may cross more than one vertebra on their ascending course, but most commonly they proceed to the next higher one. Found along the entire length of the vertebral column, they are distinguishable as developed muscles only in the thoracic area.

71. ROTATORES CERVICIS

Origin:

Transverse process of one cervical vertebra

Insertion:

Base of spine of next highest vertebra

Description:

This muscle lies deep in the Multifidus and cannot be readily separated from its deepest fibers. They are irregular and variable.

Function:

Extension of the cervical spine (assist)
Rotation of spine to opposite side

Innervation:

C3–C8 spinal nerves (dorsal rami)

MUSCLES OF CAPITAL FLEXION

The primary capital flexors are the short recti that lie between the atlas and the skull and the Longus capitis. Reinforcing these muscles are the suprahyoid muscles from the mandibular area.

72. Rectus capitis anterior
73. Rectus capitis lateralis
74. Longus capitis

Suprahyoids:

75. Mylohyoid
76. Stylohyoid
77. Geniohyoid
78. Digastricus

72. RECTUS CAPITIS ANTERIOR

Origin:

Atlas (C1) vertebra (anterior surface of lateral side)

Insertion:

Occiput (inferior surface of basilar part)

Description:

Short, flat muscle found immediately behind Longus capitis. Passes upward and medially.

Function:

Forward flexion of head (capital flexion)
Stabilization of atlanto-occipital joint

Innervation:

C1–C2 spinal nerves (ventral rami)

73. RECTUS CAPITIS LATERALIS

Origin:

Atlas (C1) (transverse process, upper surface)

Insertion:

Occiput (jugular process)

Description:

Short, flat muscle; courses upward and laterally.

Function:

Lateral bending of head
Assists head rotation (obliquity of muscle)
Stabilizes atlanto-occipital joint (assists)

Innervation:

C1–C2 spinal nerves (ventral rami)

74. LONGUS CAPITIS

Origin:

C3–C6 vertebrae (transverse processes, anterior tubercles)

Insertion:

Occiput (inferior basilar part)

Description:

Starting as four tendinous slips, muscle merges and becomes broader and thicker as it rises, converging medially toward its contralateral counterpart.

Function:

Capital flexion
Rotation of head to same side

Innervation:

C1–C3 spinal nerves (ventral rami)

75. MYLOHYOID

Origin:

Mandible (whole length of mylohyoid line from symphysis in front to last molar behind)

Insertion:

Hyoid bone (body)

Description:

Flat triangular muscle; the muscles from the two sides form a floor for the cavity of the mouth.

Function:

Raises hyoid bone and tongue for swallowing
Depresses the mandible when fixed
Capital flexion (weak accessory)

Innervation:

Trigeminal (V) nerve (inferior alveolar branch of mandibular division)

76. STYLOHYOID

Origin:

Temporal bone, styloid process (posterolateral surface)

Insertion:

Hyoid bone (body at junction with greater horn)

Description:

Slim muscle passes downward and forward and is perforated by Digastricus near its distal attachment. Muscle occasionally is absent.

Function:

Hyoid bone drawn upward and backward (in swallowing)
Capital flexion (weak accessory)
Assists in opening mouth
Participation in mastication and speech (roles not clear)

Innervation:

Facial (VII) nerve

77. GENIOHYOID

Origin:

Mandible (symphysis menti, inferior inner surface)

Insertion:

Hyoid bone (body, anterior surface)

Description:

Narrow muscle lying superficial to the Mylohyoid, it runs backward and somewhat downward. It is in contact (or may fuse) with its contralateral counterpart at the midline.

Function:

Elevation and protraction of hyoid bone
Moves tongue forward

Capital flexion (weak accessory)
Assists in depressing mandible

Innervation:

C1–C2 spinal nerves (fibers carried by hypoglossal [XII] nerve)

78. DIGASTRICUS

Origin:

Posterior belly: temporal bone (mastoid notch)
Anterior belly: mandible (lingual side of lower border)

Insertion:

Intermediate tendon and from there to hyoid bone via a fibrous sling

Description:

Consists of two bellies united by a rounded intermediate tendon. Lies below the mandible and extends as a sling from the mastoid to the symphysis menti, perforating the Stylohyoid, where the two bellies are joined by the intermediate tendon.

Function:

Mandibular depression (muscles on both sides)
Elevation of hyoid bone (in swallowing)
Anterior belly: Draws hyoid forward
Posterior belly: Draws hyoid backward
Capital flexion (weak synergist)
On electromyography, bilateral muscles always work together.

Innervation:

Anterior belly: trigeminal (V) nerve (inferior alveolar nerve)
Posterior belly: facial (VII) nerve

CERVICAL SPINE FLEXORS

The primary cervical spine flexors are the Longus colli (a prevertebral mass), the three Scalene muscles, and the Sternocleidomastoid. Superficial accessory muscles are the Infrahyoid muscles and the Platysma.

79. Longus colli
80. Scalenus anterior
81. Scalenus medius
82. Scalenus posterior
83. Sternocleidomastoid

Infrahyoids:

84. Sternothyroid
85. Thyrohyoid
86. Sternohyoid
87. Omohyoid

88. Platysma

79. LONGUS COLLI

Three Heads

Superior oblique:

Origin:

C3–C5 vertebrae (anterior tubercles of transverse processes)

Insertion:

Atlas (tubercle of anterior arch)

Inferior oblique:

Origin:

T1–T3 vertebrae (variable) (anterior bodies)

Insertion:

C5–C6 vertebrae (anterior tubercles of transverse processes)

Vertical Portion:

Origin:

T1–T3 and C5–C7 vertebrae (anterolateral bodies)

Insertion:

C2–C4 vertebrae (anterior bodies)

Description:

Situated on the anterior surface of the vertebral column from the atlas to the T3 vertebra. It is cylindrical and tapers at each end.

Function:

Cervical flexion (weak)
Cervical rotation to opposite side (inferior oblique heads)
Lateral bending (superior and inferior oblique heads)

Innervation:

C2–C6 spinal nerves (ventral rami)

THE SCALENES

These muscles are highly variable in their specific anatomy, and this possibly leads to disputes about minor functions.

80. SCALENUS ANTERIOR

Origin:

C3–C6 vertebrae (anterior tubercles of transverse processes)

Insertion:

1st rib (scalene tubercle on inner border and ridge on upper surface)

Description:

Lying deep at the side of the neck under the Sternocleidomastoid, it descends vertically. Attachments are highly variable.

Function:

Flexion of cervical spine
Elevation of 1st rib in inspiration
Rotation of cervical spine to opposite side
Lateral bending of neck to same side

Innervation:

C4–C6 cervical nerves (ventral rami)

81. SCALENUS MEDIUS

Origin:

C2–C7 (posterior tubercles of transverse processes)
Axis (sometimes)

Insertion:

1st rib (widely over superior surface)

Description:

Longest and largest of the scalenes. Descends vertically alongside of vertebrae.

Function:

Cervical flexion (weak)
Lateral bending of cervical spine to same side
Elevation of 1st rib in inspiration
Cervical rotation to opposite side

Innervation:

C3–C8 cervical nerves (ventral rami)

82. SCALENUS POSTERIOR

Origin:

C4–C6 vertebrae (variable; posterior tubercles of transverse processes)

Insertion:

2nd rib (outer surface)

Description:

Smallest and deepest lying of the scalene muscles. Attachments are highly variable. Often not separable from Scalenus medius.

Function:

Cervical flexion (weak)
Elevation of 2nd rib in inspiration
Lateral bending of cervical spine (accessory)
Cervical spine rotation to opposite side

Innervation:

C6–C8 cervical nerves (ventral rami)

83. STERNOCLEIDOMASTOID

Origin:

Sternal (medial) head: Sternum, manubrium (ventral surface)
Clavicular (lateral) head: Clavicle (superior and anterior surface of medial 1/3)

Insertion:

Temporal bone, mastoid process (lateral surface)
Occiput (lateral half of superior nuchal line)

Description:

The two heads of origin gradually merge in the neck as the muscle rises upward laterally and posteriorly.

Function:

Flexion of cervical spine (both muscles)
Lateral bending of cervical spine to same side
Rotation of head to opposite side
Capital extension (posterior fibers)
Raises sternum in forced inspiration

Innervation:

Accessory (XI) nerve (spinal part)
C2–C3 cervical nerves (ventral rami)

84. STERNOTHYROID

Origin:

Sternum, manubrium (posterior surface)
1st rib (cartilage)

Insertion:

Thyroid cartilage (oblique line)

Description:

A deep lying, somewhat broad muscle rising vertically and slightly laterally just lateral to the thyroid gland.

Function:

Cervical flexion (weak)
Depression of hyoid, mandible, and tongue (after
   elevation)
Draws larynx down after swallowing

Innervation:

C1–C3 cervical nerves (branch of ansa cervicalis)

85. THYROHYOID

Origin:

Thyroid cartilage (oblique line)

Insertion:

Hyoid bone (inferior border of greater horn)

Description:

Appears as an upward extension of Sternothyroid.
   Small, rectangular muscle lateral to thyroid cartilage.

Function:

Cervical flexion (weak)
Draws hyoid bone downward
Elevates larynx and thyroid cartilage

Innervation:

C1 cervical nerve (which runs in hypoglossal
   nerve)

86. STERNOHYOID

Origin:

Clavicle (medial end, posterior surface)
Sternum, manubrium (posterior)
Sternoclavicular ligament

Insertion:

Hyoid bone (body, lower border)

Description:

Thin strap muscle that ascends slightly medially
   from clavicle to hyoid bone.

Function:

Cervical flexion (weak)
Depresses hyoid bone after swallowing

Innervation:

C1–C3 cervical nerves (branch of ansa cervicalis)

87. OMOHYOID

Inferior belly:

Origin:

Scapula (superior margin; extent variable)
Superior transverse ligament

Insertion:

Intermediate tendon of Omohyoid under
   Sternocleidomastoid
Clavicle by fibrous expansion

Superior belly:

Origin:

Intermediate tendon of Omohyoid

Insertion:

Hyoid bone (lower border of body)

Description:

Muscle consists of two fleshy bellies united by a
   central tendon. The inferior belly is a narrow
   band that courses forward and slightly upward
   across the lower front of the neck. The superior belly rises vertically and lateral to the
   Sternohyoid.

Function:

Depression of hyoid after elevation
Mandible depression (assists)

Innervation:

C1–C3 cervical nerves (branch from ansa cervicalis)

88. PLATYSMA

Origin:

Fascia covering upper Pectoralis major and
   Deltoid

Insertion:

Mandible (below the oblique line)
Modiolus
Skin and subcutaneous tissue of lower lip

Description:

A broad sheet of muscle, it rises from the shoulder, crosses the clavicle, and rises obliquely upward and medially.

Function:

Draws lower lip downward and backward (expression of surprise or horror) and assists with
   jaw opening.

Cervical flexion (weak). Electromyogram shows great activity in extreme effort and in sudden deep inspiration.

Can pull skin up from clavicular region, increasing diameter of neck. Wrinkles skin of nuchal area obliquely, thereby decreasing concavity of neck.

Innervation:

Facial (VII) nerve (cervical branch)

Muscles of the Trunk

Back
Thorax (respiration)
Abdomen
Perineum and anus

DEEP MUSCLES OF THE BACK

These muscles consist of groups of serially arranged muscles ranging from the occiput to the sacrum. There are four subgroups plus the Quadratus lumborum.

In this section readers will note that the cervical portions of each muscle group are not included. These muscles are described as part of the neck muscles because their functions involve capital and cervical motions. They are, however, mentioned in the identification of each group for a complete overview.

Splenius (in neck only)
Erector spinae
Transversospinalis group
Interspinal-intertransverse group

Quadratus lumborum

Erector Spinae

The muscles in this group are arranged alongside the vertebral column in a large musculotendinous mass that is covered by the thoracodorsal fascia and Serratus posterior inferior below and the Rhomboid and Splenius muscles above. They vary in size and composition at different levels.

Sacral region: narrow, tendinous, and strong
Lumbar region: expands into thick muscular mass (palpable); visible surface groove on lateral side
Thoracic region: surface groove follows costal angles until covered by scapula

From all aspects of the tendinous insertion the muscles rise to form a large mass that divides in the upper lumbar region into three longitudinal columns.

Lateral column of muscle
    66. Iliocostalis cervicis (see *Muscles of the Neck*)
    89. Iliocostalis thoracis
    90. Iliocostalis lumborum

Intermediate column of muscle
    60. Longissimus capitis (see *Muscles of the Neck*)
    64. Longissimus cervicis (see *Muscles of the Neck*)
    91. Longissimus thoracis

Medial column of muscle
    63. Spinalis capitis (see *Muscles of the Neck*)
    68. Spinalis cervicis (see *Muscles of the Neck*)
    92. Spinalis thoracis

Common Origin of Erector Spinae

Sacrum (median and lateral crests, anterior surface); L1–L5 and T12 vertebrae (spinous processes); supraspinous, sacrotuberous, and sacroiliac ligaments; iliac crests (inner aspect of dorsal part)

The Iliocostalis Column (Lateral)

89. ILIOCOSTALIS THORACIS

Origin:

Ribs 12 up to 7 (upper borders at angles)

Insertion:

Ribs 1 down to 6 (at angles)
C7 (transverse process, dorsum)

90. ILIOCOSTALIS LUMBORUM

Origin:

Iliac crest (external lip)
Sacrum (posterior surface)

Insertion:

Ribs 5 or 6 to 12 (angles on inferior border)

Description (All):

This is the most lateral column of the Erector spinae. The muscles lie in a groove lateral to the vertebral column on each side. The lumbar muscle is the largest, and it subdivides as it ascends.

Function:

Extension of spine
Lateral bending of spine
Depression of ribs (lumborum)

Innervation:

Spinal nerves (T7–L2 distribution variable) (dorsal rami)

The Longissimus Column (Intermediate)

91. LONGISSIMUS THORACIS

Origin:

Sacrum
L1–L5 vertebrae (transverse processes)

Insertion:

L1–L3 vertebrae (accessory processes)
T1–T12 vertebrae (transverse processes)
Ribs 2 to 12 (between tubercles and angles)

Description (All Longissimi):

These are the intermediate Erector spinae. They lie between the Iliocostales (laterally) and the Spinales (medially). The Longissimi are the largest and longest of the three columns that compose the Erector spinae. The Longissimus thoracis splits off the Iliocostalis lumborum and Spinalis thoracis in the upper lumbar region.

Function (Longissimus thoracis):

Extension of the spine
Lateral bending of spine to same side
Depression of ribs

Innervation:

Lower cervical thoracic and lumbar spinal nerves (dorsal rami)

The Spinalis Column (Medial)

92. SPINALIS THORACIS

Origin:

T11–T12 and L1–L2 vertebrae (spinous processes)

Insertion:

T1–T4 or T8 vertebrae (spinous processes)

Description:

The smallest and thinnest of the Erector spinae, they lie closest to the vertebral column. The Spinales are inconstant, and are difficult to separate.

Function:

Extension of vertebral column

Innervation:

Spinal nerves (dorsal rami)

Transversospinales Group

Muscles of this group lie deep to the Erector spinae, filling the concave region between the spinous and transverse processes of the vertebrae. They ascend obliquely and medially from the vertebral transverse processes to adjacent and sometimes more remote vertebrae. A span over four to six vertebrae is not uncommon.

62. Semispinalis capitis (see *Muscles of the Neck*)
65. Semispinalis cervicis (see *Muscles of the Neck*)
93. Semispinalis thoracis

94. Multifidi
71. Rotatores cervicis (see *Muscles of the Neck*)
95. Rotatores thoracis
96. Rotatores lumborum

93. SEMISPINALIS THORACIS

Origin:

T6–T10 vertebrae (transverse processes)

Insertion:

C6–C7 and T1–T4 vertebrae (spinous processes)

Description:

This group is found only in the thoracic and cervical regions, extending to the head. They lie deep to the Spinalis and Longissimus columns of the Erector spinae.

Function:

Extension of thoracic spine
Rotation of spine to opposite side

Innervation:

Thoracic spinal nerves (dorsal rami)

94. MULTIFIDI

Origin:

Sacrum (as low as the S4 foramen)
Erector spinae aponeurosis
Ilium (posterior superior iliac spine) and adjacent crest
Sacroiliac ligaments (posterior)
T1–T12 vertebrae (mamillary and transverse processes)
C4–C7 vertebrae (articular processes)

Insertion:

A higher vertebra (spinous process)—may span two to four vertebrae before inserting

Description:

These muscles fill the grooves on both sides of the spinous processes of the vertebrae from the sacrum to the axis. They lie deep to the Erector spinae in the lumbar region and deep to the Semispinales above. Each fasciculus ascends obliquely, traversing over two to four vertebrae as it moves toward the midline to insert in the spinous process of a higher vertebra.

Function:

Extension of spine (muscles on both sides)
Lateral bending of spine
Rotation to opposite side (one)

Innervation:

Spinal nerves, segmentally (dorsal rami)

The Rotatores

These are the deepest muscles of the Transversospinalis group, lying as 11 pairs of very short muscles beneath the Multifidi. The fibers run obliquely upward and medially or almost horizontal. They may cross more than one vertebra on their ascending course, but most commonly they proceed to the next higher one. Found along the entire length of the vertebral column, they are distinguishable as developed muscle only in the thoracic area.

95. ROTATORES THORACIS

Origin:

Transverse process of one thoracic vertebra

Insertion:

Base of spine of next highest vertebra

Function:

Extension of thoracic spine

Innervation:

Spinal nerves

96. ROTATORES LUMBORUM

The Rotatores are highly variable and irregular in these regions.

Description:

This muscle lies deep to the Multifidi and cannot be readily separated from its deepest fibers.

Function:

Extension of spine
Rotation of spine to opposite side

Innervation:

Spinal nerves (dorsal rami)

Interspinal-Intertransverse Group

69. Interspinalis cervicis (see under *Muscles of the Neck*)
97. Interspinalis thoracis
98. Interspinalis lumborum

70. Intertransversarii cervicis, anterior and posterior (see under *Muscles of the Neck*)
99. Intertransversarii thoracis
99. Intertransversarii lumborum, medial and lateral

The Interspinales

The muscles in this group pass segmentally from one vertebra to the next. They are short, paired fasciculi lying between the spinous processes of contiguous vertebrae. They are most highly developed in the cervical region.

97. INTERSPINALES THORACIS

Origin and Insertion:

Between spinous processes of contiguous vertebrae
Three pairs: (1) between 1st and 2nd thoracic vertebrae; (2) between 2nd and 3rd thoracic vertebrae (variable); (3) between 11th and 12th thoracic vertebrae

Function:

Extension of spine

Innervation:

Spinal nerves (dorsal rami)

98. INTERSPINALES LUMBORUM

Origin and Insertion:

There are four pairs lying between the five lumbar vertebrae.

Function:

Extension of spine

Innervation:

Spinal nerves (dorsal rami)

The Intertransversarii

These are small fasciculi lying between the transverse processes of contiguous vertebrae. They are most developed in the cervical spine.

99. INTERTRANSVERSARII THORACIS AND LUMBORUM

In the thoracic area these short muscles consist of single slips lying between the transverse processes of the last three thoracic vertebrae and the first lumbar vertebra.

In the lumbar region the muscles are again paired on each side of the spinal column as medial and lateral muscles. The medial muscles connect the accessory process of one vertebra with the mamillary process of the next vertebra below. The lateral intertransversarii fill the space between transverse processes of contiguous vertebrae.

Function:

Extension of spine (muscles on both sides)
Lateral bending to same side (muscles on one side)

Innervation:

Spinal nerves
Medial muscles (dorsal rami)
Lateral muscles (ventral rami)

100. QUADRATUS LUMBORUM

Origin:

Ilium (crest, inner lip)
Iliolumbar ligament

Insertion:

12th rib (lower border)
L1–L4 vertebrae (apices of transverse processes)
T12 vertebral body (occasionally)

Description:

An irregular quadrilateral muscle located against the posterior (dorsal) abdominal wall, this muscle is encased by layers of the thoracolumbar fascia. It fills the space between the 12th rib and the iliac crest. Its fibers run obliquely upward and medially from the iliac crest to the inferior border of the 12th rib and transverse processes of the lumbar vertebrae. The muscle is variable in size and occurrence.

Function:

Extension of lumbar spine (muscles on both sides)

Lateral bending of lumbar spine to same side (pelvis fixed)
Fixation and depression of 12th rib

Innervation:

T12 and L1–L3 spinal nerves (ventral rami)

MUSCLES OF THE THORAX FOR RESPIRATION

101. Diaphragm
102. Intercostales externi
103. Intercostales interni
104. Intercostales intimi
105. Subcostales
106. Transversus thoracis
107. Levatores costarum
108. Serratus posterior superior
109. Serratus posterior inferior

101. DIAPHRAGM

Origin:

Muscle fibers take origin from the circumference of the thoracic outlet in three groups:
Sternal: Xiphoid (posterior surface)
Costal: Ribs 7 to 12 (bilaterally; inner surfaces of the cartilage and the deep surfaces on each side)
Lumbar: L1–L3 vertebrae (from the medial and lateral arcuate ligaments and from bodies of the vertebrae by two muscular crura)

Insertion:

Central tendon of diaphragm
Immediately below the pericardium and blending with it. The central tendon has no bony attachments. It has three divisions called leaflets (because of its clover-leaf pattern) in an otherwise continuous sheet of muscle, which affords the muscle great strength.

Description:

This half-dome shaped muscle of contractile and fibrous structure forms the floor of the thorax (convex upper surface) and the roof of the abdomen (concave inferior surface) (Fig. 9–4).

The diaphragm is muscular on the periphery and its central area is tendinous. It closes the opening of the thoracic outlet and forms a convex floor for the thoracic cavity. The muscle is flatter centrally than at the periphery and higher on the right (reaching rib 5) than on the left (reaching rib 6). From the peak on each side, the diaphragm abruptly descends to its costal and vertebral attachments. This descending slope is much more precipitous and longer posteriorly.

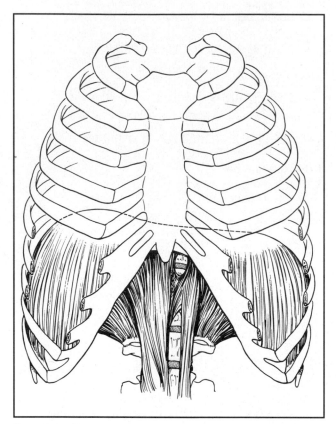

Figure 9-4. The diaphragm.

Function:

Inspiration: Contraction of the diaphragm draws the central tendon downward and forward during inspiration. This increases the vertical thoracic dimensions and pushes the abdominal viscera downward. It also decreases the pressure within the thoracic cavity, forcing air into the lungs through the open glottis by the higher pressure of the atmospheric air.

These events occur along with intercostal muscle action, which elevates the ribs, sternum, and vertebrae, increasing the anteroposterior and transverse thoracic dimensions for the inspiratory effort.

Expiration: Passive relaxation allows the half-dome to ascend, thus decreasing thoracic cavity volume and increasing its pressure.

Innervation:

Phrenic nerve, C4 (with contributions from C3 and C5)

The Intercostals

The intercostal muscles are slim layers of muscle and tendon occupying each of the intercostal spaces; the externals are the most superficial, the internals come next, and deepest are the intimi.

102. INTERCOSTALES EXTERNI

Origin:

Ribs 1 to 11 (lower borders and costal tubercles)

Insertion:

Ribs 2 to 12 (upper margins)
Aponeurotic external intercostal membrane, which continues to sternum

Description:

There are 11 of these muscles on each side of the chest. Each arises from the inferior margin of one rib and inserts on the superior margin of the rib below. They extend in the intercostal spaces from the tubercles of the ribs dorsally to the cartilages of the ribs ventrally.

The muscle fibers run obliquely inferolaterally on the dorsal thorax; they run inferomedially and somewhat ventrally on the anterior thorax (down and toward the sternum)

The externi are the thickest of the three intercostal muscles. In appearance they may seem to be continuations of the external oblique abdominal muscles.

Function:

Elevation of ribs in inspiration. There are data to support this claim for the upper four or five muscles, but the more dorsal and lateral fibers of the same muscles also are active in early expiration. It is possible that the activity of the intercostals during respiration varies with the depth of breathing.[13,14]

Depression of the ribs in expiration (supporting data sparse).

Rotation of thoracic spine to opposite side (unilateral).

Stabilization of rib cage.

Innervation:

Intercostal nerves, T1–T11. These nerves are numbered sequentially according to interspace—e.g., the 5th intercostal nerve innervates muscle occupying the 5th intercostal space between the 5th and 6th ribs.

103. INTERCOSTALES INTERNI

Origin:

Ribs 1 to 11 (ridge on inner surface, then passing down and toward spine)
Costal cartilage of same rib

Insertion:

Ribs 2 to 12 (upper border of next rib below)

Description:

There are 11 pairs of these muscles. They extend from the sternal end of the ribs anteriorly to the angle of the ribs posteriorly. The fibers run obliquely upward but at a 90-degree angle to the external intercostals.

Function:

Elevation of ribs in inspiration. This may be true at least for the 1st to 5th muscles. The more lateral muscles, whose fibers run more obliquely inferior and posterior, are active in expiration.[15] Stabilization of rib cage.

Innervation:

Intercostal nerves, T1–T11

104. INTERCOSTALES INTIMI

Origin:

Costal groove of rib above in lower costal interspaces, but no consistent evidence in upper five to six interspaces

Insertion:

Upper margin of rib below in lower costal interspaces

Description:

There is dispute about whether this is a separate muscle or just a part of the internal intercostals. It is a thin sheet lying deep to the internal intercostals, but arguments in favor of a separate muscle are not convincing. If they are separate, there may be five to six pairs, with no consistent presence in the upper costal interspaces.

Function:

Presumed to be identical to Intercostales interni.

Innervation:

Intercostal nerves

105. SUBCOSTALES

Origin:

Lower ribs (variable) on inner surface near angle

Insertion:

Inner surface of two or three ribs below rib of origin

Description:

Lying on the dorsal thoracic wall, these muscles are discretely developed only in the lower thorax. Fibers run in the same direction as those of the Intercostales interni.

Function:

Draws adjacent ribs together or depresses ribs (no supporting data).

Innervation:

Intercostal nerves

106. TRANSVERSUS THORACIS

Origin:

Sternum (caudal 1/3 and xiphoid)
Ribs 3 to 6 (costal cartilages, inner side)

Insertion:

Ribs 2 to 5 (costal cartilages, caudal borders)

Description:

A thin plane on the inner surface of the anterior wall of the thorax. The fibers pass obliquely up and laterally, diverging more as they insert. The lowest fibers are horizontal and are continuous with the Transversus abdominis. The highest fibers are almost vertical. Attachments vary from side to side in the same person and among different persons.

Function:

Draws ribs downward
Active in forced expiration

Innervation:

Intercostal nerves

107. LEVATORES COSTARUM

Origin:

C7 and T1–T11 vertebrae (transverse processes)

Insertion:

Rib immediately below transverse process of the vertebra from which muscle takes origin (on outer surface between tubercle and angle).

Description:

There are 12 pairs of these muscles on either side of the thorax on its posterior wall. Fibers run obliquely inferolaterally, like those of the external intercostal muscles. The most inferior fibers divide into two fasciculi, one of which inserts as described; the other descends to the 2nd rib below its origin.

Function:

Elevation of ribs in inspiration
Lateral bending of spine

Innervation:

Intercostal nerves, T1–T12

108. SERRATUS POSTERIOR SUPERIOR

Origin:

C7 and T1–T3 vertebrae (spinous processes)
Ligamentum nuchae
Supraspinal ligament

Insertion:

Ribs 2 to 5 (upper borders, lateral to angles)

Description:

Muscle lies on the upper dorsal thorax, over the Erector spinae and under the Rhomboids. Fibers run inferolaterally.

Function:

Elevates upper ribs
Presumably increases thoracic volume (function uncertain)

Innervation:

Intercostal nerves, T1–T4 (ventral rami)

109. SERRATUS POSTERIOR INFERIOR

Origin:

T11–T12 and L1–L2 vertebrae (spinous processes via thoracolumbar fascia)

Insertion:

Ribs 9 to 12 (inferior borders, lateral to angles)

Description:

A thin muscle, composed of four digitations, lying at the border between the thoracic and lumbar regions. It is much broader than the Serratus posterior superior and lies four ribs below it. It lies over the Erector spinae and under the external obliques. The muscle may have fewer than four digitations or may be absent.

Function:

Depresses lower ribs and moves them dorsally
Has an uncertain role in respiration

Innervation:

Intercostal nerves, T9–T12

MUSCLES OF THE ABDOMEN

Anterolateral Walls

110. Obliquus externus abdominis
111. Obliquus internus abdominis
112. Transversus abdominis
113. Rectus abdominis
114. Pyramidalis

110. OBLIQUUS EXTERNUS ABDOMINIS

Origin:

Ribs 4 to 12 (by digitations that attach to the external and inferior surfaces and alternate with digitations of the Serratus anterior)

Insertion:

Iliac crest (anterior half of outer lip)
Aponeurosis from the prominence of 9th costal cartilage to ASIS; both sides meet at the linea alba.

Description:

The largest and most superficial flat, thin muscle of the abdomen curves around the anterior and lateral walls. Its muscular fibers lie on the lateral wall while its aponeurosis traverses the anterior wall, meeting its opposite number to form the linea alba. The digitations form an oblique line that runs down and backward. The linea alba extends from the xiphoid process to the symphysis pubis.
The upper (superior) five digitations increase in size as they descend and alternate with the corresponding digitations of the Serratus anterior. The distal three digitations decrease in size as they descend and alternate with digitations of the Latissimus dorsi. The superior fibers travel inferomedially; the posterior fibers pass more vertically.

Function:

Flexion of trunk (bilateral muscles)
Tilt pelvis posteriorly
Rotation of trunk (unilateral)
Lateral bending of trunk (unilateral)
Support and compression of abdominal viscera, counteracting effect of gravity on abdominal contents
Assist defecation, micturition, emesis, and parturition (i.e., expulsion of contents of abdominal viscera and air from lungs)
Important accessory muscle of forced expiration

(during expiration it forces the viscera upward to elevate the diaphragm)

Innervation:

T7–T12 spinal nerves (ventral rami)

111. OBLIQUUS INTERNUS ABDOMINIS

Origin:

Thoracolumbar fascia
Inguinal ligament (lateral 2/3)
Iliac crest (anterior 2/3 of intermediate lip)

Insertion:

Ribs 9 to 12 (inferior borders by digitations that appear continuous with internal intercostals)
Aponeurosis that fuses with that of the external oblique and into the linea alba
Cartilages of ribs 7 to 9 (via an aponeurosis)
Pubis (pectineal line, with tendinous sheath from Transversus abdominis)

Description:

This muscle is smaller and thinner than the external oblique under which it lies on the lateral and ventral abdominal wall. The fibers from the iliac crest pass upward and medially to ribs 9 to 12 and the aponeurosis; the more lateral the fibers lie, the more they run toward the vertical. The lowest fibers pass almost horizontally on the lower abdomen.

Function:

Rotation of spine
Flexion of spine (bilateral)
Lateral bending of spine (unilateral)
Increases abdominal pressure to assist in defecation
Forces viscera upward during expiration to elevate diaphragm

Innervation:

T8–T12 spinal nerves
L1 spinal nerve (iliohypogastric and ilioinguinal branches)

112. TRANSVERSUS ABDOMINIS

Origin:

Inguinal ligament (lateral 1/3)
Iliac crest (anterior 2/3 of inner lip)
Thoracolumbar fascia (between iliac crest and 12th rib)
Ribs 7 to 12 (costal cartilages)

Insertion:

Linea alba (upper and middle fibers pass medially to blend with the broad aponeurosis)

Pubis, crest and pecten (lower fibers: via a bilateral aponeurosis that bends inferomedially as the falx inguinalis; blends with Obliquus internus)

Description:

The innermost of the flat abdominal muscles, the Transversus abdominis lies under the internal oblique. Its name derives from the direction of its fibers, which pass horizontally across the lateral abdomen to an aponeurosis and the linea alba. The length of the fibers varies considerably depending on the insertion site, the most inferior to the pubis being the longest. At its origin on ribs 7 to 12, the muscle interdigitates with similar diaphragmatic digitations separated by a narrow raphe.

Function:

Constricts (flattens) abdomen, compressing the abdominal viscera and assisting in expelling their contents
Forced expiration

Innervation:

T7–T12 spinal nerves
L1 spinal nerve (iliohypogastric and ilioinguinal branches)

113. RECTUS ABDOMINIS

Origin:

Ribs 5 to 7 (costal cartilages by three fascicles of differing size)
Sternum (costoxiphoid ligaments)

Insertion:

By two tendons inferiorly:
    Pubis (tubercle on crest)
    Ligaments covering front of symphysis pubis

Description:

A long muscular strap extending from the ventral lower sternum to the pubis. Its vertical fibers lie centrally along the abdomen, each separated from its contralateral partner by the linea alba. The muscle is interrupted (but not all the way through) by three (or more) fibrous bands called the *tendinous intersections,* which pass transversely across the muscle in a zigzag fashion. The most superior intersection generally is at the level of the xiphoid; the lowest is at the level of the umbilicus, and the second intersection is midway between the two. These are readily visible on body builders or others with well-developed musculature.

Function:

Flexion of spine (draws symphysis and sternum toward each other)

Posterior tilt of pelvis

With other abdominal muscles, compresses abdominal contents

Innervation:

T7–T12 spinal nerves

T7 innervates fibers above the superior tendinous intersection; T8 innervates fibers between the superior and middle intersections; T9 innervates fibers between the middle and distal intersections.

114. PYRAMIDALIS

Origin:

Pubis (front of body)

Insertion:

Linea alba (midway between umbilicus and pubis)

Description:

A small triangular-shaped muscle located in the extreme distal portion of the abdominal wall and lying anterior to the lower Rectus abdominis. Its origin on the pubis is wide, and it narrows as it rises to a pointed insertion. The muscle varies considerably from side to side, and may be present or absent.

Function:

Tenses the linea alba

Innervation:

T12 spinal nerve (ventral ramus)

MUSCLES OF THE PERINEUM

115. Levator ani
116. Coccygeus
117. Cremaster

118. Transversus perinei superficialis
119. Transversus perinei profundus
120. Bulbospongiosus
121. Ischiocavernosus
122. Sphincter urethrae
123. Spincter ani externus

Corrugator cutis ani (involuntary muscle, not described)

Internal anal sphincter (involuntary muscle, not described)

115. LEVATOR ANI

Origin:

Pubococcygeus part: Pubis (inner surface of superior ramus)

Iliococcygeus part: Ischium (inner surface of spine)

Obturator fascia

Insertion:

Coccyx (last two segments)

Anococcygeal raphe

Sphincter ani externus

Description:

This broad, thin sheet of muscle unites with its contralateral partner to form the pelvic floor. Anteriorly it is attached to the pubis lateral to the symphysis, posteriorly to the ischial spine, and between these to the obturator fascia. The fibers course medially with varying obliquity.

Function:

Constriction of rectum and vagina

Along with the coccygei, the levator forms a muscular pelvic diaphragm that supports the pelvic viscera and opposes sudden increases in intra-abdominal pressure, as in forced expiration, or the Valsalva maneuver.

Innervation:

S4 spinal nerve, pudendal nerve (sometimes S3 and S5)

116. COCCYGEUS

Origin:

Ischium (spine)

Sacrospinous ligament

Insertion:

Coccyx (lateral margins)

Sacrum (last segment, side)

Description:

The paired muscle lies posterior and superior to the Levator ani and contiguous with it in the same plane. The muscle occasionally is absent. It is considered the pelvic aspect of the sacrospinous ligament.

Function:

The coccygei pull the coccyx forward and support it after it has been pushed back for defecation or parturition.

With the Levatores ani and Piriformis, compresses

the posterior pelvic cavity and outlet ("the birth canal")

Innervation:

S4–S5 spinal nerves (pudendal plexus)

117. CREMASTER

Origin:

Inguinal ligament (continuous with Internal oblique). Technically this is an abdominal muscle.

Insertion:

Pubis (tubercle and crest)
Sheath of Rectus abdominis and Transversus abdominis

Description:

Consists of loose fasciculi lying along the spermatic cord and held together by areolar tissue to form the cremasteric fascia around the cord and testis. Often said to be continuous with the internal oblique abdominal muscle or with the Transversus abdominis. After passage through the superficial inguinal ring, the muscle spreads into loops of varying length over the spermatic cord.
Although the muscle fibers are striated, this is not usually a voluntary muscle. Stimulation of the skin on the medial thigh evokes a reflex response, the *cremasteric reflex.*
Found as a vestige in women.

Function:

Elevation of testis toward superficial inguinal ring
Thermoregulation of testes by adjusting position

Innervation:

L1–L2 spinal nerve (genitofemoral nerve)

118. TRANSVERSUS PERINEI SUPERFICIALIS

Origin:

Ischial tuberosity (inner and anterior part)

Insertion:

Perineal body (a centrally placed, modiolar-like structure on which perineal muscles and fascia converge)

Description:

A narrow slip of muscle in both the male and female perineum, it courses almost transversely across the perineal area in front of the anus. It is joined on the perineal body by the muscle

from the opposite side. The muscle is sometimes absent or may be doubled.

Function:

Bilateral action serves to fix the centrally located perineal body.

Innervation:

S2–S4 spinal nerve (pudendal)

119. TRANSVERSUS PERINEI PROFUNDUS

Origin:

Ischium (inferior ramus)

Insertion:

Male: Perineal body
Female: Vagina (side); perineal body

Description:

Small deep muscle with similar structure and function in both male and female. The bilateral muscles meet at the midline on the perineal body. This muscle is in the same plane as the Sphincter urethrae, and together they form most of the bulk of the urogenital diaphragm. (Together they used to be called the Constrictor urethrae).

Function:

Fixation of perineal body (uncertain)

Innervation:

S2–S4 spinal nerves (pudendal)

120. BULBOSPONGIOSUS

Formerly called:
    Male: Bulbocavernosus; Accelerator urinae
    Female: Sphincter vaginae

In the Female:

Origin:

Perineal body
Blending with Sphincter ani externus

Insertion:

Corpora cavernosus clitoridis

Description:

Surrounds the orifice of the vagina and covers the lateral parts of the vestibular bulb. The fibers run anteriorly on each side of the vagina and send a slip to cover the clitoral body.

Function:

Constriction of vaginal orifice
Constriction of deep dorsal vein, of clitoris by anterior fibers, contributing to erection of clitoris

Innervation:

S2–S4 spinal nerves (pudendal)

In the Male:

Origin:

Perineal body and its ventral extension into the median raphe

Insertion:

Urogenital diaphragm (inferior fascia)
Aponeurosis over corpus spongiosum penis
Body of penis anterior to Ischiocavernosus
Tendinous expansion over dorsal vessels of penis

Description:

Located in the midline of the perineum anterior to the anus and consisting of two symmetrical parts united by a tendinous raphe. Its fibers divide like the halves of a feather. The posterior fibers disperse on the inferior fascia of the urogenital diaphragm; the middle fibers encircle the penile bulb and the corpus spongiosum and form a strong aponeurosis with fibers from the opposite side; and the anterior fibers spread out over the corpora cavernosa.

Function:

Empties urethra at end of micturition (is capable of arresting urination).
Middle fibers assist in penis erection by compressing the bulbar erectile tissue; anterior fibers assist by constricting the deep dorsal vein.
Contracts repeatedly in ejaculation.

Innervation:

S2–S4 spinal nerves (pudendal)

121. ISCHIOCAVERNOSUS

Formerly called:
    Female: Erector clitoridis
    Male: Erector penis

In the Female:

Origin:

Ischium (tuberosity [inner surface] and ramus)
Crus clitoridis (surface)

Insertion:

Aponeurosis inserting into sides and inferior surface of crus clitoridis

Description:

Covers the unattached surface of crus clitoridis. Muscle is smaller than the male counterpart.

Function:

Compresses crus clitoridis, retarding venous return and thus assisting erection.

Innervation:

S2–S4 (pudendal)

In the Male:

Origin:

Ischium (tuberosity dorsal to crus penis and ramus)
Pubis (ramus)

Insertion:

Aponeurosis into the sides and undersurface of the body of the penis

Description:

The muscle is paired and covers the crus of the penis.

Function:

Compression of crus penis, maintaining erection by retarding return of blood through the veins

Innervation:

S2–S4 spinal nerves (pudendal)

122. SPHINCTER URETHRAE

In the Female:

Origin:

Pubis (inferior ramus on each side)
Transverse perineal ligament

Insertion:

Blend with fibers from opposite muscle posterior to urethra.

Description:

Has both external and internal fibers. The external fibers arise on the pubis and course across the pubic arch in front of the urethra to circle around it. The internal fibers encircle the lower end of the urethra.

Function:

Constricts urethra

Innervation:

S2–S4 spinal nerves (pudendal)

In the Male:

Origin:

Ischiopubic ramus (superior fibers)
Transverse perineal ligament (inferior fibers)

Insertion:

Converges with muscles from other side on perineal body

Description:

Surrounds entire length of membranous portion of urethra and is enclosed in the urogenital diaphragm fascia.

Function:

Compression of urethra (bilateral action)
Active in ejaculation
Relaxes during micturition

Innervation:

S2–S4 spinal nerves (pudendal)

123. SPHINCTER ANI EXTERNUS

Origin:

Skin surrounding margin of anus
Coccyx (via anococcygeal ligament)

Insertion:

Perineal body
Blends with many other muscles in area

Description:

Surrounds entire anal canal. Consists of three parts, all skeletal muscle:
1. Subcutaneous: around lower anal canal; fibers course horizontally beneath the skin at the anal orifice.
2. Superficial: is the only part attached to bone (terminal coccygeal segment); perineal body.
3. Deep part: thick band around the upper internal sphincter with fibers blending with the Puborectalis of the Levator ani and fascia.

Function:

Keeps anal orifice closed. It is always in a state of tonic contraction and has no antagonist. Muscle relaxes during defecation, allowing orifice to open. The muscle can be voluntarily contracted to close the orifice more tightly as in forced expiration or the Valsalva maneuver.

Innervation:

S2–S3, spinal nerve (pudendal)
S4 spinal nerve

Muscles of the Upper Extremity (Shoulder Girdle, Elbow, Forearm, Wrist, Fingers, Thumb)

MUSCLES OF THE SHOULDER GIRDLE ACTING ON THE SCAPULA

124. Trapezius
125. Rhomboid major
126. Rhomboid minor
127. Levator scapulae

128. Serratus anterior
129. Pectoralis minor

124. TRAPEZIUS

Origin:

Upper:
Occiput (external protuberance and medial 1/3 of superior nuchal line)
Ligamentum nuchae
C7 vertebrae (spinous process)
Middle
T1–T6 vertebrae (spinous processes and supraspinous ligaments)
Lower
T7–T12 vertebrae (spinous processes)

Insertion:

Upper
Clavicle (posterior surface, lateral 1/3)
Scapula (Anterior acromion process)
Middle
Scapula (medial margin of acromion and superior lip of posterior border of spine)
Lower
Scapula (aponeurosis at root of spine, then into tubercle at apex of smooth triangular surface at root)

Description:

A flat, triangular muscle lying over the posterior neck, shoulder, and upper thorax. The upper trapezius fibers course down and laterally from the occiput; the middle fibers are horizontal, and the lower fibers move upward and laterally from the vertebrae to the scapular spine. The name of the muscle is derived from the shape of the muscle with its contralateral partner: a diamond-shaped quadrilateral figure, or trapezoid.

Function:

Upper and lower:
> Rotation of scapula so glenoid faces up (inferior angle moves laterally and forward)

Upper:
> Elevation of scapula and shoulder ("shrugging")
> Rotation of head to opposite side (one)
> Capital extension (both)
> Cervical extension (both)

Middle:
> Scapular adduction (retraction)

Lower:
> Scapular adduction, depression, and upward rotation

Innervation:

Accessory (XI) nerve

125. RHOMBOID MAJOR

Origin:

T2–T5 vertebrae (spinous processes)
Supraspinal ligament

Insertion:

Scapula (medial border between root of spine above and inferior angle below)

Description:

Fibers of the muscle run slightly inferolaterally between the thoracic spine and the medial border of the scapula.

Function:

Scapular adduction
Downward rotation of scapula (glenoid faces down)
Scapular elevation

Innervation:

Dorsal scapular nerve, C5

126. RHOMBOID MINOR

Origin:

C7–T1 vertebrae (spinous processes)
Ligamentum nuchae

Insertion:

Scapula (root of spine on medial border)

Description:

Lies just superior to Rhomboid major, and its fibers run parallel to the larger muscle.

Function:

Scapular adduction
Scapular downward rotation (glenoid faces down)
Scapular elevation

Innervation:

Dorsal scapular nerve, C5

127. LEVATOR SCAPULAE

Origin:

C1 (atlas), C2 (axis) (transverse processes)
C3–C4 vertebrae (transverse processes and posterior tubercles)

Insertion:

Scapula (vertebral border between superior angle and root of spine)

Description:

Lies on the dorsolateral neck and descends deep to the Sternocleidomastoid on the floor of the posterior triangle of the neck.

Function:

Elevates and adducts scapula
Scapular rotation (glenoid faces down)
Lateral bending of cervical spine to same side (one)
Cervical rotation to same side (one)
Cervical extension (both assist)

Innervation:

C3–C4 spinal nerves
Dorsal scapular nerve (to lower fibers), C5

128. SERRATUS ANTERIOR

Origin:

Ribs 1 to 8 by digitations (superior and outer surfaces). Each digitation (except first) arises from the corresponding numbered rib.
Aponeurosis of intercostal muscles

Insertion:

Scapula (ventral surface of whole vertebral border)

First digitation: superior angle of scapula on anterior aspect

Second and third digitations: anterior surface of whole vertebral border

Fourth to eighth digitations: inferior angle of scapula

Description:

This is a large sheet of muscle curving around the thorax and arising in multiple digitations from the upper 8 or 9 ribs. On the posterior thorax it lies between the ribs and the scapula and reaches around and under to insert on the medial border of the scapula. The lower attachments interdigitate with the upper slips of the Obliquus externus abdominis.

Function:

Scapular abduction

Upward rotation of the scapula (glenoid faces up)

Medial border of scapula drawn anteriorly close to the thoracic wall (preventing "winging")

Functional Relationships:

The Serratus works with the Trapezius in a force couple to rotate the scapula upwardly (glenoid up), allowing the arm to be elevated fully (150° to 180°). Three component forces act around a center of rotation located in the center of the scapula: (1) upward pull on the acromial end of the spine of the scapula by the upper Trapezius; (2) downward pull on the base of the spine of the scapula by the lower Trapezius; (3) lateral and anterior pull on the inferior angle by the inferior fibers of the Serratus.[16, 17]

The reader is referred to comprehensive texts on kinesiology for further detail.

Innervation:

Long thoracic nerve, C5–C7

129. PECTORALIS MINOR

Origin:

Ribs 3 to 5 (upper and outer surfaces near the cartilages)

Aponeurosis of intercostal muscles

Insertion:

Scapula (coracoid process, medial border, and superior surface)

Description:

This muscle, broader at its origins, lies on the upper thorax directly under the Pectoralis major. It forms part of the anterior wall of the axilla

(along with the Pectoralis major). The fibers pass upward and laterally and converge in a flat tendon.

Function:

Scapular protraction (abduction) (scapula moves forward around the chest wall with a downward tilt)

Elevation of ribs in forced inspiration when scapula is fixed by the Levator scapulae

Innervation:

Medial and lateral pectoral nerves, C8–T1

VERTEBROHUMERAL MUSCLES

130. Latissimus dorsi
131. Pectoralis major

130. LATISSIMUS DORSI

Origin:

T6–T12 vertebrae (spinous processes by way of the thoracolumbar vertebrae)

L1–L5 and sacral vertebrae (spinous processes)

Ribs 9 to 12

Scapula (inferior angle)

Supraspinal ligament

Ilium (posterior 1/3 of iliac crest)

Insertion:

Humerus (intertubercular groove, distal)

Deep fascia of arm

Description:

A broad sheet of muscle that covers the lumbar and lower portion of the posterior thorax. From this wide origin, the muscle fibers converge on the proximal humerus. The superior fibers are almost horizontal, passing over the inferior angle of the scapula, whereas the lowest fibers are almost vertical. The muscle spirals around the inferior border of the Teres major and becomes twisted on itself so that the superior fibers course first posteriorly and then inferiorly; similarly, the most inferior vertically oriented fibers twist anteriorly and finally become superior.

Function:

Extension, adduction, and internal rotation of shoulder

Hyperextension of spine (muscles on both sides), as in lifting

The muscle is most powerful in overhead activities such as swimming (downstroke), climbing, crutch walking (elevation of trunk to arms, i.e., shoulder depression) or swinging.[18,19] It is very active in strong expiration, as in coughing and sneezing, and in deep inspiration.

Innervation:

Thoracodorsal nerve, C6–C8 (ventral rami)

131. PECTORALIS MAJOR

Origin:

Clavicular (upper) portion:
 Clavicle, anterior surface of sternal half
Sternocostal portion:
 Sternum: Half the breadth of anterior surface down to attachment of 6th rib
 Ribs (cartilage of all true ribs except rib 1 and sometimes rib 7)
 Aponeurosis of Obliquus externus abdominis

Insertion:

Humerus (sulcus intertubercularis, lateral border)

Description:

This muscle is a large, thick, fan-shaped muscle covering the anterior and superior surfaces of the thorax. The Pectoralis major forms part of the anterior wall of the axilla (the anterior axillary fold, conspicuous in abduction). The muscle is divided into two portions that converge toward the axilla.

The *clavicular* fibers pass downward and laterally toward the humeral insertion. The *sternocostal* fibers pass horizontally from midsternum and upward and laterally from the rib attachments. The lower fibers rise almost vertically toward the axilla. Both parts unite in a common tendon of insertion to the humerus.

Function:

Adduction of shoulder (glenohumeral) joint (whole muscle, proximal attachment fixed)
Internal rotation of shoulder
Elevation of thorax in forced inspiration (with both upper extremities fixed)
Clavicular fibers
 Internal rotation of shoulder
 Flexion of shoulder
 Horizontal shoulder adduction
Sternocostal fibers
 Horizontal shoulder adduction
 Extension of shoulder (with gravity and with Latissimus dorsi and Teres major)
 Draws trunk upward and forward in climbing

Innervation:

Clavicular fibers: Lateral pectoral nerve, C5–C7
Sternocostal fibers: Medial and Lateral pectoral nerves, C8–T1

SCAPULOHUMERAL MUSCLES

There are six shoulder muscles, which extend from the scapula to the humerus. Also included here are the Subclavius and the Coracobrachialis.

132. Subclavius
133. Deltoid
135. Supraspinatus
136. Infraspinatus
134. Subscapularis
138. Teres major
137. Teres minor
139. Coracobrachialis

All act on the shoulder (glenohumeral) joint. The largest of the muscles (Deltoid) also attaches to the clavicle and overlies the remaining muscles.

132. SUBCLAVIUS

Origin:

Rib 1 and its cartilage (at their junction)

Insertion:

Clavicle (inferior surface, groove in middle 1/3)

Description:

A small elongated muscle lying under the clavicle between it and the 1st rib. The fibers run upward and laterally following the contour of the clavicle.

Function:

Shoulder depression (assist)
Depresses and moves clavicle forward, thus stabilizing it during shoulder motion

Innervation:

C5–C6 (subclavian branch off brachial plexus)

133. DELTOID

Origin:

Anterior fibers: Clavicle (lateral 1/3 of front)
Middle fibers: Scapula (acromion, lateral margin, and superior surface)
Posterior fibers: Scapula (spine on lower lip of posterior border)

Insertion:

Humerus (deltoid tuberosity on lateral midshaft)

Description:

This large multipennate, triangular muscle covers the shoulder anteriorly, posteriorly, and laterally. From a wide origin on the scapula and clavicle, all fibers converge on the humeral insertion, where it gives off an expansion to the deep fascia of the arm. The anterior fibers descend obliquely backward and laterally; the middle fibers descend vertically; the posterior fibers descend obliquely forward and laterally.

Function:

Abduction of shoulder (glenohumeral joint): primarily the acromial middle fibers. The anterior and posterior fibers in this motion stabilize the limb in its cantilever position.
Flexion and internal rotation of arm
Extension and external rotation: posterior fibers
The Deltoid tends to displace the humeral head upward.

Innervation:

Axillary nerve, C5–C6

134. SUBSCAPULARIS

Origin:

Scapula (subscapular fossa and groove along axillary margin)
Aponeurosis separating this muscle from the Teres major and Triceps brachii (long head)
Tendinous laminae

Insertion:

Humerus (lesser tubercle and anterior capsule of glenohumeral joint)

Description:

This is one of the rotator cuff muscles. It is a large triangular muscle that fills the subscapular fossa of the scapula. The tendon of insertion is separated from the scapular neck by a large bursa, which is really a protrusion of the synovial lining of the joint. Variations are rare.

Function:

Internal rotation of shoulder joint
Stabilization of glenohumeral joint by humeral depression (keeps humeral head in glenoid fossa)

Innervation:

Subscapular nerves (upper and lower), C5–C6

135. SUPRASPINATUS

Origin:

Scapula (supraspinous fossa, medial 2/3)
Supraspinatus fascia

Insertion:

Humerus (greater tubercle, highest facet)
Articular capsule of glenohumeral joint

Description:

This is one of the four rotator cuff muscles. Occupying all of the supraspinous fossa, the muscle fibers converge to form a flat tendon that crosses above the glenohumeral joint on its way to a humeral insertion. This tendon is the most commonly ruptured element of the rotator cuff mechanism around the joint.

Function:

Maintains humeral head in glenoid fossa (with other rotator cuff muscles)
Abduction of shoulder
External rotation of shoulder

Innervation:

Suprascapular nerve, C5–C6

136. INFRASPINATUS

Origin:

Scapula (infraspinous fossa, medial 2/3)
Infraspinous fascia

Insertion:

Humerus (greater tubercle, middle facet)

Description:

Occupies most of the infraspinous fossa. The muscle fibers converge to form the tendon of insertion, which glides over the lateral border of the scapular spine and then passes across the posterior aspect of the articular capsule to insert on the humerus. This is the third of the rotator cuff muscles.

Function:

Stabilizes shoulder joint by depressing humeral head in glenoid fossa
External rotation of shoulder

Innervation:

Suprascapular nerve, C5–C6

137. TERES MINOR

Origin:

Scapula (proximal 2/3 of flat surface on dorsal aspect of axillary margin)
Aponeurotic laminae (two such), one of which separates it from the Teres major, the other from the Infraspinatus

Insertion:

Humerus: upper fibers to greater tubercle, most inferior of the facets; and lower fibers to humeral body below the facet

Description:

A somewhat cylindrical and elongated muscle, the Teres minor ascends laterally and upward from its origin to form a tendon that inserts on the greater tubercle of the humerus. It lies inferior to the Infraspinatus, and its fibers lie in parallel with that muscle. It is one of the rotator cuff muscles.

Function:

Maintains humeral head in glenoid fossa, thus stabilizing the shoulder joint
External rotation of shoulder
Adduction of shoulder (weak)

Innervation:

Axillary nerve, C5–C6

138. TERES MAJOR

Origin:

Scapula (dorsal surface near the inferior scapular angle on its lateral margin)
Fibrous septa between this muscle and the Teres minor and Infraspinatus

Insertion:

Humerus (intertubercular sulcus, medial lip)

Description:

The Teres major is a flattened but thick muscle that ascends laterally and upward to the humerus. Its tendon lies behind that of the Latissimus dorsi, and they generally unite for a short distance.

Function:

Adduction and extension of shoulder
Extension of shoulder from a flexed position
Internal rotation of shoulder

Innervation:

Subscapular nerve (lower), C5–C6

139. CORACOBRACHIALIS

Origin:

Scapula, coracoid process (apex)
Intermuscular septum

Insertion:

Humerus (midway along medial border of shaft)

Description:

The smallest of the muscles of the arm, it lies along the upper medial portion of the arm, appearing as a small rounded ridge. The muscle fibers lie along the axis of the humerus.

Function:

Flexion of arm
Adduction of arm

Innervation:

Musculocutaneous nerve, C6–C7

MUSCLES ACTING ON THE ELBOW

140. Biceps brachii
141. Brachialis
142. Triceps brachii
143. Brachioradialis
144. Anconeus

140. BICEPS BRACHII

Origin:

Short head: Scapula (apex of coracoid process)
Long head: Scapula: Supraglenoid tubercle

Insertion:

Radius (rough surface of radial tuberosity)
Broad bicipital aponeurosis fusing with deep fascia over forearm flexors

Description:

Long muscle on the anterior surface of the arm consisting of two heads. The tendon of origin of the short head is thick and flat; the tendon of the long head is long and narrow, curving up, over, and down the humeral head before giving way to the muscle belly. The muscle fibers of both heads lie fairly parallel to the axis of the humerus. The heads can be readily separated except for the distal portion near the elbow joint, where they join before ending in a flat tendon.
The distal tendon spirals so that its anterior surface becomes lateral at the point of insertion.

Function:

Both heads:
    Flexion of elbow
    Supination of forearm (powerful)
    Flexion of shoulder (weak)

Long head: Stabilizes and depresses humeral head in glenoid fossa during Deltoid activity.

Innervation:

Musculocutaneous nerve, C5–C6

141. BRACHIALIS

Origin:

Humerus (distal 2/3 of anterior shaft)
Intermuscular septum

Insertion:

Ulna (ulnar tuberosity and rough surface of coronoid process)
Anterior ligament of elbow joint

Description:

Positioned over the distal half of the humerus and the anterior aspect of the elbow joint

Function:

Flexion of elbow

Innervation:

Musculocutaneous nerve, C5–C6

142. TRICEPS BRACHII

Origin:

Long head: Scapula (infraglenoid tuberosity)
Lateral head: Humerus (on narrow, linear posterior ridge of shaft); Intermuscular septum (lateral)
Medial head: Humerus (posterior surface of shaft distal to radial groove down almost to trochlea); Intermuscular septum

Insertion:

Ulna (olecranon, proximal posterior surface)
Antebrachial fascia

Description:

Located along the entire dorsal aspect of the arm in the extensor compartment. It is a large muscle arising in three heads—long, lateral, and medial. All heads join in a common tendon of insertion, which begins at the midpoint of the muscle. A fourth head is not uncommon.

Function:

Extension of elbow
Long and lateral heads: Especially active in resisted extension
Long head: Extension and adduction of shoulder (assist)

Innervation:

Radial nerve, C7–C8

143. BRACHIORADIALIS

Origin:

Humerus (anterior aspect of proximal 2/3 of lateral supracondylar ridge
Lateral intermuscular septum

Insertion:

Radius (lateral side of base just proximal to styloid process)

Description:

The most superficial muscle on the radial side of the forearm, it forms the lateral side of the cubital fossa. It has a rather thin belly, which descends to the midforearm, where its long flat tendon begins and continues to the distal radius.

Function:

Flexion of elbow
Note: this muscle evolved with the extensor muscles and is innervated by the radial nerve, but its action is that of a forearm flexor.
Muscle is less active when the forearm is fully supinated because it crosses the joint laterally rather than anteriorly.

Innervation:

Radial nerve, C5–C6 (C7 innervation sometimes cited)

144. ANCONEUS

Origin:

Humerus (lateral epicondyle by a separate tendon)
Capsule of elbow joint (dorsal part)

Insertion:

Ulna (olecranon, lateral aspect, and dorsal surface of upper 1/4 of body)

Description:

A small triangular muscle on the dorsum of the elbow whose fibers descend medially a short distance to their ulnar insertion. Considered a continuation of the Triceps.

Function:

Elbow extension (assist)

Innervation:

Radial nerve, C7–C8

MUSCLES ACTING ON THE FOREARM

145. Supinator
140. Biceps brachii (see under *Muscles Acting on the Elbow*)
146. Pronator teres
147. Pronator quadratus

145. SUPINATOR

Origin:

Humerus, lateral epicondyle
Radial collateral ligament of elbow joint
Annular ligament of radioulnar joint
Ulna (dorsal surface of shaft, supinator crest)

Insertion:

Radius (superficial plane: lateral edge, radial tuberosity; deep plane: medial and back surfaces of the radial tuberosity and dorsal and medial surfaces of upper 1/3 of radial shaft)

Description:

Broad muscle whose fibers form two planes that curve around the upper radius. The two planes arise together from the epicondyle: the superficial plane from the tendon and the deep plane from the muscle fibers. This muscle is subject to considerable variation.

Function:

Supination of forearm

Innervation:

Radial nerve, C5–C6

146. PRONATOR TERES

Origin:

Humeral head (superficial head):
    Distal supracondylar ridge
    Medial epicondyle
    Common tendon of origin of flexor muscles
Ulnar head (deep head):
    Coronoid process of medial ulna

Insertion:

Radius (lateral surface of middle of shaft)

Description:

The humeral head is much larger; the thin ulnar head joins its companion at an acute angle,

and together they pass obliquely across the forearm to end in a flat tendon of insertion near the radius. The lateral border of this muscle is the medial limit of the cubital fossa, which lies just anterior to the elbow joint.

Function:

Pronation of forearm
Elbow flexion (accessory)

Innervation:

Median nerve, C6–C7

147. PRONATOR QUADRATUS

Origin:

Ulna (anterior and medial surfaces of distal 1/4 and oblique ulnar ridge on body)
Aponeurosis over middle 1/3 of the muscle

Insertion:

Radius (anterior surface of distal 1/4 of shaft; deeper fibers to a narrow triangular area above ulnar notch)

Description:

This small, flat quadrilateral muscle passes across the anterior aspect of the distal ulna to the distal radius. Its fibers are quite horizontal.

Function:

Pronation of forearm

Innervation:

Median nerve, C8–T1

MUSCLES ACTING AT THE WRIST

148. Extensor carpi radialis longus
149. Extensor carpi radialis brevis
150. Extensor carpi ulnaris

151. Flexor carpi radialis
152. Palmaris longus
153. Flexor carpi ulnaris

148. EXTENSOR CARPI RADIALIS LONGUS

Origin:

Humerus (distal 1/3 of lateral supracondylar ridge)
Lateral intermuscular septum
Common extensor tendon

Insertion:

2nd metacarpal (dorsal surface of base on radial side). Occasionally slips to 1st and 3rd metacarpals.

Description:

Descends lateral to Brachioradialis. Muscle fibers end at midforearm in a flat tendon.

Function:

Extension and radial deviation of wrist
Synergist for finger flexion by stabilization of wrist
Accessory in elbow flexion

Innervation:

Radial nerve, C6 (C5 and C7 have been reported also)

149. EXTENSOR CARPI RADIALIS BREVIS

Origin:

Humerus (lateral epicondyle via common extensor tendon)
Radial collateral ligament of elbow joint
Aponeurosis sheath and intermuscular septa

Insertion:

3rd metacarpal (dorsal surface of base on radial side on or just distal to styloid process)
Slips often sent to 2nd metacarpal

Description:

This short, thick muscle lies partially under the Extensor carpi radialis longus in the upper forearm. Its muscle fibers end well above the wrist in a flattened tendon, which descends alongside the Extensor carpi radialis longus tendon to the wrist.

Function:

Extension of wrist
Radial deviation of wrist (weak)
Finger flexion synergist (by stabilizing the wrist)

Innervation:

Radial nerve, C6–C7

150. EXTENSOR CARPI ULNARIS

Origin:

Humerus (lateral epicondyle via common extensor tendon)
Ulna (dorsal border with an aponeurosis common to Flexor carpi ulnaris and Flexor digitorum profundus)

Insertion:

5th metacarpal (tubercle on ulnar side of base)

Description:

Muscle fibers descend on the dorsal-ulnar side of the forearm and join a tendon located in the distal 1/3 of the forearm that is the most medial tendon on the dorsum of the hand. This tendon can be palpated lateral to the groove found just over the ulna's posterior border.

Function:

Extension of wrist
Ulnar deviation of hand

Innervation:

Radial nerve (deep branch), C6–C8

151. FLEXOR CARPI RADIALIS

Origin:

Humerus (epicondyle by common flexor tendon)
Intermuscular septa
Antebrachial fascia

Insertion:

2nd and 3rd metacarpals (bases)

Description:

A slender aponeurotic muscle at its origin, it descends in the forearm between the Pronator teres and the Palmaris longus. It increases in size as it descends to end in a tendon about halfway down the forearm.

Function:

Flexion of wrist
Radial deviation of wrist
Extends fingers (tenodesis action)
Flexion of elbow (weak assist)
Pronation of forearm (weak assist)

Innervation:

Median nerve, C6–C7

152. PALMARIS LONGUS

Origin:

Humerus (medial epicondyle via common flexor tendon)
Intermuscular septa

Insertion:

Flexor retinaculum (distal)
Palmar aponeurosis
Slip sent frequently to the short thumb muscles

Description:

A slim fusiform muscle, it ends in a long tendon midway in the forearm. Muscle is quite variable and frequently is absent.

Function:

Tension of palmar fascia
Flexion of wrist (weak)
Flexion of elbow (weak)

Innervation:

Median nerve, C6–C7

153. FLEXOR CARPI ULNARIS

Origin:

Humeral head: Humerus (medial epicondyle via common flexor tendon)
Ulnar head:
    Ulna (olecranon, medial process, and upper 2/3 of body via an aponeurosis)
    Intermuscular septa

Insertion:

Pisiform bone
Hamate bone (hamulus)
5th metacarpal

Description:

This is the most ulnar-lying of the flexors in the forearm. The humeral head is small in contrast to the extensive origin of the ulnar head. The two heads are connected by a tendinous arch under which the ulnar nerve descends. The muscle fibers end in a tendon that forms along the anterolateral border of the muscle's distal half.

Function:

Flexion of wrist
Ulnar deviation of wrist
Flexion of elbow (assist)

Innervation:

Ulnar nerve, C8–T1

MUSCLES ACTING ON THE FINGERS (Figs. 9–5 and 9–6)

154. Extensor digitorum
155. Extensor indicis

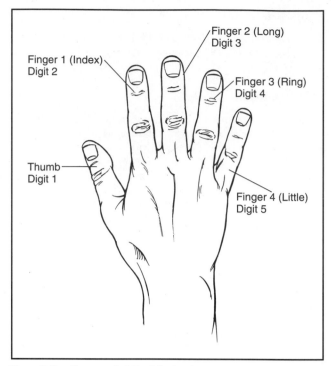

Figure 9–5.   Fingers and digits of the hand.

156. Flexor digitorum superficialis
157. Flexor digitorum profundus

154. EXTENSOR. DIGITORUM

This muscle is the common extensor of the fingers.

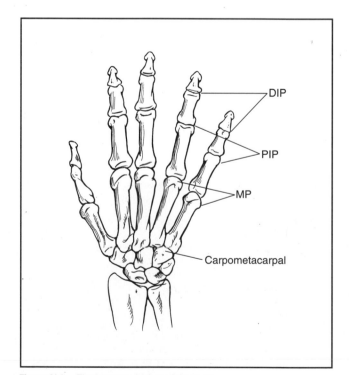

Figure 9–6.   The bones and joints of the hand.

Origin:

Humerus (lateral epicondyle via common extensor tendon)
Intermuscular septa
Antebrachial fascia

Insertion:

Divides distally into four tendons that insert in three variable intertendinous connections on digits 2 to 5:
Intermediate slip: Dorsum of base of each middle phalanx
Lateral slips (two): Dorsum of base of each distal phalanx via dorsal expansion

Description:

The muscle divides above the wrist into four distinct tendons, which traverse (with the Extensor indicis) a tunnel under the extensor retinaculum in a common sheath. Over the dorsum of the hand the four tendons diverge, one to each finger. The tendon to the index finger is accompanied by the Extensor indicis tendon.
The digital attachments are achieved by a fibrous expansion dorsal to the proximal phalanges. All of the digital extensors as well as the lumbricales and interossei are integral to this mechanism.

Function:

Extension of MP and IP joints, digits 2 to 5
    MP joints (proximal phalanges) directly
    IP joints (middle and distal phalanges) indirectly with MP joints flexed
Wrist extension (accessory)
Abduction of ring, index, and little fingers with extension but no such action on the middle finger

Innervation:

Radial nerve, C6–C8

155. EXTENSOR INDICIS

Origin:

Ulna (posterior surface of body below origin of Extensor pollicis longus
Interosseous membrane

Insertion:

2nd digit (extensor hood)

Description:

Arises just below the Extensor pollicis longus and travels adjacent with it down to the level of the wrist. After passing under the extensor retinaculum near the head of the 2nd metacarpal, it joins with the index tendon of the Extensor digitorum on its ulnar side and then inserts into the extensor hood of the 2nd digit.

Function:

Extension of MP joint of index finger
Extension of IP joints (with intrinsics)
Adduction of index finger (accessory)
Wrist extension (accessory)

Innervation:

Radial nerve, C6–C7

156. FLEXOR DIGITORUM SUPERFICIALIS

Origin:

Humeral-ulnar head:
Humerus (medial epicondyle via the common flexor tendon)
Ulna (ulnar collateral ligament of elbow joint)
Intermuscular septa
Radial head:
Radius (oblique line on anterior surface of shaft)

Insertion:

Middle phalanges (digits 2 to 5): via 4 tendons into the sides

Description:

Lies deep to the other forearm flexors but is the largest superficial flexor. The muscle separates into two planes of fibers, superficial and deep. The superficial plane (joined by radial head) divides into two tendons for the middle and ring fingers. The deep plane fibers divide and join the tendons to the index and little fingers.
The four tendons sweep under the flexor retinaculum arranged in pairs (for the middle and ring fingers, and for the index and little fingers). The tendons diverge again in the palm, and at the base of the proximal phalanges each divides into two slips to permit passage of the Flexor digitorum profundus to each finger. The slips reunite and then divide *again* for a final time to insert on both sides of each middle phalanx.
The radial head may be absent.

Function:

Flexion of PIP joints of digits 2 to 5
Flexion of MP joints of digits 2 to 5 (assist)
Flexion of wrist (accessory, especially in forceful grasp)

Innervation:

Median nerve, C7–C8

157. FLEXOR DIGITORUM PROFUNDUS

Origin:

Ulna (upper 3/4 of anterior and medial surfaces of body and also coronoid process)
Interosseous membrane

Insertion:

Splits into four tendons for digits 2 to 5
Distal phalanges and base of digits 2 to 5
Each tendon passes through openings in the continuous tendon of the Flexor digitorum superficialis and inserts on the base of the distal phalanx of each finger.

Description:

Lying deep to the superficial flexors, the profundus is situated on the ulnar side of the forearm. The muscle fibers end in four tendons below the midforearm; the tendons pass into the hand under the transverse carpal ligament. The tendon for the index finger remains distinct, but the tendons for the other fingers are interwoven with areolar tissue and tendinous slips well down into the palm.
After passing through the tendons of the Flexor digitorum superficialis they move to their insertions. The four lumbrical muscles arise with the profundus in the palm.

Function:

Flexion of DIP joints of digits 2 to 5
Flexion of MP and PIP joints of digits 2 to 5 (assist)
Flexion of wrist (accessory)

Innervation:

Median nerve, C8–T1, for digits 2 and 3
Ulnar nerve, C8–T1, for digits 4 and 5

MUSCLES ACTING ON THE LITTLE FINGER (AND HYPOTHENAR MUSCLES)

158. Extensor digiti minimi
159. Abductor digiti minimi
160. Flexor digiti minimi brevis
161. Opponens digiti minimi
162. Palmaris brevis

158. EXTENSOR DIGITI MINIMI

Origin:

Humerus (via common extensor tendon)
Intermuscular septa

Insertion:

Proximal phalanx of digit 5 (little finger) on the radial side, proximally divided into two tendons
to join the extensor expansion (hood) and the tendon of the Extensor digitorum.

Description:

A slim extensor muscle that lies medial to the Extensor digitorum. It descends in the forearm (between the Extensor digitorum and the Extensor carpi ulnaris, passes under the extensor retinaculum in its own compartment, and then divides into two tendons. The lateral tendon joins directly with the tendon of the Extensor digitorum; all three join the extensor expansion, and all insert on the proximal phalanx of digit 5.

Function:

Extension of MP and IP joints of digit 5 (little finger)
Wrist extension (accessory)
Abduction of digit 5 (accessory)

Innervation:

Radial nerve (deep branch), C6–C8

159. ABDUCTOR DIGITI MINIMI

Origin:

Pisiform bone
Tendon of Flexor carpi ulnaris

Insertion:

5th digit (base of proximal phalanx on ulnar side)
Into dorsal digital expansion of Extensor digiti minimi

Description:

Located on the ulnar border of the palm

Function:

Abduction of 5th digit away from ring finger
Flexion of proximal phalanx of 5th digit at the MP joint
Opposition of 5th digit (assist)

Innervation:

Ulnar nerve, C8–T1

160. FLEXOR DIGITI MINIMI BREVIS

Origin:

Hamate bone (hamulus)
Flexor retinaculum

Insertion:

5th digit (base of proximal phalanx on ulnar side)

Description:

This short flexor of the little finger lies in the same plane as the Abductor digiti minimi on its radial side.

Function:

Flexion of little finger at the MP joint
Opposition (assist)

Innervation:

Ulnar nerve, C8–T1

161. OPPONENS DIGITI MINIMI

Origin:

Hamate (hamulus)
Flexor retinaculum

Insertion:

5th metacarpal (entire length of ulnar margin)

Description:

A triangular muscle lying deep to the abductor and the flexor. It commonly is blended with its neighbors.

Function:

Opposition of little finger (abduction, flexion and lateral rotation, deepening the palmar hollow) to the thumb

Innervation:

Ulnar nerve, C8–T1

162. PALMARIS BREVIS

Origin:

Flexor retinaculum and palmar aponeurosis

Insertion:

Skin on ulnar border of palm

Description:

A thin superficial muscle whose fibers run directly laterally across the hypothenar eminence

Function:

Draws the skin of the ulnar side of the hand toward the palm. This increases the height of the hypothenar eminence, possibly assisting in grasp.

Innervation:

Ulnar nerve, C8–T1

INTRINSIC MUSCLES OF THE HAND

163. Lumbricales
164. Interossei (dorsal)
165. Interossei (palmar)

163. LUMBRICALES (Fig. 9–7)

Origin:

Flexor digitorum profundus tendons
1st and 2nd Lumbricales: Arise by single heads from the radial sides and palmar surface of the Flexor digitorum profundus tendons to digits 2 and 3.
3rd and 4th Lumbricales: Arise by double heads from adjacent (contiguous) sides of tendons of Flexor digitorum profundus bound for digits 3, 4, and 5.

Insertion:

Extensor digitorum (tendinous expansion). Each muscle extends distally to the radial side of its corresponding digit. Variable.

Description:

These four small muscles arise from the tendon of the Flexor digitorum profundus over the

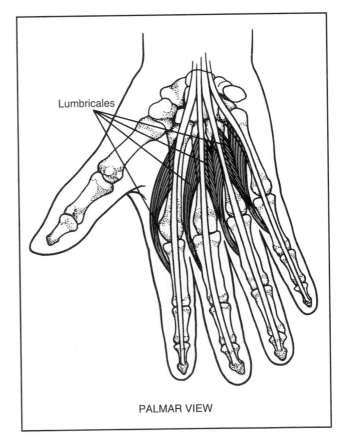

Lumbricales

PALMAR VIEW

Figure 9–7. The lumbricales.

metacarpals. They may be unipennate or bipennate. They extend to the middle phalanges of digits 2 to 5 (fingers 1 to 4), where they join the dorsal extensor expansion on the radial side of each digit (Fig. 9–7). Essentially, they link the flexor to the extension tendon systems in the hand. The exact attachments are quite variable. This gives rise to complexity of movement and differences in description.[19]

Function:

Flexion of MP joints (proximal phalanges) of digits 2 to 5 and simultaneous extension of the IP joints

Innervation:

1st and 2nd Lumbricales: Median nerve, C8–T1
3rd and 4th Lumbricales: Ulnar nerve, C8–T1
Note: The 3rd Lumbrical may receive innervation from both the ulnar and median nerves or all from the median.

164. DORSAL INTEROSSEI (Fig. 9–8)

1st Dorsal Interosseus (also called Abductor indicis):

Origin:

Lateral head: 1st metacarpal (thumb), proximal half of ulnar border
Medial head: 2nd metacarpal (index finger), entire radial border

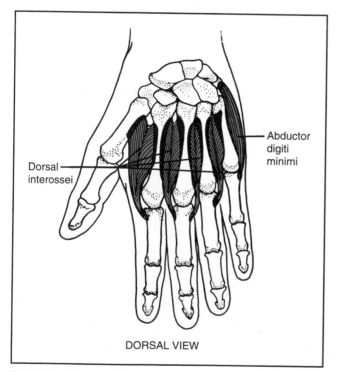

DORSAL VIEW

Figure 9–8. The dorsal interossei.

Insertion:

Index finger (digit 2) (proximal phalanx, base on radial side)
Capsule of 2nd MP joint

2nd Dorsal Interosseus:

Origin:

Adjacent sides of metacarpals of index and long fingers

Insertion:

Middle finger (radial side of proximal phalanx)

3rd Dorsal Interosseus:

Origin:

Adjacent sides of metacarpals of long and ring fingers

Insertion:

Middle finger (ulnar side of proximal phalanx)
Dorsal extensor expansion

4th Dorsal Interosseus:

Origin:

Adjacent sides of metacarpals of ring and little fingers

Insertion:

Ring finger (ulnar side of proximal phalanx)
Dorsal extensor expansion

Description:

This group comprises four bipennate muscles (Fig. 9–8). In general, they originate via two heads from the adjacent metacarpal but more so from the metacarpal of the digit where they will insert distally. They insert into the bases of the proximal phalanges.

Function:

Abduction of fingers away from an axis drawn through the center of the middle finger
Flexion of fingers at MP joints (assist)
Extension of fingers at IP joints (assist)
Thumb adduction (assist)

Innervation:

Ulnar nerve, C8–T1

165. PALMAR (VOLAR) INTEROSSEI (Fig. 9–9)

1st Palmar Interosseus:

Origin:

2nd metacarpal (entire ulnar side)

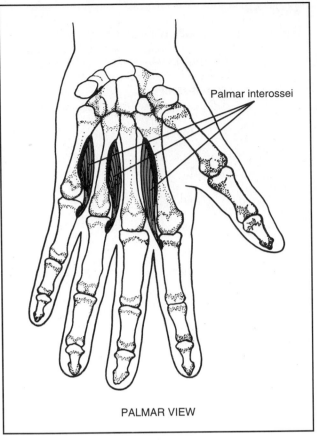

Palmar interossei

PALMAR VIEW

Figure 9–9. Palmar (volar) interossei.

Description:

The palmar Interossei are smaller than their dorsal counterparts. They are found on the palmar surface of the hand at the metacarpal bones. There are three very distinct volar Interossei (Fig. 9–9), and some authors describe a fourth Interosseus, to which they give the number 1 for its attachment on the thumb. When this is found as a discrete muscle, the other palmar Interossei become numbers 2, 3, and 4, respectively. When the thumb Interosseus exists it is on the ulnar side of the metacarpal and proximal phalanx. Some authors (including us) consider the Interosseus of the thumb part of the Adductor pollicis.

The middle finger has no interosseous muscle.

Function:

Adduction of fingers (index, ring, and little) toward an axis drawn through the center of the middle finger

Flexion of MP joints (assist)

Extension of IP joints (assist)

Innervation:

Ulnar nerve, C8–T1

MUSCLES ACTING ON THE THUMB

166. Abductor pollicis longus
167. Extensor pollicis longus
168. Extensor pollicis brevis
169. Flexor pollicis longus
171. Abductor pollicis brevis
172. Opponens pollicis
170. Flexor pollicis brevis
173. Adductor pollicis

166. ABDUCTOR POLLICIS LONGUS

Origin:

Ulna (posterior surface of shaft)
Radius (middle 1/3 of posterior surface of body)

Insertion:

1st metacarpal (radial side of base)
Trapezium

Description:

Lies immediately below the Supinator and sometimes is fused with that muscle. Traverses obliquely down and lateral to end in a tendon at the wrist. The tendon passes through a groove on the lateral side of the distal radius along with the tendon of the Extensor pollicis brevis. Its tendon is commonly split; one slip

Insertion:

Index finger (base of proximal phalanx, ulnar side)
Dorsal expansion of Extensor digitorum

2nd Palmar Interosseus:

Origin:

4th metacarpal (entire radial side)

Insertion:

Ring finger (base of proximal phalanx, radial side)
Dorsal expansion of Extensor digitorum

3rd Palmar Interosseus:

Origin:

5th metacarpal (entire radial side)

Insertion:

Little finger (base of proximal phalanx, radial side)
Dorsal expansion of Extensor digitorum

attaches to the radial side of the 1st metacarpal and the other to the trapezium.

Function:

Abduction and extension of thumb at carpometacarpal joint (CMC)
Radial deviation of wrist
Wrist flexion (weak)

Innervation:

Radial nerve, C6–C7; sometimes C8 (posterior interosseous nerve)

167. EXTENSOR POLLICIS LONGUS

Origin:

Ulna (posterior-lateral surface of middle shaft)
Interosseous membrane

Insertion:

Thumb (base of dorsal distal phalanx)

Description:

The muscle rises distal to the Abductor pollicis longus and courses down and lateral into a tendon over the distal radius, which tendon lies in a narrow oblique groove on the dorsal radius. It descends obliquely over the tendons of the carpal extensors. It separates from the Extensor pollicis brevis and can be seen during thumb extension as the radial margin of a triangular depression called the anatomic snuff box. This is a larger muscle than the Extensor pollicis brevis.

Function:

Extension of the thumb (at IP, MP, and CMC joints)
Radial deviation of wrist (accessory)

Innervation:

Radial nerve, C6–C8 (posterior interosseous nerve)

168. EXTENSOR POLLICIS BREVIS

Origin:

Radius (posterior surface of shaft)
Interosseous membrane

Insertion:

Thumb (proximal phalanx base on dorsal surface)
Attachment to distal phalanx via tendon of Extensor pollicis longus is common.

Description:

Arises distal to and lies medial to the Abductor pollicis longus and descends with it so that the tendons of the two muscles pass through the same groove on the lateral side of the distal radius. The muscle often is actually connected with the abductor. It also may be absent.

Function:

Extension of MP joint of thumb
Extension and abduction of 1st CMC joint of thumb
Radial deviation of wrist (accessory)

Innervation:

Radial nerve, C6–C7, sometimes C8

169. FLEXOR POLLICIS LONGUS

Origin:

Radius (grooved anterior surface of middle 1/2 of body)
Interosseous membrane
Ulna (coronoid process), variable
Humerus (medial epicondyle), variable

Insertion:

Thumb (base of distal phalanx, palmar surface)

Description:

Descends on the radial side of the forearm in the same plane as, but lateral to, the Flexor digitorum profundus

Function:

Flexion of IP joint of thumb
Flexion of the MP and CMC joints of thumb (accessory)
Flexion of wrist (accessory)

Innervation:

Median nerve (anterior interosseous branch), C8–T1

170. FLEXOR POLLICIS BREVIS

Origin:

Superficial head:
    Flexor retinaculum (distal border)
    Trapezium (tubercle)
Deep head:
    Trapezoid
    Capitate

Insertion:

Thumb (base of proximal phalanx on radial side)
Extensor expansion

Description:

The superficial head runs more laterally and accompanies the Flexor pollicis longus. Its tendon of insertion contains the radial sesamoid bone at a point where it unites with the tendon of the deep head. The deep head is sometimes absent.

Function:

Flexion of the MP and CMC joints of the thumb
Opposition of thumb (assist)

Innervation:

Median nerve, C8–T1 (superficial head)
Ulnar nerve, C8–T1 (deep head)

171. ABDUCTOR POLLICIS BREVIS

Origin:

Flexor retinaculum
Scaphoid (tubercle)
Trapezium (tubercle)

Insertion:

Thumb (proximal phalanx, radial side of base)
Extensor expansion

Description:

The most superficial muscle on the radial side of the thenar eminence

Function:

Abduction at CMC and MP joints (in a plane 90° from the palm)
Opposition of thumb (accessory)
Extension of IP joint (accessory)

Innervation:

Median nerve, C8–T1

172. OPPONENS POLLICIS

Origin:

Trapezium (tubercle)
Flexor retinaculum

Insertion:

1st metacarpal (along entire length of radial side of shaft)

Description:

A small triangular muscle lying deep to the abductor

Function:

Flexion of CMC joint medially across the palm
Abduction of CMC joint
Medial rotation of CMC joint

These motions occur simultaneously in the motion called opposition, which brings the thumb into contact with any of the other fingers on their palmar digital aspect (pads)

Innervation:

Median nerve, C8–T1

173. ADDUCTOR POLLICIS

Origin:

Oblique head:
  Capitate
  2nd and 3rd metacarpals (bases)
  Palmar carpal ligaments
  Tendon sheath of Flexor carpi radialis
  Flexor retinaculum (small slip)
Transverse head:
  3rd metacarpal (distal 2/3 of palmar surface)

Insertion (both heads):

Thumb (proximal phalanx, ulnar side of base)

Description:

The fibers of the oblique head pass obliquely downward (across the web of the thumb) and converge into a tendon that contains a sesamoid bone before it unites with the transverse head. The deeper transverse head (deepest of the thenar muscles) is triangular in shape. The sizes of the two heads vary considerably.

Function:

Adduction of CMC joint of thumb (approximates the pollex to the palm)
Adduction and flexion of MP joint (assist)

Innervation:

Ulnar nerve, C8–T1

Muscles of the Lower Extremity (Hip, Knee, Ankle, Toes, Hallux)

MUSCLES OF THE HIP

174. Psoas major
175. Psoas minor

176. Iliacus
177. Pectineus
178. Gracilis
179. Adductor longus
180. Adductor brevis
181. Adductor magnus

182. Gluteus maximus
183. Gluteus medius
184. Gluteus minimus
185. Tensor fasciae latae
186. Piriformis

187. Obturator internus
188. Obturator externus
189. Gemellus superior
190. Gemellus inferior
191. Quadratus femoris

192. Biceps femoris (long head)
193. Semitendinosus
194. Semimembranosus

195. Sartorius

174. PSOAS MAJOR

Origin:

L1–L5 vertebrae (transverse processes, inferior border)
T12–L5 vertebral bodies and intervertebral discs between them
Tendinous arches across the lumbar vertebral bodies

Insertion:

Femur (lesser trochanter)

Description:

A long muscle lying next to the lumbar spine, its fibers descend downward and laterally. It decreases in size as it descends along the pelvic brim. It passes anterior to the hip joint and joins in a tendon with the Iliacus to insert on the lesser trochanter.
This muscle often is referred to as the Iliopsoas because of its blending with the Iliacus.

Function:

Hip flexion with origin fixed[21]
Trunk flexion (sit-up) with insertion fixed
    (These two functions occur in conjunction with the Iliacus)
Hip external (lateral) rotation
Flexion of lumbar spine (both)
Lateral bending of lumbar spine to same side (muscle on one side)

Innervation:

Lumbar plexus with fibers from L2–L4 (ventral rami)

175. PSOAS MINOR

Origin:

T12–L1 vertebral bodies (sides) and the intervertebral discs between them

Insertion:

Ilium (iliopectineal eminence and linea terminalis [pectineal line] of inner surface of the pelvis)
Iliac fascia

Description:

Lying anterior to the Psoas major, this is a long slender muscle whose belly lies entirely within the abdomen along its posterior wall, but its long flat tendon descends to the ilium. The muscle frequently is absent.

Function:

Flexion of trunk and lumbar spine (both; weak)

Innervation:

L1 spinal nerve

176. ILIACUS

Origin:

Ilium (superior 2/3 of iliac fossa)
Iliac crest (inner lip)
Anterior sacroiliac and iliolumbar ligaments

Insertion:

Femur (lesser trochanter via insertion on tendon of the Psoas major and shaft below lesser trochanter)

Description:

A broad flat muscle, it fills the iliac fossa and descends along the fossa, converging laterally with the tendon of the Psoas major.

Function:

Hip flexion
Trunk flexion
    (These functions occur in conjunction with Psoas major)
Hip external rotation

Innervation:

Femoral nerve, L2–L3

177. PECTINEUS

Origin:

Pubis (pecten, between iliopectineal eminence and tubercle)
Anterior fascia

Insertion:

Femur (pectineal line on posterior surface)

Description:

A flat muscle forming part of the wall of the femoral triangle in the upper medial aspect of the thigh. It descends posteriorly and laterally on the medial thigh.

Function:

Hip adduction
Hip flexion (accessory)

Innervation:

Femoral nerve, L2–L4
Accessory obturator nerve when present

178. GRACILIS

Origin:

Pubis (inferior ramus near symphysis)

Insertion:

Tibia (medial surface of shaft below tibial condyle)

Description:

Lies most superficially on the medial thigh as a thin and broad muscle that tapers and narrows distally. The fibers are directed vertically and join a tendon that curves around the medial condyle of the femur and then around the medial condyle of the tibia. Its tendon is one of three (along with those of the Sartorius and Semitendinosus) that unite to form the *pes anserinus.*

Function:

Hip adduction
Knee flexion
Internal (medial) rotation of knee (accessory)

Innervation:

Obturator nerve (anterior division), L2–L3

179. ADDUCTOR LONGUS

Origin:

Pubis (anterior at the angle where the crest meets the symphysis)

Insertion:

Femur (by an aponeurosis on the middle 1/3 of the linea aspera on its medial lip)

Description:

The most anterior of the adductor muscles arises in a narrow tendon and widens into a broad muscle belly as it descends backward and laterally to insert on the femur.

Function:

Hip adduction
Hip flexion (accessory)
Hip rotation (depends on position of thigh)
Hip external (lateral) rotation (when hip is in extension; accessory)

Innervation:

Obturator (anterior division), L2–L4

180. ADDUCTOR BREVIS

Origin:

Pubis (inferior ramus)

Insertion:

Femur (proximal 1/3 of medial lip of linea aspera and distal pectineal line)

Description:

The muscle lies under the Pectineus and Adductor longus with its fibers coursing laterally and posteriorly as it broadens and descends

Function:

Hip adduction
Hip flexion

Innervation:

Obturator nerve, L2–L4

181. ADDUCTOR MAGNUS

Origin:

Pubis (inferior ramus)
Ischium (inferior ramus and inferior and lateral aspect of ischial tuberosity)

Insertion:

Femur (whole length of linea aspera by an aponeurosis; medial supracondylar line; adductor tubercle on medial condyle)
Occasionally the fibers that arise from the ramus of the pubis are inserted into a line from the greater trochanter to the linea aspera and seem to form a distinct separate muscle. When this occurs the muscle is called the *Adductor minimis.*

Description:

The largest of the adductor group, this muscle is located on the medial thigh and appears as three distinct bundles. The superior fibers from the pubic ramus are short and horizontal. The medial fibers move down and lateral. The most distal bundle descends almost vertically to a tendon on the distal 1/3 of the thigh.

Function:

Hip adduction
Hip extension (inferior fibers)
Hip flexion (superior fibers; weak)
The role of the Adductor magnus in rotation of the hip is dependent on the position of the thigh.

Innervation:

Superior and middle fibers: Obturator nerve (posterior division), L2–L4
Inferior fibers: Sciatic nerve, L4–S1

182. GLUTEUS MAXIMUS

Origin:

Ilium (posterior gluteal line and crest)
Sacrum (posterior surface)
Coccyx (posterior)
Erector spinae aponeurosis
Sacrotuberous ligament

Insertion:

Iliotibial band of fascia lata
Femur (gluteal tuberosity)

Description:

The maximus is the largest and most superficial of the gluteal muscles forming the prominence of the buttocks. The fibers descend laterally, inserting widely on the thick tendinous iliotibial tract.

Function:

Hip extension (powerful)
Hip external (lateral) rotation
Trunk extension (when insertion is fixed)
Hip abduction (upper fibers)
Hip adduction (lower fibers)
Through its insertion into the iliotibial band it stabilizes the knee.

Innervation:

Inferior gluteal nerve, L5–S2

183. GLUTEUS MEDIUS

Origin:

Ilium (outer surface between the anterior and posterior gluteal lines)
Gluteal aponeurosis

Insertion:

Femur (greater trochanter, oblique ridge on lateral surface)

Description:

The posterior fibers of the medius lie deep to the maximus; its anterior 2/3 are covered by fascia (gluteal aponeurosis). It lies on the outer surface of the pelvis.

Function:

Hip abduction (in all positions)
Hip internal rotation (anterior fibers)
Hip external (lateral) rotation (posterior fibers)
Hip flexion (anterior fibers) and hip extension (posterior fibers) as accessory functions
The Gluteus medius helps to maintain erect posture in walking. During single limb stance when the swing limb is raised from the ground, all body weight is placed on the opposite (stance) limb, which should result in a noted sagging of the pelvis of the swing limb. The action of the Gluteus medius on the stance limb prevents such a tilt or sag. When the Gluteus medius is weak, the trunk tilts (lateral lean) to the weak side with each step in an attempt to maintain balance (this is the deliberate compensation for the positive Trendelenburg sign). It is called a Gluteus medius sign or gait.
The uncompensated positive Trendelenburg results in a pelvic drop of the contralateral side. This is the so-called Trendelenburg gait.

Innervation:

Superior gluteal nerve, L4–S1

184. GLUTEUS MINIMUS

Origin:

Ilium (outer surface between the anterior and inferior gluteal lines; greater sciatic notch)

Insertion:

Femur (greater trochanter, anterior border)
Expansion to capsule of hip joint

Description:

The minimus is the smallest of the gluteal muscles and lies immediately under the medius. Its fibers pass obliquely lateral and down, forming a fan-shaped muscle that converges on the great femoral trochanter.

Function:

Hip abduction
Hip internal (medial) rotation

Innervation:

Superior gluteal nerve, L4–S1

185. TENSOR FASCIAE LATAE

Origin:

Ilium (Iliac crest; anterior part of outer lip; anterior superior iliac spine)
Fascia lata (deep surface)

Insertion:

Iliotibial band

Description:

The Tensor descends between and is attached to the deep and superficial layers of the iliotibial band. The muscle belly is highly variable in length. The muscle lies superficially on the border between the anterior and lateral thigh.

Function:

Hip flexion
Hip abduction
Hip internal (medial) rotation
Knee flexion (accessory via iliotibial band) once the knee is flexed beyond 30°
Knee external (lateral) rotation (assist)

Innervation:

Superior gluteal nerve, L4–S1

186. PIRIFORMIS

Origin:

Sacrum (anterior digitations attached to the portions of bone between the 1st, 2nd, 3rd, and 4th anterior sacral foramina)
Ilium (upper margin of great sciatic notch)
Sacrotuberous ligament (pelvic surface)

Insertion:

Femur (greater trochanter of femur; superior border of medial aspect)

Description:

Runs parallel to the posterior margin of the Gluteus medius posterior to the hip joint. It lies against the posterior wall on the interior of the pelvis. The broad muscle belly narrows to exit through the great sciatic foramen and converge on the greater trochanter. The insertion tendon often is partly blended with the common tendon of the Obturator internus and Gemelli.

Function:

Hip external (lateral) rotation

Innervation:

S1–S2 spinal nerves

187. OBTURATOR INTERNUS

Origin:

Pelvis (obturator foramen, around most of its margin; from pelvic brim to greater sciatic foramen above and obturator foramen below)
Ischium (ramus)
Pubis (inferior ramus)
Obturator membrane (pelvic surface)
Obturator fascia

Insertion:

Femur (greater trochanter, medial surface proximal to the trochanteric fossa)

Description:

Muscle lies internal in the osteoligamentous pelvis and also external behind the hip joint. The fibers converge toward the lesser sciatic foramen and hook around the body of the ischium, which acts as a pulley; it exits the pelvis via the lesser sciatic foramen, crosses the capsule of the hip joint, and proceeds to the greater trochanter.

Function:

Hip external (lateral) rotation
Abduction of flexed hip (assist)

Innervation:

Nerve to Obturator internus, L5–S2

188. OBTURATOR EXTERNUS

Origin:

Pubis and ischium (medial margin of obturator foramen, formed by rami of pubis and ischium)
Obturator membrane (outer surface)

Insertion:

Femur (trochanteric fossa)

Description:

This flat, triangular muscle covers the external aspect of the anterior pelvic wall from a very broad origin on the medial margin of the obturator foramen. Its fibers pass posteriorly and laterally in a spiral to a tendon that passes behind the femoral neck to insert in the trochanteric fossa.

This muscle, along with the other small lateral rotators, may serve more postural functions than prime movement.

Function:

Hip external (lateral) rotation
Hip adduction (assist)

Innervation:

Obturator nerve, L3–L4

189. GEMELLUS SUPERIOR

Origin:

Ischial spine (gluteal surface)

Insertion:

Femur (greater trochanter)

Description:

Muscle lies in parallel with and superior to the tendon of the Obturator internus, which it joins. This muscle may be absent.

Function:

Hip external (lateral) rotation
Hip abduction (accessory)

Innervation:

Nerve to Obturator internus, L5–S2

190. GEMELLUS INFERIOR

Origin:

Ischium (tuberosity, superior surface)

Insertion:

Femur (greater trochanter)

Description:

This small muscle parallels and joins the tendon of the Obturator internus on its inferior side.

Function:

Hip external (lateral) rotation

Innervation:

Nerve to Quadratus femoris, L5–S1

191. QUADRATUS FEMORIS

Origin:

Ischium (tuberosity, upper external border)

Insertion:

Femur (quadrate tubercle on posterior aspect)

Description:

This flat quadrilateral muscle lies between the Gemellus inferior and the Adductor longus. Its fibers pass almost horizontally, posterior to the hip joint and femoral neck.

Function:

Hip external (lateral) rotation

Innervation:

Nerve to Quadratus femoris, L5–S1

192. BICEPS FEMORIS

Origin:

Long head:
    Ischium: tuberosity, inferior and medial aspects, in common with tendon of Semitendinosus
    Sacrotuberous ligament
Short head:
    Femur: linea aspera, entire length of lateral lip; lateral supracondylar line (proximal 2/3)
    Lateral intermuscular septum

Insertion:

Fibula (head, lateral aspect)
Tibia (lateral condyle)
Aponeurosis over belly of muscle
Fascia on lateral leg

Description:

This lateral hamstring muscle is a two-head muscle on the posterolateral thigh. Its long head is a two-joint muscle. The muscle fibers of the long head descend laterally, ending in an aponeurosis that covers the posterior surface of the muscle. Fibers from the short head also converge into the same aponeurosis, which narrows into the lateral hamstring tendon of insertion. At the insertion the tendon divides into two slips to embrace the fibular collateral ligament. The short head is sometimes absent.

Function:

Knee flexion (only the short head is a pure knee flexor)
Knee external rotation
Hip extension and external (lateral) rotation (long head)

Innervation:

Long head: Sciatic nerve (tibial portion), L5–S3
Short head: Sciatic nerve (common peroneal portion), L5–S2

193. SEMITENDINOSUS

Origin:

Ischium (tuberosity, inferior medial aspect)
Aponeurosis to share tendon with Biceps femoris (long head)

Insertion:

Tibia (shaft on proximal medial side)
Deep fascia of leg

Description:

A muscle on the posteromedial thigh known for its long, round tendon, which extends from midthigh to the tibia. The Semitendinosus unites with the tendons of the Sartorius and the Gracilis to form a flattened aponeurosis called the *pes anserinus.*

Function:

Knee flexion
Knee internal rotation
Hip extension
Hip internal (medial) rotation (accessory)

Innervation:

Sciatic nerve (tibial portion), L5–S2

194. SEMIMEMBRANOSUS

Origin:

Ischium (tuberosity, superior and lateral facets)
Complex proximal tendon and aponeurosis

Insertion:

Tibia (medial condyle on posterior medial aspect)
Femur (lateral condyle, posterior aspect via fibrous expansion, which forms part of oblique popliteal ligament)
Aponeurosis over distal part of muscle (variable)

Description:

One of two medial hamstrings, the Semimembranosus derives its name from its flat, membranous tendon of origin. Its fibers descend, sometimes intermingling with those of the Semitendinosus and the Biceps femoris. The Semitendinosus overlaps this muscle throughout its extent.

Function:

Knee flexion
Knee internal rotation
Hip extension
Hip internal (medial) rotation (accessory)

Innervation:

Sciatic nerve (tibial portion), L5–S2

195. SARTORIUS

Origin:

Ilium (anterior superior iliac spine [ASIS]; notch below ASIS)

Insertion:

Tibia (medial surface of the shaft distal to the tibial condyle)

Description:

The longest muscle in the body, its parallel fibers form a narrow, thin muscle. It descends obliquely from lateral to medial to just above the knee, where it turns abruptly downward and passes posterior to the medial condyle of the femur. It expands into a broad aponeurosis before inserting on the medial surface of the tibia. The Sartorius is the most superficial of the anterior thigh muscles.

Function:

Hip external rotation, abduction, and flexion
Knee flexion
Knee internal rotation
Assists in "tailor sitting"

Innervation:

Femoral nerve (two branches usually), L2–L3

MUSCLES OF THE KNEE

196–200. Quadriceps femoris
201. Articularis genus
192. Biceps femoris (See *Muscles Acting on the Hip*)
193. Semitendinosus (See *Muscles Acting on the Hip*)
194. Semimembranosus (See *Muscles Acting on the Hip*)
202. Popliteus

196–200. QUADRICEPS FEMORIS

This muscular mass on the anterior thigh has five component muscles (or heads), which together make this the most powerful muscle group in the human body.

196. Rectus femoris
197. Vastus lateralis
198. Vastus intermedius
199. Vastus medialis longus
200. Vastus medialis oblique

These are the great extensors of the knee.

196. RECTUS FEMORIS

Origin:

Arises by two tendons, which conjoin to form an aponeurosis from which the muscle fibers arise:
Ilium (anterior inferior iliac spine)
Acetabulum (groove above posterior rim and superior margin of labrum)

Insertion:

Patella (from an aponeurosis which gradually narrows into a tendon which inserts into the center portion of the quadriceps tendon)

Description:

This most anterior of the Quadriceps lies 6° medial to the axis of the femur. Its superficial fibers are bipennate, but the deep fibers are parallel. It traverses a vertical course along the thigh.

197. VASTUS LATERALIS

Origin:

Arises via a broad aponeurosis from
Femur: linea aspera, lateral lip as far proximally as the greater trochanter; greater trochanter, anterior and inferior borders; proximal intertrochanteric line
Lateral intermuscular septum

Insertion:

Patella, into an underlying aponeurosis over the deep surface of the muscle, which narrows and attaches to the lateral border of the quadriceps tendon; to a lateral expansion, which blends with the capsule of the knee and the iliotibial band

Description:

The lateralis is the largest of the quadriceps group and, as its name suggests, it forms the bulk of the lateral thigh musculature. Its fibers run at an angle of 17° to the axis of the femur. It descends to the thigh under the iliotibial band. It is the muscle of choice for biopsy in the lower extremity.

198. VASTUS INTERMEDIUS

Origin:

Femur (anterior and lateral surfaces of upper 2/3 of shaft)
Lateral intermuscular septum (lower part)

Insertion:

Patella, into an aponeurosis on the anterior surface of the muscle which attaches to the middle part of the quadriceps tendon.

Description:

The deepest of the Quadriceps muscles, this muscle lies under the Rectus femoris, the Vastus medialis, and the Vastus lateralis. It almost completely surrounds the proximal 2/3 of the shaft of the femur. A small muscle occasionally is distinguishable from the intermedius, the Articularis genus, but more commonly it is part of the intermedius.

199. VASTUS MEDIALIS LONGUS[22, 23]

Origin:

Femur (intertrochanteric line, lower half; linea aspera, medial lip, proximal portion)
Tendons of Adductors longus and magnus
Medial intermuscular septum

Insertion:

Patella, via an aponeurosis into the superior medial margin of the quadriceps tendon

Description:

The fibers of this muscle course upward at an angle of 15° to 18° to the longitudinal axis of the femur.

200. VASTUS MEDIALIS OBLIQUE[22,23]

Origin:

Femur (linea aspera, medial lip, distal portion; medial supracondylar line, proximal portion)
Tendon of Adductor magnus
Medial intermuscular septum

Insertion:

Patella
Into the medial Quadriceps tendon and along medial margin of patella
Expansion aponeurosis to capsule of knee joint

Description:

The fibers of this muscle run at an angle of 50° to 55° to the longitudinal axis of the femur. The muscle appears to bulge quickly with training and to atrophy with disuse before the other Quadriceps show changes. This is deceiving because the Medialis oblique has the most sparse and thinnest fascial investment, making changes in it more obvious to observation.

Insertion (all):

The tendons of the five heads unite at the distal thigh to form a common strong tendon (quadriceps tendon) that inserts into the proximal margin of the patella. Fibers continue across the anterior surface to become the patellar tendon, which inserts into the tuberosity of the tibia.

Function (all):

Knee extension (None of the heads function independently.)
Hip flexion (by Rectus femoris, which crosses the hip joint)

Innervation (all):

Femoral nerve, L2–L4

201. ARTICULARIS GENUS

Origin:

Femur (anterior surface of lower shaft)

Insertion:

Knee joint (synovial membrane, upper part)

Description:

A small muscle often inseparable from the Vastus intermedius

Function:

Retracts the synovial membrane during knee extension, purportedly preventing this membrane from being entrapped between the patella and the femur.

Innervation:

Femoral nerve, L2–L4

202. POPLITEUS

Origin:

Femur (lateral condyle, groove on anterior surface)
Arcuate ligament
Capsule of knee joint

Insertion:

Tibia (posterior triangular surface above soleal line)
Tendinous expansion

Description:

Sweeps across the upper leg from lateral to medial just below the knee. Forms the lower floor of the popliteal fossa.

Function:

Knee flexion
Knee internal rotation (proximal attachment fixed)
Hip external (lateral) rotation (distal attachment fixed)

Innervation:

Tibial nerve, L4–S1

MUSCLES OF THE ANKLE

203. Tibialis anterior
204. Tibialis posterior
205. Gastrocnemius
206. Soleus
207. Plantaris
208. Peroneus longus
209. Peroneus brevis
210. Peroneus tertius

203. TIBIALIS ANTERIOR

Origin:

Tibia (lateral condyle and proximal 2/3 of lateral surface)
Interosseous membrane
Deep surface of crural fascia

Insertion:

1st (medial) cuneiform (on medial and plantar surfaces)
1st metatarsal (base)

Description:

Located on the lateral aspect of the tibia, the muscle has a thick belly proximally but is tendinous distally. The fibers pass vertically downward and end in a prominent tendon on the anterior surface of the lower leg. Muscle is contained in the most medial compartments of the extensor retinacula.

Function:

Ankle dorsiflexion (talocrural joint)
Foot inversion and adduction (supination) at subtalar and midtarsal joints

Innervation:

Deep peroneal nerve, L4–S1

204. TIBIALIS POSTERIOR

Origin:

Tibia (proximal 2/3 of posterior shaft as well as distal lateral condyle)
Fibula (proximal 2/3 of shaft and posterior surface of head)
Interosseous membrane (entire posterior surface)

Insertion:

Navicular (tuberosity)
Calcaneus (sustentaculum tali)
All three cuneiform bones
2nd, 3rd, and 4th metatarsals (bases)

Description:

Most deeply placed of the flexor group, high on the posterior leg, this muscle is overlapped by both the Flexor hallucis longus and the Flexor digitorum longus. It rises by two narrow heads and descends centrally on the leg, forming its distal tendon in the distal 1/4. This tendon passes behind the medial malleolus (with the Flexor digitorum longus), enters the foot on the plantar surface (where it contains a sesamoid bone), and then divides to its several insertions.
During weight-bearing the Tibialis posterior assists in arch support and distribution of weight on the foot to maintain balance.

Function:

Foot inversion
Ankle plantar flexion (accessory)

Innervation:

Tibial nerve, L5–S1

205. GASTROCNEMIUS

Origin:

Medial head:
    Femur (medial condyle, depression on upper posterior part; popliteal surface adjacent to medial condyle)
    Capsule of knee joint
Lateral head:
    Femur (lateral condyle and posterior surface of shaft above lateral condyle)
    Capsule of knee joint

Insertion:

Calcaneus (via tendo calcaneus into middle posterior surface)
Tendinous raphe in midline of muscle

Description:

The most superficial of the calf muscles, it gives the characteristic contour to the calf. It is a two-joint muscle with two heads arising from the condyles of the femur and descending to the calcaneus. The medial head is the larger, and its fibers extend further distally before spreading into a tendinous expansion, as does the lateral head. The two heads join as the aponeurosis narrows and form the tendo calcaneus.

Function:

Ankle plantar flexion
Knee flexion (accessory)

Innervation:

Tibial nerve, S1–S2

206. SOLEUS

Origin:

Fibula (head, posterior surface; proximal 1/3 of shaft on posterior surface)
Tibia (popliteal line and middle 1/3 of medial side of shaft)
Arch of transverse intermuscular septum

Insertion:

Calcaneus (posterior surface via tendo calcaneus along with Gastrocnemius)
Aponeurosis over posterior surface of muscle, which, with tendon of Gastrocnemius, thickens to become the tendo calcaneus

Description:

This is a one-joint muscle, the largest of the Triceps surae. It is broad and flat and lies just under the Gastrocnemius. Its anterior attachment is a wide aponeurosis, and most of its fibers course obliquely to the descending tendon on its posterior side.

Function:

Ankle plantar flexion
The Soleus is constantly active in quiet stance. It responds to the forward center of mass to prevent the body from falling forward.

Innervation:

Tibial nerve, L5–S2

207. PLANTARIS

Origin:

Femur (linea aspera, lateral lip)
Oblique (popliteal) ligament of knee joint

Insertion:

Calcaneus (posterior)

Description:

This small fusiform muscle lies between the Gastrocnemius and the Soleus. It is sometimes absent; at other times it is doubled. Its short belly is followed by a long slender tendon of insertion along the medial border of the tendo calcaneus and inserts with it on the posterior calcaneum.

Function:

Ankle plantar flexion
Knee flexion (weak accessory)

Innervation:

Tibial nerve, L4–S1

208. PERONEUS LONGUS

Origin:

Fibula (head and upper 2/3 of lateral shaft)
Tibia (lateral condyle, occasionally)
Deep fascia and intermuscular septa

Insertion:

1st metatarsal (lateral plantar side of base)
1st (medial) cuneiform (lateral plantar aspect)
2nd metatarsal (occasionally by a slip)

Description:

Muscle is found proximally on the fibular side of the leg where it is superficial to the Peroneus brevis. The belly ends in a long tendon that passes behind the lateral malleolus (with the brevis) and then runs obliquely forward lateral to the calcaneus and crosses the plantar aspect of the foot to reach the 1st metatarsal.

Function:

Foot eversion
Ankle plantar flexion (assist)
Depression of 1st metatarsal
Support of transverse arch

Innervation:

Superficial peroneal nerve, L4–S1

209. PERONEUS BREVIS

Origin:

Fibula (shaft, distal 2/3 of lateral surface)
Intermuscular septa

Insertion:

5th metatarsal (tuberosity on lateral surface of base)

Description:

The Peroneus brevis lies deep to the longus and is the shorter and smaller muscle of the two. The belly fibers descend vertically to end in a tendon, which courses (with the longus) behind the lateral malleolus (the pair of muscles shares a synovial sheath). It bends forward on the lateral side of the calcaneus, passing forward to the 5th metatarsal.

Function:

Foot eversion
Ankle plantar flexion (accessory)

Innervation:

Superficial peroneal nerve, L4–S1

210. PERONEUS TERTIUS

Origin:

Fibula (distal 1/3 of medial surface)
Interosseous membrane
Intermuscular septum

Insertion:

5th metatarsal (dorsal surface of base)

Description:

This muscle is considered part of the Extensor digitorum (i.e., the fifth tendon). The muscle descends on the lateral leg, diving under the extensor retinaculum in the same passage as the Extensor digitorum longus, to insert on the 5th metatarsal.

Function:

Ankle dorsiflexion
Foot eversion (accessory)

Innervation:

Deep peroneal nerve, L5–S1

MUSCLES ACTING ON THE TOES

211. Extensor digitorum longus
212. Extensor digitorum brevis
213. Flexor digitorum longus
214. Flexor digitorum brevis
215. Abductor digiti minimi
216. Flexor digiti minimi brevis
217. Quadratus plantae

218. Lumbricales
219. Interossei (dorsal, foot)
220. Interossei (plantar)

211. EXTENSOR DIGITORUM LONGUS

Origin:

Tibia (lateral condyle on lateral side)
Fibula (shaft, upper 3/4 of anterior surface)
Interosseous membrane
Deep fascia and intermuscular septum

Insertion:

Tendon of insertion divides into four tendons to
    dorsum of foot that form an expansion over
    each toe
Toes 2 to 5:
    Proximal phalanges of the four lesser toes
        (intermediate slip to dorsum of base of
        each)
    Distal phalanges (lateral slips [two] to dorsum
        of base of each)

Description:

Muscle lies in the lateral aspect of the anterior
    leg. It descends lateral to the Tibialis anterior,
    and its distal tendon accompanies the tendon
    of the Peroneus tertius before dividing.

Function:

MP extension of four lesser toes
PIP and DIP extension (assist) of four lesser toes
Ankle dorsiflexion (accessory)
Foot eversion (accessory)

Innervation:

Deep peroneal nerve, L4–S1

212. EXTENSOR DIGITORUM BREVIS

Origin:

Calcaneus (superior proximal surface anterolateral
    to the calcaneal sulcus)
Lateral talocalcaneal ligament
Extensor retinaculum

Insertion:

Ends in four tendons:
    (1) Hallux (proximal phalanx). This tendon is
    the largest and most medial. It frequently is de-
    scribed as a separate muscle, the *Extensor hal-
    lucis brevis.*
    (2, 3, 4) Three tendons join the Extensor digito-
    rum longus (lateral surfaces).

Description:

The muscle passes medially and distally across
    the dorsum of the foot to end in four tendons,

one to the hallux and three to toes 2, 3, and 4.
    Varies considerably.

Function:

Great toe MP extension
Toes 2 to 4 MP extension
Toes 2 to 4 IP extension (assist)

Innervation:

Deep peroneal nerve, L5–S1

213. FLEXOR DIGITORUM LONGUS

Origin:

Tibia (posterior surface of middle 2/3 of shaft)
Fascia covering the Tibialis posterior

Insertion:

Toes 2 to 5 (distal phalanges at base)

Description:

Muscle lies deep on the tibial side of the leg and
    increases in size as it descends. The tendon of
    insertion extends almost the entire length of
    the muscle and is joined in the sole of the foot
    by the tendon of the Quadratus plantae. It fi-
    nally divides into four slips, which insert into
    the four lateral toes.

Function:

Toes 2 to 5 MP, PIP, and DIP flexion
Ankle plantar flexion (accessory)
Foot inversion (accessory)

Innervation:

Tibial nerve, L5–S1

214. FLEXOR DIGITORUM BREVIS

Origin:

Calcaneus (tuberosity, medial process)
Plantar aponeurosis (central part)
Intermuscular septa (adjacent)

Insertion:

Toes 2 to 5 (by four tendons to middle pha-
    langes, both sides)

Description:

This muscle is located in the middle of the sole
    of the foot immediately above the plantar
    aponeurosis. It divides into four tendons, one
    for each of the four lesser toes. At the base of
    the proximal phalanx each is divided into two
    slips, which encircle the tendon of the Flexor

digitorum longus. The tendons divide a second time and insert onto both sides of the middle phalanges.

Function:

Toes 2 to 5 MP and PIP flexion (middle phalanges)

Innervation:

Medial plantar nerve, L5–S1

215. ABDUCTOR DIGITI MINIMI (Foot)

Origin:

Calcaneus (tuberosity, medial and lateral processes)
Plantar aponeurosis and intermuscular septum

Insertion:

Toe 5 (proximal phalanx, lateral aspect of base)
The insertion is in common with that of the Flexor digiti minimi brevis

Description:

Lies along the lateral border of the foot

Function:

Toe 5 abduction
Toe 5 PIP flexion (accessory)

Innervation:

Lateral plantar nerve, S2–S3

216. FLEXOR DIGITI MINIMI BREVIS

Origin:

5th metatarsal (base, plantar surface)
Sheath of tendon of Peroneus longus

Insertion:

Toe 5 (proximal phalanx, lateral aspect of base)

Description:

Muscle lies superficial to the 5th metatarsal and looks like an interosseous muscle. Sometimes fibers are inserted into the lateral distal half of the 5th metatarsal, and these have been described as a distinct muscle called the *Opponens digiti minimi.*

Function:

Toe 5 MP flexion

Innervation:

Lateral plantar nerve, S2–S3

217. QUADRATUS PLANTAE (Flexor digitorum accessorius)

Origin:

Lateral head:
    Calcaneus (lateral border distal to lateral process, plantar aspect)
    Long plantar ligament (lateral border)
Medial head:
    Calcaneus (medial concave surface and anterior aspect of medial process)
    Long plantar ligament (medial border)

Insertion:

Tendon of Flexor digitorum longus (lateral margin)

Description:

This muscle is sometimes known as the *Flexor digitorum accessorius,* or just Flexor accessorius.
The medial head is larger and more muscular, whereas the lateral head is more tendinous. They rise from either side of the calcaneus, pass medially, and join in an acute angle at midfoot, to end in the lateral margin of the tendon of the Flexor digitorum longus.

Function:

Toes 2 to 5 DIP flexion (in synergy with the Flexor digitorum longus)
"Corrects" the diagonal vector of the Flexor digitorum longus

Innervation:

Lateral plantar nerve, S2–S3

218. LUMBRICALES (Foot)

These are four small muscles considered accessories to the Flexor digitorum longus.

Origin:

1st: Originates by a single head from the medial side of the tendon of the Flexor digitorum longus bound for toe 2.
2nd, 3rd, and 4th: Originate by double heads from adjacent sides of tendons of the Flexor digitorum longus bound for toes 2, 3, 4, and 5.

Insertion (all):

Toes 2 to 5 (proximal phalanges and dorsal expansions of the tendons of Extensor digitorum longus)

Description:

The Lumbricales are four small muscles intrinsic to the foot. They are numbered from the medial (hallux) side of the foot so that the 1st Lumbrical goes to toe 2 and the 4th Lumbrical goes to toe 5.

Function:

Toes 2 to 5 MP flexion
Toes 2 to 5 PIP and DIP extension (assist)

Innervation:

1st: Medial plantar nerve, L5–S1
2nd, 3rd, and 4th: Lateral plantar nerve, S2–S3

219. THE DORSAL INTEROSSEI (Foot)

Origin:

Metatarsal bones (each head arises from the adjacent sides of the metatarsal bones between which it lies)

Insertion:

1st: (Toe 2 proximal phalanx, medial side of base)
2nd: (Toe 2 proximal phalanx, lateral side of base)
3rd: (Toe 3 proximal phalanx, lateral side of base)
4th: (Toe 4 proximal phalanx, lateral side of base)
All: Tendons of Extensor digitorum longus

Description:

The dorsal Interossei are four bipennate muscles, each arising by two heads. They are similar to the Interossei of the hand except that their action is considered relative to the midline of the 2nd digit (the longitudinal axis of the foot).

Function:

Toes 2 to 4 abduction from longitudinal axis of foot, which lines up through toe 2
Toes 2 to 4 MP flexion (accessory)
Toes 2 to 4 IP extension (possibly)

Innervation:

Lateral plantar nerve, S2–S3

220. PLANTAR INTEROSSEI

Origin:

3rd, 4th, and 5th metatarsal bones (base and medial sides)

Insertion:

Proximal phalanges of same toe (bases and medial sides)
Dorsal aponeuroses of Extensor digitorum longus

Description:

These are three muscles that lie along the plantar surface of the metatarsals rather than between them. Each connects with only one metatarsal.

Function:

Toes 3, 4, and 5 adduction (toward the axis of toe 2)
Toes 3, 4, and 5 MP flexion
Toes 3, 4, and 5 IP extension (assist)

Innervation:

Lateral plantar nerve, S2–S3

MUSCLES ACTING ON THE GREAT TOE

221. Extensor hallucis longus
222. Flexor hallucis longus
223. Flexor hallucis brevis
224. Abductor hallucis
225. Adductor hallucis

221. EXTENSOR HALLUCIS LONGUS

Origin:

Fibula (medial surface along middle 1/2 of body)
Interosseous membrane

Insertion:

Hallux (base of distal phalanx)
Expansion to base of proximal phalanx

Description:

This thin muscle descends in the leg between and largely covered by the Tibialis anterior and the Extensor digitorum longus. Its tendon does not emerge superficially until it reaches the distal 1/3 of the leg.

Function:

Hallux MP and IP extension
Ankle dorsiflexion (accessory)
Foot inversion (accessory)

Innervation:

Deep peroneal nerve, L4–S1

222. FLEXOR HALLUCIS LONGUS

Origin:

Fibula (inferior 2/3 of posterior surface of shaft)
Interosseous membrane and intermuscular septum

Insertion:

Hallux (distal phalanx at base on plantar surface)

Description:

This muscle lies deep in the lateral side of the leg, its fibers passing obliquely down via a tendon that runs along the whole length of its posterior surface and then crosses over the distal end of the tibia, talus, and the inferior surface of the calcaneus to run forward on the sole of the foot to the hallux.

Function:

Hallux IP flexion
Hallux MP flexion (accessory)
Ankle plantar flexion and foot inversion (accessory)

Innervation:

Tibial nerve, L5–S2

223. FLEXOR HALLUCIS BREVIS

Origin:

Cuboid (medial part of plantar surface)
Cuneiform (lateral)
Tendon of Tibialis posterior

Insertion:

Hallux:
The tendon divides distally into medial and lateral parts, which insert into the medial and lateral aspects of the base of the proximal phalanx of the hallux.

Description:

Located adjacent to the plantar surface of the 1st metatarsal. The medial tendon is blended with the Abductor hallucis before its insertion; the lateral tendon inserts near the Adductor hallucis.

Function:

Hallux MP flexion

Innervation:

Medial plantar nerve, L5–S1

224. ABDUCTOR HALLUCIS

Origin:

Calcaneus (tuberosity, medial process)
Flexor retinaculum
Plantar aponeurosis and intermuscular septum

Insertion:

Hallux (base of proximal phalanx, medial side)

Description:

This muscle lies along the medial border of the foot. Its tendon attaches distally to the medial tendon of the Flexor hallucis brevis.

Function:

Hallux abduction (away from toe 2)
Hallux MP flexion (accessory)

Innervation:

Medial plantar nerve, S2–S3

225. ADDUCTOR HALLUCIS

Origin:

Oblique head:
    2nd, 3rd, and 4th metatarsals (bases)
    Sheath of Peroneus longus tendon
Transverse head:
    Toes 3, 4, and 5 plantar metatarsophalangeal ligaments
    Transverse metatarsal ligaments

Insertion:

Hallux (proximal phalanx, lateral aspect of base)

Description:

The two heads are unequal in size, the oblique being the larger and more muscular. The oblique head crosses the foot from center to medial on a long oblique axis; the transverse head courses transversely across the metatarsophalangeal joints.

Function:

Hallux adduction (toward toe 2)
Hallux MP flexion (accessory)
Support of transverse metatarsal arch

Innervation:

Lateral plantar nerve, S2–S3

PART IV. MOTIONS AND THEIR PARTICIPATING MUSCLES *(MOTIONS OF THE NECK, TRUNK, AND LIMBS)*

In this part of the Ready Reference chapter, each motion of the axial skeleton and trunk is listed along with the muscles that participate in that motion regardless of the extent of their contribution.

As with all aspects of human anatomy, widely different opinions about functional anatomy are cited in the literature. The authors have used the American and British (primarily) versions of *Gray's*

Anatomy as the principal references, but occasionally kinesiologic imperatives have caused them to deviate from orthodoxy for some muscles.

Motions of the Cervical Spine and Head

Capital Extension (*All muscles act bilaterally*)

56.	Rectus capitis posterior major	C1, Suboccipital
57.	Rectus capitis posterior minor	C1, Suboccipital
58.	Obliquus capitis superior	C1 (Suboccipital)
59.	Obliquus capitis inferior	C1 (Suboccipital)
60.	Longissimus capitis	C3–C8
61.	Splenius capitis	C3–C6
62.	Semispinalis capitis	C2–T1
63.	Spinalis capitis	C3–T1
83.	Sternocleidomastoid (posterior)	Accessory (XI), C2–C3
124.	Trapezius (upper)	Accessory (XI)

Capital Flexion (*All muscles act bilaterally*)

72.	Rectus capitis anterior	C1–C2
73.	Rectus capitis lateralis	C1–C2
74.	Longus capitis	C1–C3
75.	Mylohyoid	Trigeminal (V)
76.	Stylohyoid	Facial (VII)
77.	Geniohyoid	C1–C2 (with Hypoglossal XII)
78.	Digastricus	
	Anterior belly	Trigeminal (V)
	Posterior belly	Facial (VII)

Cervical Extension (*All muscles act bilaterally*)

64.	Longissimus cervicis	C3–T6
65.	Semispinalis cervicis	C2–T5
66.	Iliocostalis cervicis	C4–T6
67.	Splenius cervicis	C2–C8
69.	Interspinales cervicis	C3–C8
68.	Spinalis cervicis	C3–C8
124.	Trapezius	Accessory (XI), C2–C3
70.	Intertransversarii cervicis	C3–C8
71.	Rotatores cervicis	C3–C8
94.	Multifidus	Segmental spinal nerves

Cervical Flexion (*All muscles act bilaterally*)

79.	Longus colli	C2–C6
80.	Scalenus anterior	C4–C6
81.	Scalenus medius	C3–C8
82.	Scalenus posterior	C6–C8
83.	Sternocleidomastoid	Accessory (XI), C2–C3
84.	Sternothyroid	C1–C3
85.	Thyrohyoid	C1
86.	Sternohyoid	C1–C3
87.	Omahyoid	C1–C3
88.	Platysma	Facial (VII)

Lateral Bending (*Ear to shoulder*)

The muscles used in this movement are the capital extensors and flexors on that side, and the cervical flexors and extensors on that side.

Rotation to Same Side (*Turn face to that side*)

56.	Rectus capitis posterior major	C1 (Suboccipital)
59.	Obliquus capitis inferior	C1 (Suboccipital)
60.	Longissimus capitis	C3–C8
61.	Splenius capitis	C3–C6
67.	Splenius cervicis	C2–C8
74.	Longus capitis	C1–C3
127.	Levator scapulae	C5, Dorsal scapular

Rotation to Opposite Side

124.	Trapezius (upper)	Accessory (XI)
62.	Semispinalis capitis	C2–T1
65.	Semispinalis cervicis	C2–T5
71.	Rotatores cervicis	C3–C8
79.	Longus colli (inferior oblique)	C2–C6
80.	Scalenus anterior	C4–C6
81.	Scalenus medius	C3–C8
82.	Scalenus posterior	C6–C8
83.	Sternocleidomastoid	Accessory (XI) C2–C3
94.	Multifidus	Segmental spinal nerves

Motions of the Thoracic Spine

Thoracic Extension

89.	Iliocostalis thoracis	Thoracic spinal nerves
91.	Longissimus thoracis	Thoracic and lumbar spinal nerves
92.	Spinalis thoracis	Segmental spinal nerves
93.	Semispinalis thoracis	Thoracic spinal nerves
94.	Multifidus	Segmental spinal nerves
95.	Rotatores thoracis	Spinal nerves
97.	Interspinales thoracis	Spinal nerves

99. Intertransversarii thoracis   Spinal nerves

Motions of the Lumbar Spine and Pelvis

Lumbar Forward Flexion

110. Obliquus externus abdominis (both)	T7–T12
111. Obliquus internus abdominis (both)	T8–L1
113. Rectus abdominis (both)	T7–T12
174. Psoas major	L2–L4
175. Psoas minor	L1
176. Iliacus	L2–L3 (Femoral)

Lumbar Extension

90. Iliocostalis lumborum (both)	Segmental spinal nerves
94. Multifidus	Segmental spinal nerves
96. Rotatores lumborum (both)	Segmental spinal nerves
98. Interspinales lumborum	Segmental spinal nerves
100. Quadratus lumborum (both)	T12–L3

Lumbar Lateral Bending

90. Iliocostalis lumborum	Segmental spinal nerves
99. Intertransversarii lumborum	L1–L5
100. Quadratus lumborum	T12–L3
110. Obliquus externus abdominis	T7–T12
111. Obliquus internus abdominis	T8–L1
174. Psoas major	L2–L4

Lumbar Rotation to Same Side

111. Obliquus internus abdominis	T8–L1

Lumbar Rotation to Opposite Side

94. Multifidi	Segmental spinal nerves
96. Rotatores lumborum	Segmental spinal nerves
110. Obliquus externus abdominis	T7–T12

Motions of Respiration

Quiet Inspiration

101. Diaphragm	C4, Phrenic
102. Intercostales externi	T1–T11, Intercostal nerves
103. Intercostales interni	T1–T11, Intercostal nerves
104. Intercostales intimi	T1–T11, Intercostal nerves
107. Levatores costarum	T1–T12, Intercostal nerves
80. Scalenus anterior	C4–C6, Cervical nerves
81. Scalenus medius	C3–C8, Cervical nerves
82. Scalenus posterior	C6–C8, Cervical nerves
108. Serratus posterior superior	T1–C4, Intercostal nerves
Deep back extensors	Segmental spinal nerves

Expiration *(During exertion, coughing, Valsalva, etc.)*

110. Obliquus externus abdominis	T7–T12, Intercostal nerves
111. Obliquus internus abdominis	T8–T12, Intercostal nerves
113. Rectus abdominis	T7–T12, Spinal nerves
112. Transversus abdominis	T7–T12, Intercostal nerves
102. Intercostales externi (supporting data scarce)	T1–T11, Intercostal nerves
106. Transversus thoracis	T1–T11, Intercostal nerves
130. Latissimus dorsi	C6–C8, Thoracodorsal

Forced Inspiration

All muscles of quiet inspiration plus:

83. Sternocleidomastoid	Accessory (XI); C2–C3
88. Platysma	Facial (VII)
131. Pectoralis major	C5–T1, Medial and lateral pectorals
129. Pectoralis minor	C8–T1, Medial pectoral
130. Latissimus dorsi	C6–C8, Thoracodorsal

Upper Extremity Motions

THE SCAPULA

Scapular Elevation (Shrugging)

124. Trapezius (upper)	Accessory (XI); C3–C4
127. Levator scapulae	C5, Dorsal scapular
125. Rhomboid major	C5, Dorsal scapular

Scapular Depression

124. Trapezius (lower)   Accessory (XI)

Scapular Abduction (Protraction)

128. Serratus anterior   C5–C7, Long thoracic
129. Pectoralis minor   C8–T1, Medial pectoral

Scapular Adduction (Retraction)

124. Trapezius (middle and lower)   Accessory (XI); C3–C4
125. Rhomboid major   C5, Dorsal scapular
126. Rhomboid minor   C5, Dorsal scapular

Scapular Upward Rotation (Glenoid fossa up)

124. Trapezius (upper and lower)   Accessory (XI); C3–C4
128. Serratus anterior   C5–C7, Long thoracic

Scapular Downward Rotation (Glenoid fossa down)

125. Rhomboid major   C5, Dorsal scapular
126. Rhomboid minor   C5, Dorsal scapular
127. Levator scapulae   C3–C4, C5, Dorsal scapular
129. Pectoralis minor   C8–T1, Medial pectoral

THE SHOULDER (GLENOHUMERAL MOTIONS)

Shoulder Abduction

133. Deltoid   C5–C6, Axillary
135. Supraspinatus   C5–C6, Suprascapular

Shoulder Adduction

130. Latissimus dorsi   C6–C8, Thoracodorsal
131. Pectoralis major   C5–T1, Medial and lateral pectorals
137. Teres minor   C5–C6, Axillary
138. Teres major   C5–C6, Subscapular (lower)
139. Coracobrachialis   C6–C7, Musculocutaneous

Shoulder Internal Rotation (Medial rotation)

130. Latissimus dorsi   C6–C8, Thoracodorsal
131. Pectoralis major   C5–T1, Medial and lateral pectorals
134. Subscapularis   C5–C6, Subscapular (upper and lower)
138. Teres major   C5–C6, Subscapular (lower)
133. Deltoid (anterior)   C5–C6, Axillary

Shoulder External Rotation (Lateral rotation)

133. Deltoid (posterior)   C5–C6, Axillary
136. Infraspinatus   C5–C6, Suprascapular
137. Teres minor   C5–C6, Axillary

Shoulder Flexion

133. Deltoid (anterior)   C5–C6, Axillary
131. Pectoralis major (clavicular)   C5–C7, Lateral pectoral
139. Coracobrachialis   C6–C7, Musculocutaneous
140. Biceps brachii   C5–C6, Musculocutaneous

Shoulder Extension

130. Latissimus dorsi   C6–C8, Thoracodorsal
133. Deltoid (posterior)   C5–C6, Axillary
138. Teres major   C5–C6, Subscapular, (lower)
142. Triceps brachii (long head)   C7–C8, Radial

ELBOW AND FOREARM MOTIONS

Elbow Flexion

140. Biceps brachii   C5–C6, Musculocutaneous
141. Brachialis   C5–C6, Musculocutaneous
143. Brachioradialis   C5–C6, Radial
146. Pronator teres   C6–C7, Median
148. Extensor carpi radialis longus   C6, Radial
151. Flexor carpi radialis   C6–C7, Median
152. Palmaris longus   C6–C7, Median
153. Flexor carpi ulnaris   C8–T1, Ulnar

Elbow Extension

142.	Triceps brachii	C7–C8, Radial
144.	Anconeus	C7–C8, Radial

Forearm Pronation

146.	Pronator teres	C6–C7, Median
147.	Pronator quadratus	C8–T1, Median

Forearm Supination

145.	Supinator	C5–C6, Radial
140.	Biceps brachii	C5–C6, Musculocutaneous

WRIST AND HAND MOTIONS

Wrist Flexion

151.	Flexor carpi radialis	C6–C7, Median
153.	Flexor carpi ulnaris	C8–T1, Ulnar
152.	Palmaris longus	C6–C7, Median
166.	Abductor pollicis longus	C6–C7, Radial
156.	Flexor digitorum superficialis	C7–C8, Median
169.	Flexor pollicis longus	C8–T1, Median
157.	Flexor digitorum profundus	
	Digits 2–3	C8–T1, Median
	Digits 4–5	C8–T1, Ulnar

Wrist Extension

148.	Extensor carpi radialis longus	C6, Radial
149.	Extensor carpi radialis brevis	C6–C7, Radial
150.	Extensor carpi ulnaris	C6–C8, Radial
154.	Extensor digitorum	C6–C8, Radial
158.	Extensor digiti minimi	C6–C8, Radial
155.	Extensor indicis	C6–C8, Radial

Wrist Radial Deviation (Abduction)

148.	Extensor carpi radialis longus	C6, Radial
149.	Extensor carpi radialis brevis	C6–C7, Radial
151.	Flexor carpi radialis	C6–C7, Median
167.	Extensor pollicis longus	C6–C8, Radial

168.	Extensor pollicis brevis	C6–C7, Radial
166.	Abductor pollicis longus	C6–C7, Radial

Wrist Ulnar Deviation (Adduction)

150.	Extensor carpi ulnaris	C6–C8, Radial
153.	Flexor carpi ulnaris	C8–T1, Ulnar

THUMB MOTIONS

Thumb Flexion

Carpometacarpal (CMC)

169.	Flexor pollicis longus	C8–T1, Median
172.	Opponens pollicis	C8–T1, Median
170.	Flexor pollicis brevis	
	Superficial head	C8–T1, Median
	Deep head	C8–T1, Ulnar

Metacarpophalangeal (MP)

170.	Flexor pollicis brevis	
	Superficial head	C8–T1, Median
	Deep head	C8–T1, Ulnar
169.	Flexor pollicis longus	C8–T1, Median
173.	Adductor pollicis	C8–T1, Ulnar

Interphalangeal (IP)

169.	Flexor pollicis longus	C8–T1, Median

Thumb Extension

Carpometacarpal (CMC)

166.	Abductor pollicis longus	C6–C7, Radial
168.	Extensor pollicis brevis	C6–C7, Radial
167.	Extensor pollicus longus	C6–C8, Radial

Metacarpophalangeal (MP)

168.	Extensor pollicis brevis	C6–C7, Radial
167.	Extensor pollicis longus	C6–C8, Radial

Interphalangeal (IP)

167.	Extensor pollicis longus	C6–C8, Radial
171.	Abductor pollicis brevis	C8–T1, Median

Thumb Abduction *(Away from digit 2 [index finger])*

Carpometacarpal (CMC)

166.	Abductor pollicis longus (in plane of palm)	C6–C7, Radial
168.	Extensor pollicis brevis	C6–C7, Radial
171.	Abductor pollicis brevis (perpendicular to palm)	C8–T1, Median
172.	Opponens pollicis	C8–T1, Median

Metacarpophalangeal (MP)

171.	Abductor pollicis brevis	C8–T1, Median

Thumb Adduction *(Toward digit 2)*

Carpometacarpal (CMC)

173.	Adductor pollicis	C8–T1, Ulnar
164.	1st Dorsal Interosseous	C8–T1, Ulnar

Metacarpophalangeal (MP)

173.	Adductor pollicis	C8–T1, Ulnar

Thumb Opposition *(Combination of Internal rotation, abduction, and flexion)*

172.	Opponens pollicis	C8–T1, Median
171.	Abductor pollicis brevis	C8–T1, Median
170.	Flexor pollicis brevis	
	Superficial head	C8–T1, Median
	Deep head	C8–T1, Ulnar

DIGIT 2, 3, AND 4 MOTIONS (INDEX, LONG, AND RING FINGERS)

Finger Flexion

Metacarpophalangeal (MP)

163.	Lumbricales	
	1st and 2nd, for digits 2 and 3	C8–T1, Median

	3rd and 4th, for digits 4 and 5	C8–T1, Ulnar
165.	Palmar Interossei for digits 2, 4, and 5	C8–T1, Ulnar
164.	Dorsal interossei for digits 2, 3, and 4	C8–T1, Ulnar
156.	Flexor digitorum superficialis for digits 2 to 5	C7–C8, Median
157.	Flexor digitorum profundus	
	For digits 2 and 3	C8–T1, Median
	For digits 4 and 5	C8–T1, Ulnar

Proximal Interphalangeal (PIP)

156.	Flexor digitorum superficialis	C7–C8, Median
157.	Flexor digitorum profundus	
	For digits 2 and 3	C8–T1, Median
	For digits 4 and 5	C8–T1, Ulnar

Distal Interphalangeal (DIP)

157.	Flexor digitorum profundus	
	For digits 2 and 3	C8–T1, Median
	For digits 4 and 5	C8–T1, Ulnar

Finger Extension

Metacarpophalangeal (MP)

154.	Extensor digitorum (digits 2–5)	C6–C8, Radial
155.	Extensor indicis (digit 2)	C6–C7, Radial

Proximal and Distal Interphalangeal (PIP and DIP)

154.	Extensor digitorum (digits 2–5)	C6–C8, Radial
155.	Extensor indicis (digit 2)	C6–C7, Radial
163.	Lumbricales	
	1st and 2nd, for digits 2 and 3	C8–T1, Median
	3rd and 4th for digits 4 and 5	C8–T1, Ulnar
165.	Palmar interossei for digits 2, 4, and 5	C8–T1, Ulnar
164.	Dorsal interossei for digits 2, 3, and 4	C8–T1, Ulnar

Finger Abduction

164.	Dorsal interossei	
	1st and 2nd, for digits 2 and 3	C8–T1, Ulnar

3rd and 4th, for   C8–T1, Ulnar
digits 3 and 4
154. Extensor digitorum   C6–C8, Radial
(digits 2, 4, and 5)

Finger Adduction

165. Palmar interossei
1st, 2nd, and 3rd,   C8–T1, Ulnar
for digits 2, 4, and
5

Little Finger Flexion

Carpometacarpal (CMC)

161. Opponens digiti   C8–T1, Ulnar
minimi

Metacarpophalangeal (MP)

160. Flexor digiti minimi   C8–T1, Ulnar
brevis
159. Abductor digiti min-   C8–T1, Ulnar
imi
163. 4th Lumbrical   C8–T1, Ulnar
165. 3rd palmar   C8–T1, Ulnar
Interosseus
156. Flexor digitorum   C7–C8, Median
superficialis
157. Flexor digitorum   C8–T1, Ulnar
profundus, for
digit 5

Proximal Interphalangeal (PIP)

156. Flexor digitorum   C7–C8, Median
superficialis
157. Flexor digitorum   C8–T1, Ulnar
profundus, for
digit 5

Distal Interphalangeal (DIP)

157. Flexor digitorum   C8–T1, Ulnar
profundus, for
digit 5

Little Finger Extension

Metacarpophalangeal (MP)

154. Extensor digitorum   C6–C8, Radial
158. Extensor digiti min-   C6–C8, Radial
imi

Proximal and Distal Interphalangeal (PIP and DIP)

154. Extensor digitorum   C6–C8, Radial
158. Extensor digiti min-   C6–C8, Radial
imi

163. 4th Lumbrical   C8–T1, Ulnar
165. 3rd palmar Interos-   C8–T1, Ulnar
seus

Little Finger Abduction

158. Extensor digiti min-   C6–C8, Radial
imi
159. Abductor digiti min-   C8–T1, Ulnar
imi

Little Finger Adduction

165. 3rd palmar Inter-   C8–T1, Ulnar
osseus

Little Finger Opposition

159. Abductor digiti min-   C8–T1, Ulnar
imi
161. Opponens digiti   C8–T1, Ulnar
minimi
160. Flexor digiti minimi   C8–T1, Ulnar
brevis
163. 4th Lumbrical   C8–T1, Ulnar
165. 3rd palmar In-   C8–T1, Ulnar
terosseus

Lower Extremity Motions

HIP MOTIONS

Hip Flexion

176. Iliacus   L2–L3, Femoral
174. Psoas major   L2–L4, Spinal nerves
196. Rectus femoris   L2–L4, Femoral
195. Sartorius   L2–L3, Femoral
177. Pectineus   L2–L4, Femoral
179. Adductor longus   L2–L4, Obturator
180. Adductor brevis   L2–L4, Obturator
181. Adductor magnus   L2–L4, Obturator
(superior)
185. Tensor fasciae latae   L4–S1, Superior gluteal
183. Gluteus medius   L4–S1, Superior gluteal
(anterior)

Hip Extension

182. Gluteus maximus   L5–S2, Inferior gluteal
192. Biceps femoris   L5–S3, Sciatic (tibial)
(long head)
193. Semitendinosus   L5–S2, Sciatic (tibial)
194. Semimembranosus   L5–S2, Sciatic (tibial)
181. Adductor magnus   L4–S1, Sciatic
(inferior)

183. Gluteus medius (posterior) — L4–S1, Superior gluteal
186. Piriformis (flexed hip) — S1–S2, Sacral nerves

Hip Abduction

183. Gluteus medius — L4–S1, Superior gluteal
184. Gluteus minimus — L4–S1, Superior gluteal
185. Tensor fasciae latae — L4–S1, Superior gluteal
195. Sartorius — L2–L3, Femoral
182. Gluteus maximus (upper) — L5–S2, Inferior gluteal
186. Piriformis — S1–S2, Sacral nerves
189. Gemellus superior — L5–S2, Nerve to Obturator internus
187. Obturator internus — L5–S2, Nerve to Obturator internus

Hip Adduction

181. Adductor magnus (superior and middle) — L2–L4, Obturator
180. Adductor brevis — L2–L4, Obturator
179. Adductor longus — L2–L4, Obturator
177. Pectineus — L2–L3, Femoral
188. Obturator externus — L3–L4, Obturator
182. Gluteus maximus (lower) — L5–S2, Inferior gluteal
178. Gracilis — L2–L3, Obturator

Hip Internal Rotation (*Medial rotation*)

183. Gluteus medius (anterior) — L4–S1, Superior gluteal
184. Gluteus minimus — L4–S1, Superior gluteal
185. Tensor fasciae latae — L4–S1, Superior gluteal
194. Semimembranosus — L5–S2, Sciatic (tibial)
193. Semitendinosus — L5–S2, Sciatic (tibial)
181. Adductor magnus (position dependent) — L2–L4, Obturator; L4–S1, Sciatic

Hip External Rotation (*Lateral rotation*)

182. Gluteus maximus — L5–S2, Inferior gluteal
188. Obturator externus — L3–L4, Obturator
191. Quadratus femoris — L5–S1, Nerve to Quadratus femoris
189. Gemellus superior — L5–S1, Nerve to Obturator internus
190. Gemellus inferior — L5–S1, Nerve to Quadratus femoris
187. Obturator internus — L5–S2, Nerve to Obturator internus
186. Piriformis — S1–S2, Sacral nerves

195. Sartorius — L2–L3, Femoral
192. Biceps femoris (long head) — L5–S3, Sciatic (tibial)
183. Gluteus medius (posterior) — L4–S1, Superior gluteal
174. Psoas major — L2–L4, Spinal nerves

KNEE MOTIONS

Knee Flexion

194. Semimembranosus — L5–S2, Sciatic (tibial)
193. Semitendinosus — L5–S2, Sciatic (tibial)
192. Biceps femoris
   Long head — L5–S3, Sciatic (tibial)
   Short head — L5–S2, Sciatic (common peroneal)
178. Gracilis — L2–L3, Obturator
195. Sartorius — L2–L3, Femoral
202. Popliteus — L4–S1, Tibial
185. Tensor fasciae latae (via iliotibial band) — L4–S1, Superior gluteal
207. Plantaris — L4–S1, Tibial
205. Gastrocnemius — S1–S2, Tibial

Knee Extension

196–200. Quadriceps (all five heads) — L2–L4, Femoral
   196. Rectus femoris
   197. Vastus lateralis
   198. Vastus intermedius
   199. Vastus medialis longus
   200. Vastus medialis oblique

Knee Internal Rotation (Knee Flexed)

194. Semimembranosus — L5–S2, Sciatic (tibial)
193. Semitendinosus — L5–S2, Sciatic (tibial)
195. Sartorius — L2–L3, Femoral
178. Gracilis — L2–L3, Obturator
202. Popliteus — L4–S1, Tibial

Knee External Rotation (Knee Flexed)

192. Biceps femoris
   Long head — L5–S3, Sciatic (tibial)
   Short head — S1–S3, Sciatic (common peroneal)
185. Tensor fasciae latae — L4–S1, Superior gluteal

ANKLE AND FOOT MOTIONS

Ankle Plantar Flexion

205.	Gastrocnemius	S1–S2, Tibial
206.	Soleus	L5–S2, Tibial
204.	Tibialis posterior	L5–S1, Tibial
208.	Peroneus longus	L4–S1, Superficial peroneal
209.	Peroneus brevis	L4–S1, Superficial peroneal
207.	Plantaris	L4–S1, Tibial
222.	Flexor hallucis longus	L5–S2, Tibial
213.	Flexor digitorum longus	L5–S1, Tibial

Ankle Dorsiflexion

203.	Tibialis anterior	L4–S1, Deep peroneal
210.	Peroneus tertius	L5–S1, Deep peroneal
221.	Extensor hallucis longus	L4–S1, Deep peroneal
211.	Extensor digitorum longus	L4–S1, Deep peroneal

Foot Inversion

204.	Tibialis posterior	L5–S1, Tibial
203.	Tibialis anterior	L4–S1, Deep peroneal
221.	Extensor hallucis longus	L4–S1, Deep peroneal
222.	Flexor hallucis longus	L5–S2, Tibial
213.	Flexor digitorum longus	L5–S1, Tibial

Foot Eversion

208.	Peroneus longus	L4–S1, Superficial peroneal
209.	Peroneus brevis	L4–S1, Superficial peroneal
210.	Peroneus tertius	L5–S1, Deep peroneal
211.	Extensor digitorum longus	L4–S1, Deep peroneal

MOTIONS OF THE HALLUX

Great Toe Flexion

Proximal Joint (MP)

223.	Flexor hallucis brevis	S2–S3, Medial plantar
222.	Flexor hallucis longus	L5–S2, Tibial

224.	Abductor hallucis	S2–S3, Medial plantar
225.	Adductor hallucis	S2–S3, Lateral plantar

Distal Joint (IP)

222.	Flexor hallucis longus	L5–S2, Tibial

Great Toe Extension (Toe 1)

Proximal Joint (MP)

212.	Extensor digitorum brevis	L5–S1, Deep peroneal
221.	Extensor hallucis longus	L4–S1, Deep peroneal

Distal Joint (IP)

221.	Extensor hallucis longus	L4–S1, Deep peroneal

Great Toe Abduction (Away from toe 2)

224.	Abductor hallucis	S2–S3, Medial plantar

Great Toe Adduction (Toward toe 2)

225.	Adductor hallucis	S2–S3, Lateral plantar

MOTIONS OF TOES 2, 3, AND 4

Toe Flexion

MP Joints

218.	Lumbricales 1st (for digit 2)	L5–S1, Medial plantar
	2nd, 3rd, and 4th (for digits 3, 4, and 5)	S2–S3, Lateral plantar
220.	Plantar interossei 1st, 2nd, and 3rd (for digits 3, 4, and 5)	S2–S3, Lateral plantar
219.	Dorsal interossei 1st, 2nd, 3rd, and 4th (for digits 2–5)	S2–S3, Lateral plantar
214.	Flexor digitorum brevis	L5–S1, Medial plantar
213.	Flexor digitorum longus	L5–S1, Tibial

PIP Joints

214.	Flexor digitorum brevis	L4–S1, Medial plantar
213.	Flexor digitorum longus	L5–S1, Tibial

DIP Joints

213.	Flexor digitorum longus	L5–S1, Tibial
217.	Quadratus plantae	S2–S3, Lateral plantar

Toe Extension

Proximal Joints (MP)

211.	Extensor digitorum longus	L4–S1, Deep peroneal
212.	Extensor digitorum brevis	L5–S1, Deep peroneal

Middle and Distal Joints (PIP and DIP)

211.	Extensor digitorum longus	L4–S1, Deep peroneal
212.	Extensor digitorum brevis	L4–S1, Deep peroneal
218.	Lumbricales	
	1st (for digit 2)	L5–S1, Medial plantar
	2nd, 3rd, and 4th (for digits 3, 4, and 5)	S2–S3, Lateral plantar
220.	Plantar interossei 1st, 2nd, and 3rd (for digits 3, 4, and 5)	S2–S3, Lateral plantar
219.	Dorsal interossei 1st, 2nd, 3rd, and 4th (for digits 2–5)	S2–S3, Lateral plantar

Toe Abduction (*Away from axial line through digit 2*)

219.	Dorsal interossei 2nd, 3rd, and 4th (for digits 2, 3, and 4)	S2–S2, Lateral plantar

Toe Adduction (*Toward axial line through 2nd digit*)

220.	Plantar interossei 1st, 2nd, and 3rd (for digits 3, 4, and 5)	S2–S3, Lateral plantar

MOTIONS OF THE LITTLE TOE

Little Toe Flexion

MP Joint

216.	Flexor digiti minimi brevis	S2–S3, Lateral plantar
218.	Lumbrical, 4th	S1–S2, Lateral plantar

220.	Interosseus, 3rd plantar	S2–S3, Lateral plantar
214.	Flexor digitorum brevis	L5–S1, Medial plantar
213.	Flexor digitorum longus	L5–S1, Tibial

PIP Joint

214.	Flexor digitorum brevis	L5–S1, Medial plantar
213.	Flexor digitorum longus	L5–S1, Tibial
215.	Abductor digiti minimi	S2–S3, Lateral plantar

DIP Joint

213.	Flexor digitorum longus	L5–S1, Tibial
217.	Quadratus plantae	S2–S3, Lateral plantar

Little Toe Extension

Proximal Joint (MP)

211.	Extensor digitorum longus	L4–S1, Deep peroneal

Middle and Distal Joints (PIP and DIP)

211.	Extensor digitorum longus	L4–S1, Deep peroneal
218.	Lumbrical, 4th	S2–S3, Lateral plantar
220.	Interosseus, 3rd plantar	S2–S3, Lateral plantar

Little Toe Abduction (*Away from digit 4*)

215.	Abductor digiti minimi	S2–S3, Lateral plantar

Little Toe Adduction (*Toward digit 4*)

220.	Interosseus, 3rd plantar	S2–S3, Lateral plantar

PART V. CRANIAL AND PERIPHERAL NERVES AND THE MUSCLES THEY INNERVATE

Cranial Nerves and the Muscles They Innervate

OCULOMOTOR (III)

3.	Levator palpebrae superioris
6.	Rectus superior

7. Rectus inferior
8. Rectus medialis
11. Obliquus inferior

TROCHLEAR (IV)

10. Obliquus superior

TRIGEMINAL (V) *(largest of the cranial nerves)*

28. Masseter (mandibular division)
29. Temporalis (mandibular division)
30. Lateral pterygoid (mandibular division)
31. Medial pterygoid (mandibular division)
78. Digastricus, anterior belly (mandibular nerve)
75. Mylohyoid (lingual nerve)
46. Tensor veli palatini

ABDUCENS (VI)

9. Rectus lateralis

FACIAL (VII)

1. Occipitofrontalis
   Frontalis (temporal branch)
   Occipitalis (posterior auricular branch)
2. Temporoparietalis (temporal branch)
4. Orbicularis oculi (temporal and zygomatic branches)
5. Corrugator supercilii (temporal branch)
12. Procerus (buccal branch)
13. Nasalis (buccal branch)
14. Depressor septi (buccal branch)
15. Levator labii superioris (buccal branch)
16. Levator labii superioris alaeque nasi (buccal branch)
17. Levator anguli oris (buccal branch)
18. Zygomaticus major (buccal branch)
19. Zygomaticus minor (buccal branch)
20. Risorius (buccal branch)
21. Mentalis (mandibular branch)
22. Transversus menti (mandibular branch)
23. Depressor anguli oris (mandibular branch)
24. Depressor labii inferioris (mandibular branch)
25. Orbicularis oris (buccal branch)
26. Buccinator (buccal branch)
27. Auriculares (posterior auricular branch)
78. Digastricus, posterior belly (posterior auricular nerve)
76. Stylohyoid (posterior auricular nerve)
88. Platysma

GLOSSOPHARYNGEAL (IX)

44. Stylopharyngeus

VAGUS (X)

41. Inferior pharyngeal constrictor (via pharyngeal plexus)
42. Middle pharyngeal constrictor (via pharyngeal plexus)
43. Superior pharyngeal constrictor (via pharyngeal plexus)
45. Salpingopharyngeus (via pharyngeal plexus)
49. Palatopharyngeus (via pharyngeal plexus)
46. Levator veli palatini (via pharyngeal plexus)
48. Musculus uvulae (via pharyngeal plexus)
50. Cricothyroid (superior laryngeal nerve)
51. Posterior cricoarytenoid (recurrent laryngeal nerve)
52. Lateral cricoarytenoid (recurrent laryngeal nerve)
53. Transverse arytenoid (recurrent laryngeal nerve)
54. Oblique arytenoid (recurrent laryngeal nerve)
55. Thyroarytenoid (recurrent laryngeal nerve)

ACCESSORY (XI) *(with the vagus forms the pharyngeal plexus)*

48. Musculus uvulae (via pharyngeal plexus)
46. Levator veli palatini (via pharyngeal plexus)
43. Superior pharyngeal constrictor (via pharyngeal plexus)
42. Middle pharyngeal constrictor (via pharyngeal plexus)
41. Inferior pharyngeal constrictor (via pharyngeal plexus)
45. Salpingopharyngeus (via pharyngeal plexus)
83. Sternocleidomastoid (spinal part and communication with C2–C3)
124. Trapezius
49. Palatopharyngeus (via pharyngeal plexus)

HYPOGLOSSAL (XII) *(motor nerve to the tongue)*

32. Genioglossus
33. Hyoglossus
34. Chondroglossus
35. Styloglossus
36. Palatoglossus
37. Superior longitudinal
38. Inferior longitudinal
39. Transverse lingual
40. Vertical lingual
77. Geniohyoid (with fibers from 1st cervical nerve)
85. Thyrohyoid

Peripheral Nerves

NERVES FROM CERVICAL AND BRACHIAL PLEXUSES (UPPER EXTREMITY MUSCLE INNERVATION)

CERVICAL PLEXUS (FIG. 9–10)

1. Ventral primary divisions of first four cervical nerves (C1–C4)

2. C2, C3, and C4 divide into superior and inferior branches.

3. The cervical plexus communicates with three motor cranial nerves (vagus, hypoglossal, accessory).

4. Special nerves often leave both the cervical and brachial plexuses and supply motor innervation to individual muscles. These special nerves, when named, are usually named for the muscle they supply (e.g., nerve to Rectus capitis anterior). These nerves are listed under the appropriate spinal nerves (myotomes) in Part IV of this Ready Reference section.

BRACHIAL PLEXUS (FIG. 9–11)

1. Comprises the ventral primary divisions of the last four cervical (C5–C8) and the first thoracic (T1) nerves.

2. Supplies the nerves to the upper extremity.

The Cervical Plexus

Suboccipital Nerve (C1)

56. Rectus capitis posterior major
57. Rectus capitis posterior minor
58. Obliquus capitis superior
59. Obliquus capitis inferior

Phrenic Nerve (C4; Contributions from C3 and C5)

101. Diaphragm

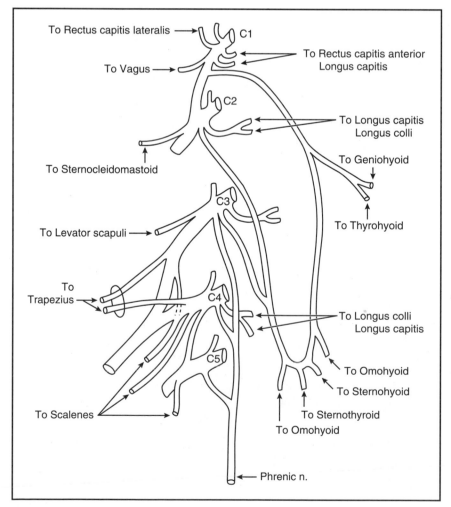

Figure 9–10. Cervical plexus.

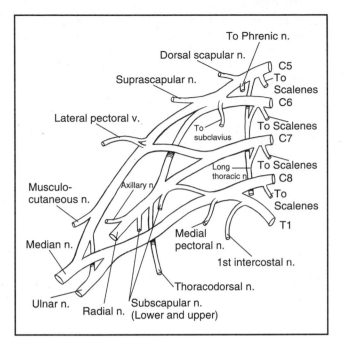

Figure 9–11. Brachial plexus.

The Brachial Plexus

Dorsal Scapular Nerve (C5)

127. Levator scapulae
125. Rhomboid major
126. Rhomboid minor

Long Thoracic Nerve (C5–C7)

128. Serratus anterior

Suprascapular Nerve (C5, C6)

135. Supraspinatus
136. Infraspinatus

Lateral Pectoral Nerve (C5–C7)

131. Pectoralis major
    (clavicular)

Medial Pectoral Nerve (C8–T1)

131. Pectoralis major
    (sternocostal)
129. Pectoralis minor

Subscapular Nerve (Upper and Lower) (C5, C6)

134. Subscapularis   C5–C6 (upper and lower)
138. Teres major   C5–C6 (lower)

Thoracodorsal Nerve (C6–C8)

130. Latissimus dorsi   C6–C8

Musculocutaneous Nerve (C5–C7)

140. Biceps brachii   C5–C6
141. Brachialis   C5–C6
139. Coracobrachialis   C6–C7

Axillary Nerve (C5–C6)

137. Teres minor
133. Deltoid

Median Nerve (C6–T1)

Supplies most of the flexor muscles of the forearm and the thenar muscles of the hand. The nerve has no branches above the elbow except on occasion when the nerve to the Pronator teres arises there.

Muscular Branches in Forearm:

151. Flexor carpi radialis   C6–C7
146. Pronator teres   C6–C7
152. Palmaris longus   C6–C7
156. Flexor digitorum   C7–C8
    superficialis

Anterior Interosseus Nerve:

169. Flexor pollicis   C8–T1
    longus
157. Flexor digitorum   C8–T1
    profundus (digits
    2, 3)
147. Pronator quadratus   C8–T1

Muscular Branch in Hand:

171. Abductor pollicis   C8–T1
    brevis
172. Opponens pollicis   C8–T1
170. Flexor pollicis bre-   C8–T1
    vis (superficial)

1st Common Palmar Digital Nerve:

163. Lumbrical, 1st   C8–T1

2nd Common Palmar Digital Nerve:

163. Lumbrical, 2nd   C8–T1

Radial Nerve (C5–C8)

Supplies the extensor muscles of the arm and forearm.

142. Triceps brachii   C7–C8
144. Anconeus   C7–C8

| 143. Brachoradialis | C5–C6 |
| 148. Extensor carpi radialis longus | C6 (with C5 and C7) |

Deep Branch (of Radial Nerve)

145. Supinator	C5–C6
149. Extensor carpi radialis brevis	C6–C7
158. Extensor digiti minimi	C6–C8
150. Extensor carpi ulnaris	C6–C8
154. Extensor digitorum	C6–C8
155. Extensor indicis	C6–C7
167. Extensor pollicis longus	C6–C8
168. Extensor pollicis brevis	C6–C7
166. Abductor pollicis longus	C6–C7

Ulnar Nerve (C8–T1)

Supplies the muscles on the ulnar side of the forearm and hand (all are C8–T1)

173. Adductor pollicis
159. Abductor digiti minimi
161. Opponens digiti minimi
160. Flexor digiti minimi brevis
157. Flexor digitorum profundus (digits 4, 5)
163. Lumbricales, 3rd and 4th
153. Flexor carpi ulnaris
162. Palmaris brevis
164. Dorsal interossei
165. Palmar interossei
170. Flexor pollicis brevis (deep head)

NERVES OF THE THORACIC REGION

Superior Thoracic Nerves (T1–T6) (Thoracic Intercostal Nerves)

102. Intercostales interni	T1–T11
103. Intercostales externi	T1–T11
104. Intercostales intimi	T1–T11
105. Subcostales	T1–T11
107. Levatores costarum	T1–T12
108. Serratus posterior superior	T1–T4
106. Transversus thoracis	T1–T11

Lower Thoracic Nerves (T7–T12) (Thoracoabdominal Intercostal Nerves)

102. Intercostales interni	T1–T11
103. Intercostales externi	T1–T11
104. Intercostales intimi	T1–T11
110. Obliquus externus abdominis	T7–T12
111. Obliquus internus abdominis	T8–T12
112. Transversus abdominis	T7–T12
113. Rectus abdominis	T7–T12
109. Serratus posterior inferior	T9–T12

Subcostal Nerve (T12)

114. Pyramidalis
112. Transversus abdominis

NERVES FROM LUMBAR AND SACRAL PLEXUSES (LOWER EXTREMITY MUSCLE INNERVATION)

Muscles Innervated Directly off Lumbar Plexus (Fig. 9–12)

100. Quadratus lumborum	T12–L3
174. Psoas major	L2–L4
175. Psoas minor	L1

Iliohypogastric (L1 [T12])

| 112. Transversus abdominis | L1 (and T7–T12) |
| 111. Obliquus internus abdominis | L1 (and T8–T12) |

Ilioinguinal (L1)

| 112. Transversus abdominis | L1 (and T7–T12) |
| 111. Obliquus internus abdominis | L1 (and T8–T12) |

Genitofemoral (L1–L2)

| 117. Cremaster | L1–L2 |

Accessory Obturator (L3–L4) When Present

| 177. Pectineus | L2–L4 (and L2–L4 femoral) |

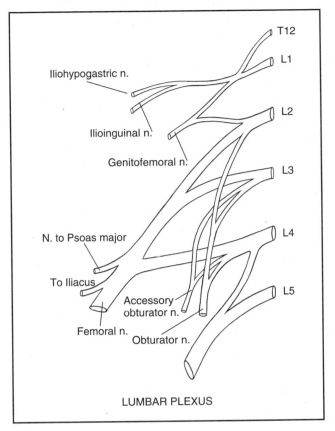

Figure 9–12

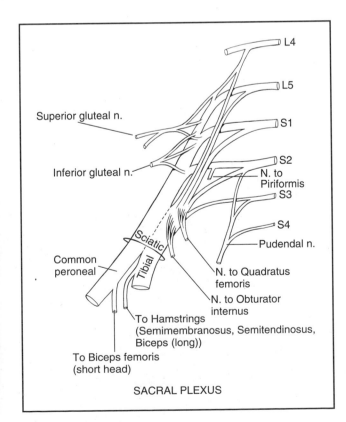

Figure 9–13

Obturator (L2–L4)

Anterior Branch

180.	Adductor brevis	L2–L4
179.	Adductor longus	L2–L4
178.	Gracilis	L2–L3

Posterior Branch

181.	Adductor magnus (superior and middle)	L2–L4
188.	Obturator externus	L3–L4

Femoral (L2–L4)

176.	Iliacus	L2–L3
177.	Pectineus	L2–L4
195.	Sartorius	L2–L3
196.	Rectus femoris	L2–4
198.	Vastus intermedius	L2–4
197.	Vastus lateralis	L2–4
199.	Vastus medialis longus	L2–4
200.	Vastus medialis oblique	L2–4
201.	Articularis genus	L2–4

Muscles Innervated Directly off Sacral Plexus

191.	Quadratus femoris	L5–S1
190.	Gemellus inferior	L5–S1
189.	Gemellus superior	L5–S2
187.	Obturator internus	L5–S2
186.	Piriformis	S1–S2

Superior Gluteal (L4–S1)

Superior Branch

184.	Gluteus minimus	L4–S1

Inferior Branch

183.	Gluteus medius	L4–S1
185.	Tensor fasciae latae	L4–S1

Inferior Gluteal (L5–S2)

182.	Gluteus maximus	L5–S2

Sciatic Nerve (L4–S3)

This is the largest nerve in the body, and it innervates the muscles of the posterior thigh and all of the muscles of the leg and foot. The sciatic trunk has a tibial component and a common peroneal component, which innervate five muscles before dividing to form the tibial and common peroneal nerves.

Common Peroneal Component (Dorsal divisions L4–L5 and S1–S2)

192. Biceps femoris   L5–S2
  (short head)

Tibial Component (Ventral divisions L4–L5 and S1–S3)

181. Adductor magnus   L4–S1
  (inferior)
192. Biceps femoris   L5–S3
  (long head)
194. Semimembranosus   L5–S2
193. Semitendinosus   L5–S2

Tibial Division (Medial Popliteal Nerve) (L4–S3)

This larger of the two main divisions of the sciatic nerve sends branches high in the leg to the posterior leg muscles (Triceps surae) and Popliteus. Lower branches supply motor innervation to the more distal posterior muscles. Its named branches are the lateral and medial plantar nerves.

High Branches

205. Gastrocnemius   S1–S2
  (both heads)
207. Plantaris   L4–S1
206. Soleus   L5–S2
202. Popliteus   L4–S1

Low Branches

206. Soleus   L5–S2
204. Tibialis posterior   L5–S1
213. Flexor digitorum   L5–S1
  longus
222. Flexor hallucis   L5–S2
  longus

Lateral Plantar Nerve (S2–S3)

217. Quadratus plantae   S2–S3
215. Abductor digiti   S2–S3
  minimi

Deep Branch

218. Lumbricales (2nd,   S2–S3
  3rd, 4th)
225. Adductor hallucis   S2–S3
219. Dorsal interossei   S2–S3
  (1st, 2nd, 3rd)
220. Plantar interossei   S2–S3
  (1st, 2nd, 3rd)

Superficial Branch

216. Flexor digiti minimi   S2–S3
  brevis
219. Dorsal interossei   S2–S3
  (4th)

Medial Plantar Nerve (L5–S1)

218. Lumbricales, 1st   L5–S1
  (foot)
224. Abductor hallucis   S2–S3
223. Flexor hallucis bre- S2–S3
  vis
214. Flexor digitorum   L5–S1
  brevis

Common Peroneal Division (L4–S2)

This smaller of the two divisions of the sciatic nerve divides into the deep and superficial peroneal nerves.

Deep Peroneal Nerve

203. Tibialis anterior   L4–S1
221. Extensor hallucis   L4–S1
  longus
211. Extensor digitorum   L4–S1
  longus
212. Extensor digitorum   L5–S1
  brevis
210. Peroneus tertius   L5–S1

Superficial Peroneal Nerve

208. Peroneus longus   L4–S1
209. Peroneus brevis   L4–S1

Pudendal Plexus (Muscular Branches) (S2–S4)

115. Levator ani   S4
116. Coccygeus   S4–S5
123. Sphincter ani exter- S2–S4
  nus

Pudendal (Perineal branch) S2–S4

118. Transversus perinei   S2–S4
  superficialis
119. Transversus perinei   S2–S4
  profundus
120. Bulbocavernosus   S2–S4
121. Ischiocavernosus   S2–S4
122. Sphincter urethrae   S2–S4

PART VI. MYOTOMES: THE MOTOR NERVE ROOTS AND THE MUSCLES THEY INNERVATE

In this portion of the Ready Reference chapter, the spinal roots for the axial and trunk skeletal muscles are outlined along with the muscles innervated by each root. There are many variations of these innervation patterns, but this text presents a consensus of the opinion found in classic anatomy and neurology texts.

The muscles are presented here as originating from the dorsal or ventral primary rami. Each muscle is always preceded by its reference number for cross-reference. Named peripheral nerves for individual muscles are listed parenthetically after the muscle.

The Spinal Nerve Roots and the Muscles They Innervate (Myotomes)

The spinal nerves arise in the spinal cord and exit from it via the intervertebral foramina. There are 31 pairs: cervical (8), thoracic (12), lumbar (5), sacral (5), and coccygeal (1).

Each spinal nerve has two roots that unite to form the nerve: the *ventral root* (motor), which ex- its the cord from the ventral (anterior) horn, and the *dorsal root* (sensory), which enters the cord from the dorsal (posterior) horn. This text will address only the motor (ventral) roots.

Each motor root divides into two parts:

1. *Primary ventral rami* (Fig. 9–14)
   The ventral rami supply the ventral and lateral trunk muscles and all limb muscles. The cervical, lumbar, and sacral ventral rami merge near their origin to form plexuses. The thoracic ventral rami remain individual and are distributed segmentally.
2. *Primary dorsal rami*
   The dorsal rami supply the muscles of the dorsal neck and trunk. The dorsal primary rami do not join any of the plexuses.

The major plexuses formed by the cervical, lumbar, and sacral nerves are as follows:

Cervical plexus (ventral primary rami of C1–C4 and connecting cranial nerves)
Brachial plexus (ventral primary rami of C5–T1 and connections from C4 and T2
Lumbosacral plexus (ventral primary rami of lumbar, sacral, pudendal, and coccygeal nerves)
Lumbar plexus (ventral primary rami of L1–L4 and communication from T12)
Sacral plexus (ventral primary rami of L4–L5 and S1–S3)
Pudendal plexus (ventral primary rami of S2–S4)
Coccygeal plexus (S4–S5)

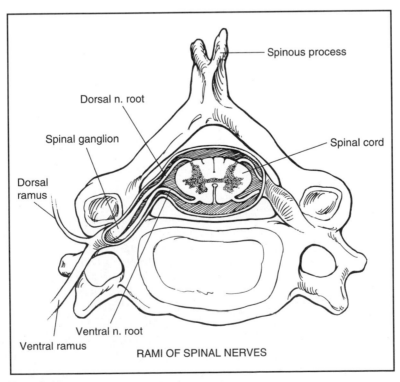

RAMI OF SPINAL NERVES

Spinous process

Dorsal n. root

Spinal ganglion

Spinal cord

Dorsal ramus

Ventral n. root

Ventral ramus

Figure 9–14

THE CERVICAL ROOTS AND NERVES

C1

Ventral Primary Ramus

73. Rectus capitis lateralis
72. Rectus capitis anterior
74. Longus capitis
77. Geniohyoid
84. Sternothyroid
85. Thyrohyoid
86. Sternohyoid
87. Omohyoid

Dorsal Primary Ramus

56. Rectus capitis posterior major
57. Rectus capitis posterior minor
58. Obliquus capitis superior
59. Obliquus capitis inferior

C2

Ventral Primary Ramus

72. Rectus capitis anterior
73. Rectus capitis lateralis
74. Longus capitis
79. Longus colli
70. Intertransversarii cervicis (anterior)
83. Sternocleidomastoid
77. Geniohyoid
84. Sternothyroid
86. Sternohyoid
87. Omohyoid

Dorsal Primary Ramus[2,3]

62. Semispinalis capitis
65. Semispinalis cervicis
67. Splenius cervicis

C3

Ventral Primary Ramus

79. Longus colli
74. Longus capitis
70. Intertransversarii cervicis (anterior)
87. Omohyoid
86. Sternohyoid
84. Sternothyroid
127. Levator scapulae
81. Scaleneus medius
83. Sternocleidomastoid
101. Diaphragm (phrenic)

Dorsal Primary Ramus

60. Longissimus capitis
61. Splenius capitis
62. Semispinalis capitis
63. Spinalis capitis
64. Longissimus cervicis
65. Semispinalis cervicis
67. Splenius cervicis
68. Spinalis cervicis
69. Interspinales cervicis
70. Intertransversarii cervicis (posterior)
71. Rotatores cervicis
94. Multifidi

C4

Ventral Primary Ramus

79. Longus colli
70. Intertransversarii cervicis (anterior)
127. Levator scapulae
80. Scalenus anterior
81. Scalenus medius
101. Diaphragm (phrenic)

Dorsal Primary Ramus

60. Longissimus capitis
61. Splenius capitis
62. Semispinalis capitis
63. Spinalis capitis
64. Longissimus cervicis
65. Semispinalis cervicis
66. Iliocostalis cervicis
67. Splenius cervicis
68. Spinalis cervicis
69. Interspinalis cervicis
70. Intertransversarii cervicis (posterior)
71. Rotatores cervicis
94. Multifidi

C5

Ventral Primary Ramus

79. Longus colli
70. Intertransversarii cervicis (anterior)
80. Scalenus anterior
81. Scalenus medius
132. Subclavius
101. Diaphragm (phrenic)

127. Levator scapulae (dorsal scapular)
125. Rhomboid major (dorsal scapular)
126. Rhomboid minor (dorsal scapular)
128. Serratus anterior (long thoracic)
131. Pectoralis major, clavicular (lateral pectoral)

135. Supraspinatus (suprascapular)
136. Infraspinatus (suprascapular)
134. Subscapularis (subscapular, upper and lower)
138. Teres major (subscapular, lower)

133. Deltoid (axillary)
137. Teres minor (axillary)

140. Biceps brachii (musculocutaneous)
141. Brachialis (musculocutaneous)
143. Brachioradialis (radial)
145. Supinator (radial)

Dorsal Primary Ramus

60. Longissimus capitis
61. Splenius capitis
62. Semispinalis capitis
63. Spinalis capitis
64. Longissimus cervicis
65. Semispinalis cervicis
66. Iliocostalis cervicis
67. Splenius cervicis
68. Spinalis cervicis
69. Interspinalis cervicis
70. Intertransversarii cervicis (posterior)
71. Rotatores cervicis
94. Multifidi

C6

Ventral Primary Ramus

79. Longus colli
70. Intertransversarii cervicis (anterior)
80. Scalenus anterior
81. Scalenus medius
82. Scalenus posterior
132. Subclavius

128. Serratus anterior (long thoracic)
131. Pectoralis major, clavicular (lateral pectoral)
136. Infraspinatus (suprascapular)
135. Supraspinatus (suprascapular)
134. Subscapularis (subscapular, upper and lower)
138. Teres major (subscapular, lower)
133. Deltoid (axillary)
137. Teres minor (axillary)

139. Coracobrachialis (musculocutaneous)
140. Biceps brachii (musculocutaneous)
141. Brachialis (musculocutaneous)
143. Brachioradialis (radial)
145. Supinator (radial)
148. Extensor carpi radialis longus (radial)
149. Extensor carpi radialis brevis (radial)
150. Extensor carpi ulnaris (radial)
154. Extensor digitorum (radial)
155. Extensor indicis (radial)
158. Extensor digiti minimi (radial)
166. Abductor pollicis longus (radial)

167. Extensor pollicis longus (radial)
168. Extensor pollicis brevis (radial)

146. Pronator teres (median)
151. Flexor carpi radialis (median)
152. Palmaris longus (median)

130. Latissimus dorsi (thoracodorsal)

Dorsal Primary Ramus

60. Longissimus capitis
61. Splenius capitis
62. Semispinalis capitis
63. Spinalis capitis
64. Longissimus cervicis
65. Semispinalis cervicis
66. Iliocostalis cervicis
67. Splenius cervicis
68. Spinalis cervicis
69. Interspinales cervicis
70. Intertransversarii cervicis (posterior)
71. Rotatores cervicis
94. Multifidi

C7

Ventral Primary Ramus

70. Intertransversarii cervicis (anterior)
81. Scalenus medius
82. Scalenus posterior
128. Serratus anterior (long thoracic)
130. Latissimus dorsi (thoracodorsal)
131. Pectoralis major, clavicular (lateral pectoral)
139. Coracobrachialis (musculocutaneous)

142. Triceps brachii (radial)
144. Anconeus (radial)
148. Extensor carpi radialis longus (radial)
149. Extensor carpi radialis brevis (radial)
150. Extensor carpi ulnaris (radial)
154. Extensor digitorum (radial)
155. Extensor indicis (radial)
158. Extensor digiti minimi (radial)
166. Abductor pollicis longus (radial)
167. Extensor pollicis longus (radial)
168. Extensor pollicis brevis (radial)

146. Pronator teres (median)
151. Flexor carpi radialis (median)
152. Palmaris longus (median)
156. Flexor digitorum superficialis (median)

Dorsal Primary Ramus

60. Longissimus capitis
62. Semispinalis capitis
63. Spinalis capitis
64. Longissimus cervicis

65. Semispinalis cervicis
66. Iliocostalis cervicis
67. Splenius cervicis
68. Spinalis cervicis
69. Interspinales cervicis
70. Intertransversarii cervicis (posterior)
71. Rotatores cervicis
94. Multifidi

C8

Ventral Primary Ramus

70. Intertransversarii cervicis (anterior)
81. Scalenus medius
82. Scalenus posterior
130. Latissimus dorsi (thoracodorsal)
131. Pectoralis major, sternocostal (medial pectoral)
129. Pectoralis minor (medial pectoral)

142. Triceps brachii (radial)
144. Anconeus (radial)
150. Extensor carpi ulnaris (radial)
154. Extensor digitorum (radial)
158. Extensor digiti minimi (radial)
166. Abductor pollicis longus (radial)
167. Extensor pollicis longus (radial)
168. Extensor pollicis brevis (radial)

147. Pronator quadratus (median)
156. Flexor digitorum superficialis (median)
157. Flexor digitorum profundus, 2nd and 3rd (median)
163. Lumbricales, 1st and 2nd (median)
169. Flexor pollicis longus (median)
170. Flexor pollicis brevis, superficial (median)
171. Abductor pollicis brevis (median)
172. Opponens pollicis (median)

170. Flexor pollicis brevis, deep (ulnar)
173. Adductor pollicis (ulnar)
153. Flexor carpi ulnaris (ulnar)
157. Flexor digitorum profundus, 4th and 5th (ulnar)
163. Lumbricales, 3rd and 4th (ulnar)
164. Interossei, dorsal (ulnar)
165. Interossei, palmar (ulnar)
159. Abductor digiti minimi (ulnar)
161. Opponens digiti minimi (ulnar)
160. Flexor digiti minimi brevis (ulnar)
162. Palmaris brevis (ulnar)

Dorsal Primary Ramus

60. Longissimus capitis
62. Semispinalis capitis
63. Spinalis capitis
64. Longissimus cervicis
65. Semispinalis cervicis
66. Iliocostalis cervicis

67. Splenius cervicis
68. Spinalis cervicis
69. Interspinales cervicis
70. Intertransversarii cervicis (posterior)
71. Rotatores cervicis
94. Multifidi

THE THORACIC ROOTS AND NERVES

There are 12 pairs of thoracic nerves arising from the ventral primary rami: T1 to T11 are called the intercostales and T12 is called the subcostal nerve. These nerves are not part of a plexus. T1 and T2 innervate the upper extremity as well as the thorax; T3–T6 innervate only the thoracic muscles; the lower thoracic nerves innervate the thoracic and abdominal muscles.

T1

Ventral Primary Ramus

107. Levatores costarum
102. Intercostales externi
103. Intercostales interni
104. Intercostales intimi
108. Serratus posterior superior
106. Transversus thoracis

131. Pectoralis major (sternocostal) (medial pectoral)
129. Pectoralis minor (medial pectoral)

147. Pronator quadratus (median)
157. Flexor digitorum profundus
     Digits 2 and 3 (median)
163. Lumbricales, 1st and 2nd (median)
169. Flexor pollicis longus (median)
170. Flexor pollicis brevis, superficial (median)
171. Abductor pollicis brevis (median)
172. Opponens pollicis (median)

153. Flexor carpi ulnaris (ulnar)
157. Flexor digitorum profundus
     Digits 4 and 5 (ulnar)
159. Abductor digiti minimi (ulnar)
160. Flexor digiti minimi brevis (ulnar)
161. Opponens digiti minimi (ulnar)
162. Palmaris brevis (ulnar)

163. Lumbricales, 3rd and 4th (ulnar)
164. Interossei, dorsal (ulnar)
165. Interossei, palmar (ulnar)
170. Flexor pollicis brevis, deep head (ulnar)
173. Adductor pollicis (ulnar)

Dorsal Primary Ramus

62. Semispinalis cervicis
93. Semispinalis thoracis
64. Longissimus cervicis

91. Longissimus thoracis
63. Spinalis capitis
92. Spinalis thoracis
89. Iliocostalis thoracis
66. Iliocostalis cervicis
94. Multifidi
99. Intertransversarii thoracis
95. Rotatores thoracis
97. Interspinales thoracis

T2

Ventral Primary Ramus

107. Levatores costarum
102. Intercostales externi
103. Intercostales interni
104. Intercostales intimi
105. Subcostales
108. Serratus posterior superior
106. Transversus thoracis

Dorsal Primary Ramus

93. Semispinalis thoracis
64. Longissimus cervicis
91. Longissimus thoracis
92. Spinalis thoracis
66. Iliocostalis cervicis
89. Iliocostalis thoracis
94. Multifidi
95. Rotatores thoracis
97. Interspinales thoracis
99. Intertransversarii thoracis

T3

Ventral Primary Ramus

107. Levatores costarum
102. Intercostales externi
103. Intercostales interni
104. Intercostales intimi
105. Subcostales
108. Serratus posterior superior
106. Transversus thoracis

Dorsal Primary Ramus

93. Semispinalis thoracis
64. Longissimus cervicis
66. Iliocostalis cervicis
91. Longissimus thoracis
92. Spinalis thoracis
89. Iliocostalis thoracis
94. Multifidi
95. Rotatores thoracis

97. Interspinales thoracis
99. Intertransversarii thoracis

T4, T5, T6

Ventral Primary Ramus

107. Levatores costarum
102. Intercostales externi
103. Intercostales interni
104. Intercostales intimi
105. Subcostales
106. Transversus thoracis

Dorsal Primary Ramus

93. Semispinalis thoracis
91. Longissimus thoracis
92. Spinalis thoracis
89. Iliocostalis thoracis
94. Multifidi
95. Rotatores thoracis
97. Interspinales thoracis
99. Intertransversarii thoracis

T7

Ventral Primary Ramus

107. Levatores costarum
103. Intercostales interni
102. Intercostales externi
104. Intercostales intimi
106. Transversus thoracis
105. Subcostales
110. Obliquus externus abdominis
112. Transversus abdominis
113. Rectus abdominis

Dorsal Primary Ramus

93. Semispinalis thoracis
91. Longissimus thoracis
92. Spinalis thoracis
89. Iliocostalis thoracis
94. Multifidi
95. Rotatores thoracis
97. Interspinales thoracis
99. Intertransversarii thoracis

T8

Ventral Primary Ramus

107. Levatores costarum
103. Intercostales interni

102. Intercostales externi
104. Intercostales intimi
106. Transversus thoracis
105. Subcostales
110. Obliquus externus abdominis
111. Obliquus internus abdominis
112. Transversus abdominis
113. Rectus abdominis

Dorsal Primary Ramus

93. Semispinalis thoracis
91. Longissimus thoracis
92. Spinalis thoracis
89. Iliocostalis thoracis
94. Multifidi
95. Rotatores thoracis
97. Interspinales thoracis
99. Intertransversarii thoracis

T9, T10, T11

Ventral Primary Ramus

107. Levatores costarum
103. Intercostales interni
102. Intercostales externi
104. Intercostales intimi
106. Transversus thoracis
105. Subcostales
109. Serratus posterior inferior
110. Obliquus externus abdominis
111. Obliquus internus abdominis
112. Transversus abdominis
113. Rectus abdominis

Dorsal Primary Ramus

93. Semispinalis thoracis
91. Longissimus thoracis
92. Spinalis thoracis
89. Iliocostalis thoracis
94. Multifidi
95. Rotatores thoracis
97. Interspinales thoracis
99. Intertransversarii thoracis

T12

Ventral Primary Ramus

100. Quadratus lumbarum
107. Levatores costarum
112. Transversus abdominis
109. Serratus posterior inferior
110. Obliquus externus abdominis
111. Obliquus internus abdominis

113. Rectus abdominis
114. Pyramidalis

Dorsal Primary Ramus

93. Semispinalis thoracis
91. Longissimus thoracis
92. Spinalis thoracis
89. Iliocostalis thoracis
94. Multifidi
95. Rotatores thoracis
97. Interspinales thoracis
99. Intertransversarii thoracis

THE LUMBAR ROOTS AND NERVES

The lumbar plexus is formed by the first four lumbar nerves and a communicating branch from T12. The fourth lumbar nerve gives its largest part to the lumbar plexus and a smaller part to the sacral plexus. The fifth lumbar nerve and the small segment of the fourth lumbar nerve form the lumbosacral trunk, which is part of the sacral plexus.

L1

Ventral Primary Ramus

100. Quadratus lumborum
175. Psoas minor
112. Transversus abdominis
111. Obliquus internus abdominis
117. Cremaster (genitofemoral)

Dorsal Primary Ramus

90. Iliocostalis lumborum
91. Longissimus thoracis
96. Rotatores lumborum
94. Multifidi
98. Interspinales lumborum
99. Intertransversarii lumborum

L2

Ventral Primary Ramus

100. Quadratus lumborum
174. Psoas major
176. Iliacus
117. Cremaster (genitofemoral)
177. Pectineus (femoral)
178. Gracilis (obturator)
179. Adductor longus (obturator)
180. Adductor brevis (obturator)

181. Adductor magnus, superior and middle fibers (obturator)
195. Sartorius (femoral)
196–200. Quadriceps femoris (femoral)
   196. Rectus femoris
   197. Vastus intermedius
   198. Vastus lateralis
   199. Vastus medialis longus
   200. Vastus medialis oblique
201. Articularis genus

Dorsal Primary Ramus

90. Iliocostalis lumborum
96. Rotatores lumborum
94. Multifidi
98. Interspinales lumborum
99. Intertransversarii lumborum

L3

Ventral Primary Ramus

100. Quadratus lumborum
174. Psoas major
176. Iliacus (femoral)
177. Pectineus (femoral)
178. Gracilis (obturator)
179. Adductor longus (obturator)
180. Adductor brevis (obturator)
181. Adductor magnus, superior and medial fibers (obturator)
188. Obturator externus (obturator)
195. Sartorius (femoral)
196–200. Quadriceps femoris (femoral)
   196. Rectus femoris
   197. Vastus intermedius
   198. Vastus lateralis
   199. Vastus medialis longus
   200. Vastus medialis oblique
201. Articularis genus

Dorsal Primary Ramus

90. Iliocostalis lumborum
96. Rotatores lumborum
94. Multifidi
98. Interspinales lumborum
99. Intertransversarii lumborum

L4

Ventral Primary Ramus

175. Psoas major
177. Pectineus (femoral)
179. Adductor longus (obturator)

180. Adductor brevis (obturator)
181. Adductor magnus
   Superior and middle (obturator)
   Lower (sciatic, tibial)
183. Gluteus medius (superior gluteal)
184. Gluteus minimus (superior gluteal)
185. Tensor fasciae latae (superior gluteal)
188. Obturator externus (obturator)

196–200. Quadriceps femoris (femoral)
   196. Rectus femoris
   197. Vastus lateralis
   198. Vastus intermedius
   199. Vastus medialis longus
   200. Vastus medialis oblique
201. Articularis genus (femoral)
207. Plantaris (tibial)
202. Popliteus (tibial)

208. Peroneus longus (superficial peroneal)
209. Peroneus brevis (superficial peroneal)
203. Tibialis anterior (deep peroneal)
211. Extensor digitorum longus (deep peroneal)
221. Extensor hallucis longus (deep peroneal)

Dorsal Primary Ramus

90. Iliocostalis lumborum
96. Rotatores lumborum
94. Multifidi
98. Interspinales lumborum
99. Intertransversarii lumborum

L5

Ventral Primary Ramus

181. Adductor magnus, inferior fibers (sciatic, tibial division)
182. Gluteus maximus (inferior gluteal)
183. Gluteus medius (superior gluteal)
184. Gluteus minimus (superior gluteal)
185. Tensor fasciae latae (superior gluteal)

187. Obturator internus (nerve to obturator internus)
189. Gemellus superior (nerve to obturator internus)
190. Gemellus inferior (nerve to obturator internus)
191. Quadratus femoris (nerve to quadratus femoris)
192. Biceps femoris (short head) (sciatic, common peroneal division)
194. Semimembranosus (sciatic, tibial division)
193. Semitendinosus (sciatic, tibial division)
207. Plantaris (tibial)
202. Popliteus (tibial)
206. Soleus (tibial)
203. Tibialis anterior (deep peroneal)
204. Tibialis posterior (tibial)
208. Peroneus longus (superficial peroneal)
209. Peroneus brevis (superficial peroneal)
210. Peroneus tertius (deep peroneal)

211. Extensor digitorum longus (deep peroneal)
212. Extensor digitorum brevis (deep peroneal)
213. Flexor digitorum longus (tibial)
214. Flexor digitorum brevis (medial plantar)
221. Extensor hallucis longus (deep peroneal)
222. Flexor hallucis longus (tibial)
218. Lumbricales, 1st (foot) (medial plantar)

Dorsal Primary Ramus

90. Iliocostalis lumborum
96. Rotatores lumborum
94. Multifidi
98. Interspinales lumborum
99. Intertransversarii lumborum

THE LUMBOSACRAL ROOTS AND NERVES

The intermingling of the ventral primary rami of the lumbar, sacral, and coccygeal nerves is known as the lumbosacral plexus. There is uncertainty about any motor innervation from the dorsal primary rami below S3. The nerves branching off this plexus supply the lower extremity in part and also the perineum and coccygeal areas via the pudendal and coccygeal plexuses.

S1

Ventral Primary Ramus

181. Adductor magnus, inferior fibers (sciatic, tibial division)
182. Gluteus maximus (inferior gluteal)
183. Gluteus medius (superior gluteal)
184. Gluteus minimus (superior gluteal)
185. Tensor fasciae latae (superior gluteal)

186. Piriformis
187. Obturator internus (nerve to obturator internus)
189. Gemellus superior (nerve to obturator internus)
190. Gemellus inferior (nerve to obturator internus)
191. Quadratus femoris (nerve to quadratus femoris)
192. Biceps femoris
    Short head (sciatic, common peroneal division)
    Long head (sciatic, tibial division)
194. Semimembranosus (sciatic, tibial division)
193. Semitendinosus (sciatic, tibial division)
205. Gastrocnemius (tibial)
207. Plantaris (tibial)
202. Popliteus (tibial)
206. Soleus (tibial)

203. Tibialis anterior (deep peroneal)
204. Tibialis posterior (tibial)
208. Peroneus longus (superficial peroneal)
209. Peroneus brevis (superficial peroneal)
210. Peroneus tertius (deep peroneal)

211. Extensor digitorum longus (deep peroneal)
212. Extensor digitorum brevis (deep peroneal)
213. Flexor digitorum longus (tibial)
214. Flexor digitorum brevis (medial plantar)
221. Extensor hallucis longus (deep peroneal)
222. Flexor hallucis longus (tibial)
218. Lumbricales, 1st (foot) (medial plantar)

Dorsal Primary Rami

94. Multifidi

S2

Ventral Primary Ramus

182. Gluteus maximus (inferior gluteal)
186. Piriformis
187. Obturator internus (nerve to obturator internus)
189. Gemellus superior (nerve to obturator internus)
192. Biceps femoris
    Short head (sciatic, common peroneal division)
    Long head (sciatic, tibial division)
194. Semimembranosus (sciatic, tibial division)
193. Semitendinosus (sciatic, tibial division)
205. Gastrocnemius (tibial)
206. Soleus (tibial)

217. Quadratus plantae (lateral plantar)
215. Abductor digiti minimi (lateral plantar)
216. Flexor digiti minimi brevis (foot) (lateral plantar)
222. Flexor hallucis longus (tibial)
225. Adductor hallucis (lateral plantar)
218. Lumbricales 2nd, 3rd, and 4th (foot) (lateral plantar)
219. Interossei, dorsal (lateral plantar)
220. Interossei, plantar (lateral plantar)
224. Abductor hallucis (medial plantar)

118. Transversus perinei superficialis (pudendal)
119. Transversus perinei profundus (pudendal)
120. Bulbocavernosus (pudendal)
121. Ischiocavernosus (pudendal)
122. Sphincter urethrae (pudendal)
123. Sphincter ani externus (pudendal)

Dorsal Primary Rami

94. Multifidi

S3

Ventral Primary Ramus

192. Biceps femoris (long head) (sciatic, tibial division)
217. Quadratus plantae (lateral plantar)

215. Abductor digiti minimi (lateral plantar)
216. Flexor digiti minimi brevis (foot) (lateral plantar)
225. Adductor hallucis (lateral plantar)
218. Lumbricales 2nd, 3rd, and 4th (foot) (lateral plantar)
219. Interossei, dorsal (lateral plantar)
220. Interossei, plantar (lateral plantar)

118. Transversus perinei superficialis (pudendal)
119. Transversus perinei profundus (pudendal)
120. Bulbocavernosus (pudendal)
121. Ischiocavernosus (pudendal)
122. Sphincter urethrae (pudendal)
123. Sphincter ani externus (pudendal)

S4 and S5

Ventral Primary Ramus

115. Levator ani (S4)
116. Coccygeus (S4 and S5)
123. Sphincter ani externus (S4)
118. Transversus perinei superficialis (S4, pudendal)
119. Transversus perinei profundus (S4, pudendal)
120. Bulbocavernosus (S4, pudendal)
121. Ischiocavernosus (S4, pudendal)
122. Sphincter urethrae (S4, pudendal)

REFERENCES

BIBLIOGRAPHY

General Anatomy Sources

Basmajian JV, De Luca DJ. *Muscles Alive,* 5th ed. Baltimore: Williams & Wilkins, 1985.

Clemente CD. *Gray's Anatomy,* 30th Am ed. Philadelphia: Lea & Febiger, 1985.

Clemente CD. *Anatomy: A Regional Atlas of the Human Body.* Baltimore: Urban & Schwarzenberg, 1987.

Figge FHJ. *Sobotta's Atlas of Human Anatomy.* Vol. 1. *Atlas of Bones, Joints and Muscles,* 8th English ed. New York: Hafner, 1968.

Grant JCB. *An Atlas of Anatomy,* 5th ed. Baltimore: Williams & Wilkins, 1962.

Hollingshead WH. *Functional Anatomy of the Limbs and Back.* Philadelphia: W.B. Saunders, 1969.

Hoppenfeld S. *Physical Examination of the Spine and Extremities.* New York: Appleton Century Crofts, 1976.

Kendall FP, McCreary EK, Provance PG. *Muscles: Testing and Function,* 4th ed. Baltimore: Williams & Wilkins, 1993.

Long C, Brown ME. Electromyographic kinesiology of the hand: Muscles moving the long finger. J Bone Joint Surg (Am) 46:1683–1706, 1964.

Netter FH. *Atlas of Human Anatomy.* Summit, NJ: CIBA-Geigy Corp, 1989.

Pernkopf E. *Atlas of Topographical and Applied Human Anatomy.* Philadelphia: W.B. Saunders, 1980.

Williams PL, Warwick R, Dyson M, Bannister LH. *Gray's Anatomy,* 37th Br ed. London: Churchill-Livingstone, 1989.

CITED REFERENCES

Gross Anatomy

1. Clemente CD. *Gray's Anatomy,* 30th Am ed. Philadelphia: Lea & Febiger, 1985.
2. Williams PL, Warwick R, Dyson M, Bannister LH. *Gray's Anatomy,* 37th Br ed. London: Churchill-Livingstone, 1989.
3. Figge FHJ. *Sobotta's Atlas of Human Anatomy.* Vol. 1. *Atlas of Bones, Joints and Muscles,* 8th English ed. New York: Hafner, 1968.
4. Clemente CD. *Anatomy: A Regional Atlas of the Human Body.* Baltimore, Urban & Schwarzenberg, 1987.
5. Netter FH. *Atlas of Human Anatomy.* Summit, NJ: CIBA-Geigy Corp, 1989.
6. Hollingshead WH. *Functional Anatomy of the Limbs and Back.* Philadelphia: W.B. Saunders, 1969.
7. Grant JCB. *An Atlas of Anatomy,* 5th ed. Baltimore: Williams & Wilkins, 1962.
8. Moore KL. *Clinically Oriented Anatomy,* 3rd ed. Baltimore: Williams & Wilkins, 1992.
9. DuBrul EL. *Sicher and DuBrul's Oral Anatomy,* 8th ed. St. Louis: Ishiyaku EuroAmerica, 1988.
10. Nairn RI. The circumoral musculature: Structure and function, Br Dental J 138:49–56, 1975.
11. Lightoller GH. Facial muscles: The modiolus and muscles surrounding the rima oris with remarks about the panniculus adiposus. J Anat 60:1–85, 1925.
12. Perry J, Nickel VL. Total cervical-spine fusion for neck paralysis. J Bone Joint Surg 41A:37–60, 1959.
13. Hoskiko D. Electromyographic investigation of the intercostal muscles during speech. Arch Phys Med 43:115–119, 1962.
14. Basmajian JV. *Muscles Alive,* 2nd ed. Baltimore: Williams & Wilkins, 1967.
15. Jones DS, Beargie RJ, Pauly JE. An electromyographic study of some muscles of costal respiration in man. Anat Rec 117:17–24, 1953.
16. Sodeberg SL. *Kinesiology: Application to Pathologic Motion.* Baltimore: Williams & Wilkins, 1986.
17. Doody SG, Freedman L, Waterland JC. Shoulder movements during abduction in the scapular plane. Arch Phys Med Rehabil 10:595–604, 1970.
18. Perry J. Muscle control of the shoulder. In: Rowe CR (ed). *The Shoulder.* New York: Churchill-Livingstone, 1988, pp. 17–34.
19. Flatt AE. Kinesiology of the hand. Am Acad Orthop Surg Instructional Course Lectures XVIII. St. Louis: C.V. Mosby, 1961.
20. McKibben B. Action of the iliopsoas muscle in the newborn. J Bone Joint Surg (Br) 50:161–165, 1968.
21. Lieb FJ, Perry J. Quadriceps function: An anatomical and mechanical study using amputated limbs. J Bone Joint Surg (Am) 53:749–758, 1971.
22. Lieb FJ, Perry J. Quadriceps function: An electromyographic study under isometric conditions. J Bone Joint Surg (Am) 53:749–758, 1971.

Index

Note: Page numbers in *italics* refer to illustrations; page numbers followed by t refer to tables.

A

Abdomen, muscles of, 366–368
Abducens (VI cranial) nerve, 411
Abduction, finger, 142–144, 406–407
  hip, 182–185, 408
    from flexed position, 186–189
    in infants and children, 249–250
    with flexion and external rotation, 173–175
  scapular, 58–63, 404
    in infants and children, 253–254
  shoulder, 90–93, 404
    horizontal, 94–96
    in infants and children, 255
  thumb, 156–159, 406
  vocal cord, 312
Abductor digiti minimi muscle, 382
  in finger abduction, 142–144
  of foot, 399
Abductor hallucis muscle, 401
Abductor pollicis brevis muscle, 387
  in thumb abduction, 158–159
  in thumb opposition, 163t
Abductor pollicis longus muscle, 385–386
  in thumb abduction, 156–157
  in wrist flexion, 124t
Accessory (XI cranial) nerve, 411
Accessory obturator nerve, 414
Active resistance test, 2
Adduction, finger, 145–147, 407
  hip, 190–194, 408
    in infants and children, 251
  scapular, 69–72
    in infants and children, 255
    scapular depression and, 73–75
    with downward rotation, 76–80
  shoulder, 404
    horizontal, 97–101
  soft palate, 302–303
  thumb, 160–162, 406
  vocal cord, 312
Adductor brevis muscle, 389
  in hip adduction, 190–194
  in hip flexion, 169t
Adductor hallucis muscle, 401
Adductor longus muscle, 389
  in hip adduction, 190–194
  in hip flexion, 169t
Adductor magnus muscle, 389–390
  in hip adduction, 190–194
  in hip flexion, 169t
Adductor pollicis muscle, 387
  in thumb adduction, 160–162
Anconeus muscle, 377–378
Ankle, dorsiflexion of, 218–220, 409
  foot eversion with, 224–226

Ankle, dorsiflexion of *(Continued)*
  in infants and children, 252
  in upright motor control, 321
  motions of, 409
  muscles of, 395–397
  plantar flexion of, 211–217, 409
    foot eversion with, 224–226
    in infants and children, 252
    in upright motor control, 324
Articularis genus muscle, 395
Auriculares muscles, 344

B

Back, muscles of, 360
  deep, in trunk rotation, 45t
Biceps brachii muscle, 376–377
  in elbow flexion, 108–113
  in forearm supination, 118–120
  in shoulder abduction, 90t, 93
  in shoulder flexion, 81t, 83
  in shoulder scaption, 88t
Biceps femoris muscle, 392–393
  in hip extension, 176
  in knee flexion, 202, 204–205
Brachial plexus, 412, *413*
  nerves from, 413–414
Brachialis muscle, 377
  in elbow flexion, 108–113
Brachioradialis muscle, 377
  in elbow flexion, 108–113
Break test, 2
Buccinator muscle, 278, 343–344
  in cheek compression, 281
Bulbar testing. See *Muscle testing, bulbar*.
Bulbospongiosus muscle, 369–370

C

Capital extension, 12–15, 402
Capital extensor muscles, 12–14, 351–353
Capital flexion, *15*, 21–23, 402
Capital flexor muscles, 21–23, 355–357
Cervical extension, 15–18, 402
Cervical extensor muscles, 16–18, 353–355
Cervical flexion, *15*, 24–29, 402
Cervical flexor muscles, 24–27, 357–360
Cervical nerves, spinal roots of, 418–420
Cervical plexus, 412
Cervical rotation, 31, 402
Cervical spine, motions of, 402
Cheeks, compression of, 281
Chondroglossus muscle, 290t, 346
Coccygeus muscle, 368–369
Combined cervical flexion, *15*, 28–29

Combined cervical flexion *(Continued)*
  to isolate single sternocleidomastoid, 30
Combined neck extension, *15*, 19–20
Common peroneal nerve, 416
Coracobrachialis muscle, 376
  in shoulder flexion, 81–83
  in shoulder scaption, 88t
Corrugator supercilii muscle, 268t, *272*, 336
  in frowning, 273
Coughing, evaluation of, laryngeal function and, 313
  functional anatomy of, 55, 311
  to test forced expiration, 54
Cranial nerves, 410–411. See also named nerve.
Cremaster muscle, 369
Cricoarytenoid muscle, 350–351
Cricothyroid muscle, 308, 350

D

Deep peroneal nerve, 416
Deltoid muscle, 374–375
  anterior, in shoulder flexion, 81–83
    in shoulder horizontal adduction, 97t
    in shoulder internal rotation, 105t
    in shoulder scaption, 88–89
  middle, in shoulder abduction, 90–93
    in shoulder flexion, 81–83
    in shoulder scaption, 88–89
  posterior, in scapular adduction, 72
    in shoulder extension, 84–87
    in shoulder external rotation, 102t
    in shoulder horizontal abduction, 94–96
Depression, scapular, 73–75
Depressor anguli oris muscle, 278, 283, 343
Depressor labii inferioris muscle, 278, 283, 343
Depressor septi muscle, 276t, 277, 339
Deviation, of jaw, lateral, 288–289
  of tongue, 294
Diaphragm, 363–364
  in quiet inspiration, 50–53
Digastricus muscle, 357
DIP joints. See *Distal interphalangeal (DIP) joints*.
Distal interphalangeal (DIP) joints, of fingers, flexion of, 135, 138, 406, 407
  of toes, extension of, 232–233, 410
Dorsal interossei muscles, 384
  in finger abduction, 142–144
  in finger MP flexion, 132–134
  in thumb adduction, 160t
  of foot, 400

Dorsal scapular nerve, 413
Dorsiflexion, ankle, 218–220, 409
    foot eversion with, 224–226
    in infants and children, 252
    in upright motor control, 321

E
Ear, muscles of, 344
Elbow, extension of, 114–117, 405
    in infants and children, 257
    flexion of, 108–113, 404
    in infants and children, 258
    motions of, 404–405
    muscles acting on, 376–378
    supination of, in infants and children, 259
Elevation, of larynx, in swallowing, 312
    of pelvis, 38–40
    of soft palate, 302–303
    of tongue, posterior, 295–296
    scapular, 65–68, 403
Epicranius, 335
Erector spinae muscles, 360
Eversion, foot, 224–226, 409
Expiration, forced, 53–54, 403
Extension, elbow, 114–117, 405
    in infants and children, 257
    finger, 139–141, 406, 407
    hallux, 232–233, 409
    hip, 176–181, 407–408
        gluteus maximus in, test to isolate, 179–180
        hip flexion tightness in, modified tests for, 181
        in infants and children, 247–248
        in upright motor control, 322
    in upright motor control, 321–324
    knee, 207–210, 408
        in infants and children, 247–248
        in upright motor control, 323
    neck, capital, 12–14, 15, 402
        cervical, 15, 16–18, 402
        combined (capital plus cervical), 15, 19–20
        in infants and children, 237
    shoulder, 84–87
    thumb, interphalangeal, 154–155, 406
        metacarpophalangeal, 152–153, 405
    toe, 232–233, 410
    trunk, 34–37, 402–403
        in infants and children, 243–244
        lumbar, 36, 403
        lumbar and thoracic, 36–37
        thoracic, 35, 402–403
    wrist, 128–131, 405
Extensor carpi radialis brevis muscle, 379
    in wrist extension, 128–131
Extensor carpi radialis longus muscle, 378–379
    in wrist extension, 128–131
Extensor carpi ulnaris muscle, 379
    in wrist extension, 128–131
Extensor digiti minimi muscle, 382
    in finger MP extension, 139–141
    in wrist extension, 128t
Extensor digitorum brevis muscle, 398
    in toe extension, 232–233
Extensor digitorum longus muscle, 398
    in foot eversion, 224t
    in toe extension, 232–233
Extensor digitorum muscle, 380–381
    in finger MP extension, 139–141
    in wrist extension, 128t

Extensor hallucis longus muscle (Continued)
    in hallux extension, 232–233
Extensor indicis muscle, 381
    in finger MP extension, 139–141
    in wrist extension, 128t
Extensor pollicis brevis muscle, 386
    in thumb MP extension, 152–153
Extensor pollicis longus muscle, 386
    in thumb IP extension, 154–155
    in wrist extension, 128t
Extraocular muscles. See Eye(s), muscles of
Eye(s), axes of, 264
    closing of, 271
    motions of, 265–266
    muscles of, 263–267, 336–338
        actions of, 264–266
    opening of, 269
    tracking of, 266–267
Eyebrows, muscles of, 268t, 273–275, 335–336
Eyelids, muscles of, 268–271, 335–336

F
Face. See also specific structure.
    muscles of, observation of, 267
Facial (VII cranial) nerve, 411
    peripheral vs central lesions of, 269
Fair (grade 3) muscle, 5
Fair+ (grade 3+) muscle, 6
Femoral nerve, 415
Finger(s), 380
    abduction of, 142–144, 406–407
    adduction of, 145–147, 407
    extension of, 139–141, 406, 407
    flexion of, interphalangeal, distal, 135, 138, 406, 407
        proximal, 135–137, 406, 407
        metacarpophalangeal, 132–134, 406, 407
    little, muscles acting on, 382–383
    opposition of thumb to, 163–165, 406
    muscles acting on, 380–383
Flexion, ankle. See also Dorsiflexion, ankle.
    plantar, 211–217, 409
        foot eversion with, 224–226
        in infants and children, 252
        in upright motor control, 324
    elbow, 108–113, 404
        in infants and children, 258
    finger, interphalangeal, distal, 135, 138, 406, 407
        proximal, 135–137, 406, 407
        metacarpophalangeal, 132–134, 406, 407
    hallux, distal joint, 230–231, 409
        proximal joint, 227–228, 409
    hip, 169–172, 407
        in infants and children, 245–246
        in upright motor control, 321
        tightness in, modified hip extension tests for, 181
        with abduction and external rotation, 173–175
    in upright motor control, 320–321
    knee, 202–206, 408
        in infants and children, 245–246
        in upright motor control, 321
        with hip flexion, abduction, and external rotation, 173–175
    neck, capital, 15, 21–23, 402
        cervical, 15, 24–27, 402
            combined (capital and cervical), 15,

Flexion, neck, (Continued)
        to isolate single sternocleidomastoid, 30
    in infants and children, 238–239
    shoulder, 81–83
        in infants and children, 256
    thumb, interphalangeal, 148, 150, 405
        metacarpophalangeal, 148–149, 405
    toe, distal joint, 230–231, 409–410
        proximal joint, 227, 229, 409
    trunk, 41–44, 403
        in infants and children, 240–241
    wrist, 124–127, 405
Flexor carpi radialis muscle, 379
    in forearm pronation, 121t
    in wrist flexion, 124–127
Flexor carpi ulnaris muscle, 380
    in wrist flexion, 124–127
Flexor digiti minimi brevis muscle, 382–383
    of foot, 399
    in toe MP flexion, 227t
Flexor digitorum brevis muscle, 398–399
    in toe IP flexion, 230–231
    in toe MP flexion, 227t
Flexor digitorum longus muscle, 398
    in foot inversion, 221t
    in toe IP flexion, 230–231
    in toe MP flexion, 227t
Flexor digitorum profundus muscle, 382
    in finger MP flexion, 132t
    in wrist flexion, 124t
Flexor digitorum superficialis muscle, 381
    in finger MP flexion, 132t
    in finger PIP flexion, 135–137
    in wrist flexion, 124t
Flexor hallucis brevis muscle, 401
    in hallux MP flexion, 227–228
Flexor hallucis longus muscle, 400–401
    in foot inversion, 221t
    in hallux IP flexion, 230–231
Flexor pollicis brevis muscle, 386–387
    in thumb MP flexion, 148–149
Flexor pollicis longus muscle, 386
    in thumb IP flexion, 148, 150
    in thumb MP flexion, 149
    in wrist flexion, 124t
Foot. See also Ankle; Hallux; Toe(s).
    eversion of, 224–226, 409
    inversion of, 221–223, 409
        with dorsiflexion, 218–220
Forearm, motions of, 405
    muscles acting on, 378
    pronation of, 121–123, 405
    supination of, 118–120, 405
Forehead, muscles of, 268, 274–275, 335
Frowning, 273

G
Gastrocnemius muscle, 396
    in ankle plantar flexion, 211–214
    in foot inversion, 221t
Gemellus inferior muscle, 392
    in hip external rotation, 194–197
Gemellus superior muscle, 392
    in hip external rotation, 194–197
Genioglossus muscle, 290, 346
    in tongue channeling, 296
    in tongue deviation, 294
    in tongue protrusion, 294
    in tongue retraction, 295
Geniohyoid muscle, 356–357
Genitofemoral nerve, 414
Glossopharyngeal (IX cranial) nerve, 411

Gluteus maximus muscle, 390
  in hip abduction, 182t
  in hip extension, 176–181
    test to isolate, 179–180
  in hip external rotation, 194–197
Gluteus medius muscle, 390
  in hip abduction, 182–185
    from flexed position, 186t
  in hip internal rotation, 198–201
Gluteus minimus muscle, 390–391
  in hip abduction, 182–185
    from flexed position, 186t
  in hip internal rotation, 198–201
Good (grade 4) muscle, 5
Gracilis muscle, 389
  in hip adduction, 190–194

H
Hallux, extension of, 232–233, 409
  flexion of, distal joint, 230–231, 409
    proximal joint, 227–228, 409
  motions of, 409
  muscles acting on, 400–401
Hamstring muscles, in hip extension,
    176–181
  in knee flexion, 202–206
Hand. See also *Finger(s)*; *Thumb*; *Wrist*.
  bones and joints of, *380*
  intrinsic muscles of, 383–385
  motions of, 405
Head, motions of, *15*, 402. See also *Capital*
    entries.
  rotation of, 31
Hip, (see) abduction of, 182–185, 408
  from flexed position, 186–189
  in infants and children, 249–250
  (see) with flexion and external rotation,
    173–175
  adduction of, 190–194, 408
    in infants and children, 251
  extension of, 176–181, 407–408
    gluteus maximus in, test to isolate,
      179–180
    in infants and children, 247–248
    in upright motor control, 322
  external rotation of, 195–197, 408
    with flexion and abduction, 173–175
  (see) flexion of, 169–172, 407
    in infants and children, 245–246
    in upright motor control, 321
    tightness in, modified hip extension
      tests for, 181
    (see) with abduction and external rota-
      tion, 173–175
  internal rotation of, 198–201, 408
  motions of, 407–408
  muscles of, 387–393
Hip abductor muscles, in hip flexion, with
    abduction and external rotation, 173t
Hip external rotator muscles, in hip flexion,
    with abduction and external rotation,
    173t
Hip flexor muscles, in hip flexion, with ab-
    duction and external rotation, 173t
Hyoglossus muscle, 290, 346
Hypoglossal (XII cranial) nerve, 411

I
Iliacus muscle, 388
  in hip flexion, 169–172
Iliocostalis cervicis muscle, 354
  in cervical extension, 16–18

Iliocostalis cervicis muscle (Continued)
  in cervical rotation, 31
Iliocostalis lumborum muscle, 360–361
  in elevation of pelvis, 38t
  in trunk extension, 34–37
Iliocostalis thoracis muscle, 360
  in trunk extension, 34–37
Iliohypogastric nerve, 414
Iliolingual nerve, 414
Inferior gluteal nerve, 415
Inferior longitudinal muscle, of tongue, 290t,
    291, 347
  in tongue channeling, 296
  in tongue curling, 297
Inferior pharyngeal constrictor muscle, 304,
    348
Infrahyoid muscles, 308t, 358–359
Infranuclear lesions, tongue and, 293
Infraspinatus muscle, 375
  in shoulder external rotation, 102–104
  in shoulder horizontal abduction, 94t
Inspiration, forced, 403
  quiet, 50–53, 403
Intercostales externi muscles, 364
  in quiet inspiration, 50–53
Intercostales interni muscles, 364–365
  in forced expiration, 54–55
  in quiet inspiration, 50–53
Intercostales intimi muscles, 365
  in quiet inspiration, 50–53
Interphalangeal (IP) joints, of fingers, distal,
    flexion of, 135, 138, 406, 407
  proximal, flexion of, 135–137, 406, 407
  flexion of, 230–231, 409–410
Interspinales cervicis muscles, 355
Interspinales lumborum muscles, 362
Interspinales thoracis muscles, 362
Intertransversarii cervicis muscles, 355
Intertransversarii lumborum muscles, 363
Intertransversarii thoracis muscles, 363
Inversion, foot, 221–223, 409
  with dorsiflexion, 218–220
IP joints. See *Interphalangeal (IP) joints*.
Ischiocavernosus muscle, 370

J
Jaw, closure of, 287–288
  lateral deviation of, 288–289
  muscles of, 284–289, 344–346
  opening of, 286–287
  protrusion of, 289

K
Knee, extension of, 207–210, 408
  in infants and children, 247–248
  in upright motor control, 323
  flexion of, 202–206, 408
    in infants and children, 245–246
    in upright motor control, 321
    with hip flexion, abduction, and exter-
      nal rotation, 173–175
  motions of, 408
  muscles of, 393–395

L
Larynx, elevation of, in swallowing, 312
  muscles of, 308–313, 350–351

Lateral cricoarytenoid muscle, 308t, *309*, 351
  in vocal cord abduction and adduction,
    312
Lateral pectoral nerve, 413
Lateral plantar nerve, 416
Lateral pterygoid muscle, 284t, *285*, 286, 345
  in jaw opening, 286–287
  in jaw protrusion, 289
  in lateral jaw deviation, 288–289
Latissimus dorsi muscle, 373–374
  in elevation of pelvis, 38t
  in scapular adduction, 69t
    with downward rotation, 76t
  in scapular depression and adduction, 73t
  in shoulder extension, 84–87
  in shoulder internal rotation, 105t
  in trunk rotation, 45t
Levator anguli oris muscle, 278, 282,
    341–342
Levator ani muscle, 368
Levator labii superioris alaeque nasi muscle,
    278, 282, 341
Levator labii superioris muscle, 278, 282,
    340–341
Levator palpebrae superioris muscle, 268,
    336
  in eye opening, 269
Levator scapulae muscle, 372
  in scapular adduction, with downward ro-
    tation, 76t
  in scapular elevation, 65–68
Levator veli palatini muscle, 298–299, 349
  in soft palate elevation and adduction,
    302–303
Levatores costarum muscles, 365–366
Lips, closing of, 280
Long thoracic nerve, 413
Longissimus capitis muscle, 352–353
  in capital extension, 12–14
  in cervical rotation, 31
Longissimus cervicis muscle, 353–355
  in cervical extension, 16–18
Longissimus thoracis muscle, 361
  in trunk extension, 34–37
Longus capitis muscle, 356
  in capital flexion, 21–23
  in cervical rotation, 31
Longus colli muscle, 357
  in cervical rotation, 31
Lower extremity. See also specific structure.
  motions of, 407–410
  muscles of, 387–401
    innervation of, 414–416
Lower thoracic nerves, 414
Lumbar extension, 36, 403
Lumbar nerves, spinal roots of, 422–424
Lumbar plexus, *415*
  muscles innervated directly off, 414
  nerves from, 414–415
Lumbar spine. See also *Pelvis*; *Trunk*.
  extension of, 36, 403
  motions of, 403
Lumbosacral nerves, spinal roots of, 424–425
Lumbricales muscles, 383–384
  in finger MP flexion, 132–134
  of foot, 399–400
    in toe MP flexion, 227, 229

M
Manual muscle testing. See *Muscle testing,
    manual*.
Masseter muscle, 284, 286, 344–345
  in jaw closure, 287–288

Mastication, muscles of. See *Jaw, muscles of.*
Medial pectoral nerve, 413
Medial plantar nerve, 416
Medial popliteal nerve, 416
Medial pterygoid muscle, 284t, *285,* 286, 345–346
  in jaw closure, 287–288
  in jaw protrusion, 289
  in lateral jaw deviation, 288–289
Median nerve, 413
Mentalis muscle, 278, 283, 342
Metacarpophalangeal (MP) joints, extension of, 139–141, 406, 407
  flexion of, 132–134, 406, 407
Metatarsophalangeal (MP) joints, extension of, 232–233, 410
  flexion of, 227–229, 409
Middle pharyngeal constrictor muscle, 304, 348
Modiolus, 279, 340, *341*
Motor control, upright, 320–324
Motor nerve roots, 417–425
  cervical, 418–420
  lumbar, 422–424
  lumbosacral, 424–425
  thoracic, 420–422
Mouth, muscles of, 278–283, 339–344
MP joints. See *Metacarpophalangeal (MP) joints; Metatarsophalangeal (MP) joints.*
Multifidi muscles, 361–362
  in trunk extension, 34–37
Muscle(s), abductor digiti minimi, 382
  in finger abduction, 142–144
  of foot, 399
  abductor hallucis, 401
  abductor pollicis brevis, 387
    in thumb abduction, 158–159
    in thumb opposition, 163t
  abductor pollicis longus, 385–386
    in thumb abduction, 156–157
    in wrist flexion, 124t
  acting at wrist, 378–380
  acting on elbow, 376–378
  acting on fingers, 380–383
  acting on forearm, 378
  acting on great toe, 400–401
  acting on little finger, 382–383
  acting on thumb, 385–387
  acting on toes, 397–401
  adductor brevis, 389
    in hip adduction, 190–194
    in hip flexion, 169t
  adductor hallucis, 401
  adductor longus, 389
    in hip adduction, 190–194
    in hip flexion, 169t
  adductor magnus, 389–390
    in hip adduction, 190–194
    in hip flexion, 169t
  adductor pollicis, 387
    in thumb adduction, 160–162
  anconeus, 377–378
  articularis genus, 395
  auriculares, 344
  biceps brachii, 376–377
    in elbow flexion, 108–113
    in forearm supination, 118–120
    in shoulder abduction, 90t, 93
    in shoulder flexion, 81t, 83
    in shoulder scaption, 88t
  biceps femoris, 392–393
    in hip extension, 176
    in knee flexion, 202, 204–205
  brachialis, 377
    in elbow flexion, 108–113

Muscle(s) *(Continued)*
  brachioradialis, 377
    in elbow flexion, 108–113
  buccinator, 278, 343–344
    in cheek compression, 281
  bulbospongiosus, 369–370
  capital extensor, 12–14, 351–353
  capital flexor, 21–23, 355–357
  cervical extensor, 16–18, 353–355
  cervical flexor, 24–27, 357–360
  chondroglossus, 290t, 346
  coccygeus, 368–369
  coracobrachialis, 376
    in shoulder flexion, 81–83
    in shoulder scaption, 88t
  corrugator supercilii, 268t, *272,* 336
    in frowning, 273
  cremaster, 369
  cricoarytenoid, 350–351
  cricothyroid, 308, 350
  deltoid, 374–375
    anterior, in shoulder flexion, 81–83
      in shoulder horizontal adduction, 97t
      in shoulder internal rotation, 105t
      in shoulder scaption, 88–89
    middle, in shoulder abduction, 90–93
      in shoulder flexion, 81–83
      in shoulder scaption, 88–89
    posterior, in scapular adduction, 72
      in shoulder extension, 84–87
      in shoulder external rotation, 102t
      in shoulder horizontal abduction, 94–96
  depressor anguli oris, 278, 283, 343
  depressor labii inferioris, 278, 283, 343
  depressor septi, 276t, 277, 339
  digastricus, 357
  dorsal interossei, 384
    in finger abduction, 142–144
    in finger MP flexion, 132–134
    in thumb adduction, 160t
    of foot, 400
      in toe MP flexion, 227t
  erector spinae, 360
  extensor carpi radialis brevis, 379
    in wrist extension, 128–131
  extensor carpi radialis longus, 378–379
    in wrist extension, 128–131
  extensor carpi ulnaris, 379
    in wrist extension, 128–131
  extensor digiti minimi, 382
    in finger MP extension, 139–141
    in wrist extension, 128t
  extensor digitorum, 380–381
    in finger MP extension, 139–141
    in wrist extension, 128t
  extensor digitorum brevis, 398
    in toe extension, 232–233
  extensor digitorum longus, 398
    in foot eversion, 224t
    in toe extension, 232–233
  extensor hallucis longus, 400
    in hallux extension, 232–233
  extensor indicis, 381
    in finger MP extension, 139–141
    in wrist extension, 128t
  extensor pollicis brevis, 386
    in thumb MP extension, 152–153
  extensor pollicis longus, 386
    in thumb IP extension, 154–155
    in wrist extension, 128t
  extraocular. See *Muscle(s), ocular.*
  fair (grade 3), 5
  fair+ (grade 3+), 6

Muscle(s) *(Continued)*
  flexor carpi radialis, 379
    in forearm pronation, 121t
    in wrist flexion, 124–127
  flexor carpi ulnaris, 380
    in wrist flexion, 124–127
  flexor digiti minimi brevis, 382–383
    of foot, 399
      in toe MP flexion, 227t
  flexor digitorum brevis, 398–399
    in toe IP flexion, 230–231
    in toe MP flexion, 227t
  flexor digitorum longus, 398
    in foot inversion, 221t
    in toe IP flexion, 230–231
    in toe MP flexion, 227t
  flexor digitorum profundus, 382
    in finger MP flexion, 132t
    in wrist flexion, 124t
  flexor digitorum superficialis, 381
    in finger MP flexion, 132t
    in finger PIP flexion, 135–137
    in wrist flexion, 124t
  flexor hallucis brevis, 401
    in hallux MP flexion, 227–228
  flexor hallucis longus, 400–401
    in foot inversion, 221t
    in hallux IP flexion, 230–231
  flexor pollicis brevis, 386–387
    in thumb MP flexion, 148–149
  flexor pollicis longus, 386
    in thumb IP flexion, 148, 150
    in thumb MP flexion, 149
    in wrist flexion, 124t
  gastrocnemius, 396
    in ankle plantar flexion, 211–214
    in foot inversion, 221t
  gemellus inferior, 392
    in hip external rotation, 194–197
  gemellus superior, 392
    in hip external rotation, 194–197
  genioglossus, 290, 346
    in tongue channeling, 296
    in tongue deviation, 294
    in tongue protrusion, 294
    in tongue retraction, 295
  geniohyoid, 356–357
  gluteus maximus, 390
    in hip abduction, 182t
    in hip extension, 176–181
      test to isolate, 179–180
    in hip external rotation, 194–197
  gluteus medius, 390
    in hip abduction, 182–185
      from flexed position, 186t
    in hip internal rotation, 198–201
  gluteus minimus, 390–391
    in hip abduction, 182–185
      from flexed position, 186t
    in hip internal rotation, 198–201
  good (grade 4), 5
  gracilis, 389
    in hip adduction, 190–194
  hamstring, in hip extension, 176–181
    in knee flexion, 202–206
  hip abductor, in hip flexion, with abduction and external rotation, 173t
  hip external rotator, in hip flexion, with abduction and external rotation, 173t
  hip flexor, in hip flexion, with abduction and external rotation, 173t
  hyoglossus, 290, 346
  iliacus, 388
    in hip flexion, 169–172

Muscle(s) (*Continued*)
iliocostalis cervicis, 354
   in cervical extension, 16–18
   in cervical rotation, 31
iliocostalis lumborum, 360–361
   in elevation of pelvis, 38t
   in trunk extension, 34–37
iliocostalis thoracis, 360
   in trunk extension, 34–47
inferior longitudinal, of tongue, 290t, *291*, 347
   in tongue channeling, 296
   in tongue curling, 297
inferior pharyngeal constrictor, 304, 348
infrahyoid, 308t, 358–359
infraspinatus, 375
   in shoulder external rotation, 102–104
   in shoulder horizontal abduction, 94t
innervated by cranial nerves, 410–411
innervated directly off lumbar plexus, 414
innervated directly off sacral plexus, 415
innervated from cervical roots, 418–420
innervated from lumbar roots, 422–424
innervated from lumbosacral roots, 424–425
innervated from pudendal plexus, 416
innervated from thoracic roots, 420–422
intercostales externi, 364
   in quiet inspiration, 50–53
intercostales interni, 364–365
   in forced expiration, 54–55
   in quiet inspiration, 50–53
intercostales intimi, 365
   in quiet inspiration, 50–53
interspinales cervicis, 355
interspinales lumborum, 362
interspinales thoracis, 362
intertransversarii cervicis, 355
intertransversarii lumborum, 363
intertransversarii thoracis, 363
ischiocavernosus, 370
lateral cricoarytenoid, 308t, *309*, 351
   in vocal cord abduction and adduction, 312
lateral pterygoid, 284t, *285*, 286, 345
   in jaw opening, 286–287
   in jaw protrusion, 289
   in lateral jaw deviation, 288–289
latissimus dorsi, 373–374
   in elevation of pelvis, 38t
   in scapular adduction, 69t
     with downward rotation, 76t
   in scapular depression and adduction, 73t
   in shoulder extension, 84–87
   in shoulder internal rotation, 105t
   in trunk rotation, 45t
levator anguli oris, 278, 282, 341–342
levator ani, 368
levator labii superioris, 278, 282, 340–341
levator labii superioris alaeque nasi, 278, 282, 341
levator palpebrae superioris, 268, 336
   in eye opening, 269
levator scapulae, 372
   in scapular adduction, with downward rotation, 76t
   in scapular elevation, 65–68
levator veli palatini, 298–299, 349
   in soft palate elevation and adduction, 302–303
levatores costarum, 365–366
lingual, 347

Muscle(s) (*Continued*)
longissimus capitis, 352–353
   in capital extension, 12–14
   in cervical rotation, 31
longissimus cervicis, 353–355
   in cervical extension, 16–18
longissimus thoracis, 361
   in trunk extension, 34–37
longus capitis, 356
   in cervical rotation, 31
longus colli, 357
   in cervical rotation, 31
lumbricales, 383–384
   in finger MP flexion, 132–134
   of foot, 399–400
     in toe MP flexion, 227, 229
masseter, 284, 286, 344–345
   in jaw closure, 287–288
medial pterygoid, 284t, *285*, 286, 345–346
   in jaw closure, 287–288
   in jaw protrusion, 289
   in lateral jaw deviation, 288–289
mentalis, 278, 283, 342
middle pharyngeal constrictor, 304, 348
multifidi, 361–362
   in trunk extension, 34–37
mylohyoid, 356
nasalis, 276t, 277, 339
normal (grade 5), 4–5
oblique arytenoid, 308, *309*, 351
obliquus capitis inferior, 352
   in capital extension, 12–14
   in cervical rotation, 31
obliquus capitis superior, 352
   in capital extension, 12–14
obliquus externus abdominis, 366–367
   in elevation of pelvis, 38t
   in forced expiration, 54–55
   in trunk flexion, 41t
   in trunk rotation, 45–49
obliquus inferior oculi, 263, 338
   actions of, 266
obliquus internus abdominis, 367
   in elevation of pelvis, 38t
   in forced expiration, 54–55
   in trunk flexion, 41t
   in trunk rotation, 45–49
obliquus superior oculi, 263, 338
   actions of, 265
obturator externus, 391–392
   in hip external rotation, 194–197
obturator internus, 391
   in hip external rotation, 194–197
occipitofrontalis, *274*, 335
   in raising eyebrows, 275
ocular, 263–267, 336–338
   actions of, 264–266
of abdomen, 366–368
of ankle, 395–397
of back, 360
   deep, in trunk rotation, 45t
of ear, 344
of eye. See *Muscle(s), ocular.*
of eyebrows, 268t, 273–275, 335–336
of eyelids, 268–271, 335–336
of forehead, 268, 274–275, 335
of hand, intrinsic, 383–385
of hip, 387–393
of jaw, 284–289, 344–346
of knee, 393–395
of larynx, 308–313, 350–351
of lower extremity, 387–401
   innervation of, 414–416
of mouth, 278–283, 339–344

Muscle(s) (*Continued*)
of neck, 351–360
of nose, 276–277, 338–339
of palate, 298–303, 349–350
of perineum, 368–371
of pharynx, 304–307, 347–349
of shoulder girdle, 371–373
of thorax, for respiration, 363–366
of tongue, 290–297, 346–347
   extrinsic, 290, 346–347
   intrinsic, 290t, *291*, 347
of trunk, 360–371
of upper extremity, 371–387
   innervation of, 412–414
omohyoid, 359
opponens digiti minimi, 383
   in thumb opposition, 163–165
opponens pollicis, 387
   in thumb opposition, 163–165
orbicularis oculi, 268t, *270*, 336
   in eye closing, 271
orbicularis oris, 278, 343
   in lip closing, 280
palatoglossus, 290t, 347
   in soft palate elevation and adduction, 302–303
   in tongue posterior elevation, 295–296
palatopharyngeus, 298–299, 350
   in occlusion of nasopharynx, 303
palmar interossei, 384–385
   in finger adduction, 145–147
   in finger MP flexion, 132–134
palmaris brevis, 383
palmaris longus, 379–380
   in thumb abduction, 156t
   in wrist flexion, 124t
pectineus, 388–389
   in hip adduction, 190–194
   in hip flexion, 169t
pectoralis major, 374
   in scapular adduction, with downward rotation, 76t
   in scapular depression and adduction, 73t
   in shoulder flexion, 81t, 83
   in shoulder horizontal adduction, 97–101
   in shoulder internal rotation, 105t
pectoralis minor, 373
   in scapular adduction, with downward rotation, 76t
   in scapular depression and adduction, 73t
peroneus brevis, 397
   in foot eversion, 224–226
peroneus longus, 397
   in foot eversion, 224–226
   isolation of, 226
peroneus tertius, 397
   in foot eversion, 224t, 226
pharyngeal constrictor, 348
piriformis, 391
   in hip external rotation, 194–197
plantar interossei, 400
   in toe MP flexion, 227t
plantaris, 396–397
platysma, 359–360
plus and minus grade, 5–6
poor (grade 2), 5
poor- (grade 2-), 6
popliteus, 395
posterior cricoarytenoid, 308, 350–351
   in vocal cord abduction and adduction, 312

Muscle(s) *(Continued)*
  procerus, 276, 339
    in wrinkling bridge of nose, 277
  pronator quadratus, 378
    in forearm pronation, 121–123
  pronator teres, 378
    in forearm pronation, 121–123
  psoas major, 388
    in hip flexion, 169–172
    in trunk flexion, 41t
  psoas minor, 388
    in trunk flexion, 41t
  pyramidalis, 368
  quadratus femoris, 392
    in hip external rotation, 194–197
  quadratus lumborum, 363
    in elevation of pelvis, 38–40
  quadratus plantae, 399
  quadriceps femoris, 394
    in knee extension, 207–210
  rectus abdominis, 367–368
    in forced expiration, 54–55
    in trunk flexion, 41–44
    in trunk rotation, 45t
  rectus capitis anterior, 355–356
    in capital flexion, 21–23
  rectus capitis lateralis, 356
    in capital flexion, 21–23
  rectus capitis posterior major, 352
    in capital extension, 12–14
    in cervical rotation, 31
  rectus capitis posterior minor, 352
    in capital extension, 12–14
    in cervical rotation, 31
  rectus femoris, 394
    in hip flexion, 169t
    in knee extension, 207–209
  rectus inferior bulbi, 263, 336–338
    actions of, 265
  rectus lateralis bulbi, 263, 336–338
    actions of, 265
  rectus medialis bulbi, 263, 336–338
    actions of, 265
  rectus superior bulbi, 263, 336–338
    actions of, 265
  rhomboid major, 372
    in scapular adduction, 69t, 72
      with downward rotation, 76–80
    in scapular elevation, 65t, 68
  rhomboid minor, 372
    in scapular adduction, 69t, 72
      with downward rotation,
        76–80
    in scapular elevation, 65t, 68
  risorius, 342
  rotatores cervicis, 355
  rotatores lumborum, 362
    in trunk extension, 34t
  rotatores thoracis, 362
    in trunk extension, 34t
  salpingopharyngeus, 304, 349
  sartorius, 393
    in hip flexion, 169t
      with abduction and external rotation,
        173–175
  scalenus anterior, 358
    in cervical flexion, 24–27
    in cervical rotation, 31
  scalenus medius, 358
    in cervical flexion, 24–27
  scalenus posterior, 358
    in cervical flexion, 24–27
    in cervical rotation, 31
  scapulohumeral, 374–376
  semimembranosus, 393

Muscle(s) semimembranosus *(Continued)*
    in hip extension, 176
    in knee flexion, 202–203
  semispinalis capitis, 353
    in capital extension, 12–14
    in cervical rotation, 31
  semispinalis cervicis, 354
    in cervical extension, 16–18
    in cervical rotation, 31
  semispinalis thoracis, 361
    in trunk extension, *34*, 34t
  semitendinosus, 393
    in hip extension, 176
    in knee flexion, 202–203
  serratus anterior, 372–373
    in scapular abduction, 58–63
    in shoulder scaption, 88t
  serratus posterior inferior, 366
  serratus posterior superior, 366
  soleus, 396
    in ankle plantar flexion, 211–217
  sphincter ani externus, 371
  sphincter urethrae, 370–371
  spinalis capitis, 353
  spinalis cervicis, 354
  spinalis thoracis, 361
    in trunk extension, 34–37
  splenius capitis, 353
    in capital extension, 12–14
    in cervical rotation, 31
  splenius cervicis, 354
    in cervical extension, 16–18
    in cervical rotation, 31
  sternocleidomastoid, 358
    combined cervical flexion to isolate, 30
    in cervical flexion, 24–27
    in cervical rotation, 31
  sternohyoid, 359
  sternothyroid, 358–359
  styloglossus, 290, 346–347
    in tongue posterior elevation, 295–296
    in tongue retraction, 295
  stylohyoid, 356
  stylopharyngeus, 304, 348–349
  subclavius, 374
  subcostales, 365
  subscapularis, 375
    in shoulder internal rotation, 105–107
    superior longitudinal, of tongue, 290t,
      291, 347
      in tongue channeling, 296
      in tongue curling, 297
  superior pharyngeal constrictor, 304, 348
  supinator, 378
    in forearm supination, 118–120
  suprahyoid, 284t, *285*, 290t, 356–357
    in jaw opening, 286–287
  supraspinatus, 375
    in shoulder abduction, 90–93
    in shoulder flexion, 81–83
    in shoulder scaption, 88–89
  temporalis, 284, 286, 345
    in jaw closure, 287–288
  temporoparietalis, 335
  tensor fasciae latae, 391
    in hip abduction, 182t
      from flexed position, 186–189
    in hip flexion, 169t
    in hip internal rotation, 198–201
  tensor veli palatini, 298–299, 349
    in soft palate elevation and adduction,
      302–303
  teres major, 376
    in shoulder extension, 84–87
    in shoulder internal rotation, 105t

Muscle(s) *(Continued)*
  teres minor, 375–376
    in shoulder external rotation, 102–104
    in shoulder horizontal abduction, 94t
  testing of. See *Muscle testing.*
  thyroarytenoid, 308t, *309*, 351
  thyrohyoid, 359
  tibialis anterior, 395–396
    in foot dorsiflexion and inversion,
      218–220
  tibialis posterior, 396
    in foot inversion, 221–223
  trace (grade 1), 5
  transverse arytenoid, 308, *309*, 351
  transverse lingual, 290t, *291*, 347
    in tongue channeling, 296
  transversus abdominis, 367
    in forced expiration, 54–55
    in quiet inspiration, *50*
  transversus menti, 342–343
  transversus perinei profundus, 369
  transversus perinei superficialis, 369
  transversus thoracis, 365
  trapezius, 371–372
    in scapular abduction, 58t
    in scapular adduction, 69–72
    in scapular depression and adduction,
      73–75
    in scapular elevation, 65–68
    in shoulder scaption, 88t
  triceps brachii, 377
    in elbow extension, 114–117
    in shoulder extension, 84t
    in shoulder horizontal abduction, 96
  uvular (musculus uvulae), 298–299,
    349–350
    in soft palate elevation and adduction,
      302–303
  vastus intermedius, 394
    in knee extension, 207–209
  vastus lateralis, 394
    in knee extension, 207–209
  vastus medialis longus, 394
    in knee extension, 207–209
  vastus medialis oblique, 394–395
    in knee extension, 207–209
  vertical lingual, 290t, *291*, 347
    in tongue channeling, 296
  zero (grade 0), 5
  zygomaticus major, 278, 282, 342
  zygomaticus minor, 342
Muscle testing, bulbar, grading in, 262
    introduction to, 262
    patient and examiner positions in,
      262
    precautions in, 262
  manual, application of resistance in,
    2–3
    criteria for assigning grades in, 4–6
    documentation form for, *7–8*
    examiner and value of, 3–4
    grading system for, 2
    in infants and children, 236–259
    influence of patient on, 4
    preparing for, 6–9
    screening tests for, 6
  upright motor control, 320–324
Musculocutaneous nerve, 413
Musculus uvulae, 298–299, 349–350
  in soft palate elevation and adduction,
    302–303
Mylohyoid muscle, 356
Myotomes, 417–425
  cervical, 418–420
  lumbar, 422–424

Myotomes *(Continued)*
  lumbosacral, 424–425
  thoracic, 420–422

N

Nasalis muscle, 276t, 277, 339
Nasopharynx, occlusion of, 303
Neck. See also *Cervical* entries.
  extension of, capital, 12–15, 402
    cervical, 15–18, 402
    combined (capital plus cervical), *15,*
      19–20
    in infants and children, 237
  flexion of, capital, *15,* 21–23, 402
    cervical, *15,* 24–27, 402
      combined (capital and cervical), *15,*
        28–29
    in infants and children, 238–239
  motions of, *15,* 402
  muscles of, 351–360
Nerve(s), abducens (VI cranial), 411
  accessory (XI cranial), 411
  accessory obturator, 414
  axillary, 413
  cervical, spinal roots of, 418–420
  cranial, 410–411
  dorsal scapular, 413
  facial (VII cranial), 411
    peripheral vs central lesions of, 269
  femoral, 415
  from brachial plexus, 413–414
  from cervical plexus, 412
  from lumbar plexus, 414–415
  from sacral plexus, 415–416
  genitofemoral, 414
  glossopharyngeal (IX cranial), 411
  hypoglossal (XII cranial), 411
  iliohypogastric, 414
  iliolingual, 414
  inferior gluteal, 415
  lateral pectoral, 413
  lateral plantar, 416
  long thoracic, 413
  lower thoracic, 414
  lumbar, spinal roots of, 422–424
  lumbosacral, spinal roots of, 424–425
  medial pectoral, 413
  medial plantar, 416
  medial popliteal, 416
  median, 413
  musculocutaneous, 413
  obturator, 415
  oculomotor (III cranial), 410–411
  of thoracic region, 414
  peripheral, 412–416
  peroneal, 416
  phrenic, 412
  pudendal, 416
  radial, 413–414
  sciatic, 415–416
  spinal, roots of, 417–425
  subcostal, 414
  suboccipital, 412
  subscapular, 413
  superior gluteal, 415
  superior thoracic, 414
  suprascapular, 413
  thoracic, spinal roots of, 420–422
  thoracodorsal, 413
  tibial, 416
  trigeminal (V cranial), 411
  trochlear (IV cranial), 411

Nerve(s) *(Continued)*
  ulnar, 414
  vagus (X cranial), 411
Normal (grade 5) muscle, 4–5
Nose, muscles of, 276–277, 338–339
  wrinkling bridge of, 277

O

Oblique arytenoid muscles, 308, *309,* 351
Obliquus capitis inferior muscle, 352
  in capital extension, 12–14
  in cervical rotation, 31
Obliquus capitis superior muscle, 352
  in capital extension, 12–14
Obliquus externus abdominis muscle,
  366–367
  in elevation of pelvis, 38t
  in forced expiration, 54–55
  in trunk flexion, 41t
  in trunk rotation, 45–49
Obliquus inferior oculi muscle, 263, 338
  actions of, 266
Obliquus internus abdominis muscle, 367
  in elevation of pelvis, 38t
  in forced expiration, 54–55
  in trunk flexion, 41t
  in trunk rotation, 45–49
Obliquus superior oculi muscle, 263, 338
  actions of, 265
Obturator externus muscle, 391–392
  in hip external rotation, 194–197
Obturator internus muscle, 391
  in hip external rotation, 194–197
Obturator nerve, 415
Occipitofrontalis muscle, *274,* 335
  in raising eyebrows, 275
Ocular muscles, 263–267, 336–338
  actions of, 264–266
Oculomotor (III cranial) nerve, 410–411
Omohyoid muscle, 359
Opponens digiti minimi muscle, 383
  in thumb opposition, 163–165
Opponens pollicis muscle, 387
  in thumb opposition, 163–165
Opposition, thumb, 163–165, 406
Orbicularis oculi muscle, 268t, *270,* 336
  in eye closing, 271
Orbicularis oris muscle, 278, 343
  in lip closing, 280

P

Palate, description of, 301
  muscles of, 298–303, 349–350
  soft, elevation and adduction of, 302–303
Palatoglossus muscle, 290t, 347
  in soft palate elevation and adduction,
    302–303
  in tongue posterior elevation, 295–296
Palatopharyngeus muscle, 298–299, 350
  in occlusion of nasopharynx, 303
Palmar interossei muscles, 384–385
  in finger adduction, 145–147
  in finger MP flexion, 132–134
Palmaris brevis muscle, 383
Palmaris longus muscle, 379–380
  in thumb abduction, 156t
  in wrist flexion, 124t
Paresis, bilateral, tongue and, 293
Pectineus muscle, 388–389
  in hip adduction, 190–194
  in hip flexion, 169t

Pectoralis major muscle, 374
  in scapular adduction, with downward ro-
    tation, 76t
  in scapular depression and adduction, 73t
  in shoulder flexion, 81t, 83
  in shoulder horizontal adduction, 97–101
  in shoulder internal rotation, 105t
Pectoralis minor muscle, 373
  in scapular adduction, with downward ro-
    tation, 76t
  in scapular depression and adduction, 73t
Pelvis. See also *Lumbar* entries.
  elevation of, 38–40
  motions of, 403
Perineum, muscles of, 368–371
Peripheral nerves, 412–416
Peroneal nerve, 416
Peroneus brevis muscle, 397
  in foot eversion, 224–226
Peroneus longus muscle, 397
  in foot eversion, 224–226
    isolation of, 226
Peroneus tertius muscle, 397
  in foot eversion, 224t, 226
Pharyngeal reflex test, 307
Pharynx, in swallowing, 314
  muscles of, 304–307, 347–349
  posterior wall of, constriction of, 307
Phrenic nerve, 412
PIP joints. See *Proximal interphalangeal
  (PIP) joints.*
Piriformis muscle, 391
  in hip external rotation, 194–197
Plantar flexion, ankle, 211–217, 409
  foot eversion with, 224–226
  in infants and children, 252
  in upright motor control, 324
Plantar interossei muscles, 400
  in toe MP flexion, 227t
Plantaris muscle, 396–397
Platysma muscle, 359–360
Poor (grade 2) muscle, 5
Poor- (grade 2-) muscle, 6
Popliteus muscle, 395
Posterior cricoarytenoid muscle, 308,
  350–351
  in vocal cord abduction and adduction,
    312
Procerus muscle, 276, 339
  in wrinkling bridge of nose, 277
Pronation, forearm, 121–123, 405
Pronator quadratus muscle, 378
  in forearm pronation, 121–123
Pronator teres muscle, 378
  in forearm pronation, 121–123
Protrusion, of jaw, 289
  of tongue, 294
Proximal interphalangeal (PIP) joints, of fin-
  gers, flexion of, 135–137, 406, 407
  of toes, extension of, 232–233, 410
Psoas major muscle, 388
  in hip flexion, 169–172
  in trunk flexion, 41t
Psoas minor muscle, 388
  in trunk flexion, 41t
Pudendal nerve, 416
Pudendal plexus, muscles innervated from,
  416
Pyramidalis muscle, 368

Q

Quadratus femoris muscle, 392
  in hip external rotation, 194–197

Quadratus lumborum muscle, 363
  in elevation of pelvis, 38–40
Quadratus plantae muscle, 399
Quadriceps femoris muscle, 394
  in knee extension, 207–210

R
Radial nerve, 413–414
Range of motion, available, 6
Rectus abdominis muscle, 367–368
  in forced expiration, 54–55
  in trunk flexion, 41–44
  in trunk rotation, 45t
Rectus capitis anterior muscle, 355–356
  in capital flexion, 21–23
Rectus capitis lateralis muscle, 356
  in capital flexion, 21–23
Rectus capitis posterior major muscle, 352
  in capital extension, 12–14
  in cervical rotation, 31
Rectus capitis posterior minor muscle, 352
  in capital extension, 12–14
  in cervical rotation, 31
Rectus femoris muscle, 394
  in hip flexion, 169t
  in knee extension, 207–209
Rectus inferior bulbi muscle, 263, 336–338
  actions of, 265
Rectus lateralis bulbi muscle, 263, 336–338
  actions of, 265
Rectus medialis bulbi muscle, 263, 336–338
  actions of, 265
Rectus superior bulbi muscle, 263, 336–338
  actions of, 265
Resistance, application of, in manual muscle
  testing, 2–3
Respiration, motions of, 50–55, 403
  muscles of thorax for, 363–366
Rhomboid major muscle, 372
  in scapular adduction, 69t, 72
    with downward rotation, 76–80
  in scapular elevation, 65t, 68
Rhomboid minor muscle, 372
  in scapular adduction, 69t, 72
    with downward rotation, 76–80
  in scapular elevation, 65t, 68
Risorius muscle, 342
Rotation, cervical, 31, 402
  hip, external, 195–197, 408
    with flexion and external rotation,
      173–175
    internal, 198–201, 408
  scapular, downward, 76–80, 404
    upward, 58–63, 404
  shoulder, external, 102–104, 404
    internal, 105–107, 404
  trunk, 45–49, 403
    in infants and children, 242
Rotatores cervicis muscles, 355
Rotatores lumborum muscles, 362
  in trunk extension, 34t
Rotatores thoracis muscles, 362
  in trunk extension, 34t

S
Sacral plexus, 415
  muscles innervated directly off, 415
  nerves from, 415–416
Salpingopharyngeus muscle, 304, 349
Sartorius muscle, 393
  in hip flexion, 169t

Sartorius muscle (Continued)
  with abduction and external rotation,
    173–175
Scalenus anterior muscle, 358
  in cervical flexion, 24–27
  in cervical rotation, 31
Scalenus medius muscle, 358
  in cervical flexion, 24–27
Scalenus posterior muscle, 358
  in cervical flexion, 24–27
  in cervical rotation, 31
Scaption, shoulder, 88–89
Scapula, abduction of, 58–63, 404
    in infants and children, 253–254
  adduction of, 69–72, 404
    depression and, 73–75
    in infants and children, 255
    with downward rotation, 76–80
  elevation of, 65–68, 403
  motions of, 64, 403–404
Scapulohumeral muscles, 374–376
Sciatic nerve, 415–416
Screening tests, 6
Semimembranosus muscle, 393
  in hip extension, 176
  in knee flexion, 202–203
Semispinalis capitis muscle, 353
  in capital extension, 12–14
  in cervical rotation, 31
Semispinalis cervicis muscle, 354
  in cervical extension, 16–18
  in cervical rotation, 31
Semispinalis thoracis muscle, 361
  in trunk extension, 34, 34t
Semitendinosus muscle, 393
  in hip extension, 176
  in knee flexion, 202–203
Serratus anterior muscle, 372–373
  in scapular abduction, 58–63
  in shoulder scaption, 88t
Serratus posterior inferior muscle, 366
Serratus posterior superior muscle, 366
Shoulder, abduction of, 90–93, 404
    horizontal, 94–96
    in infants and children, 255
  adduction of, 404
    horizontal, 97–101
  extension of, 84–87, 404
  external rotation of, 102–104, 404
  flexion of, 81–83, 404
    in infants and children, 256
  internal rotation of, 105–107, 404
  motions of, 404
  scaption of, 88–89
Shoulder girdle, muscles of, 371–373
Soleus muscle, 396
  in ankle plantar flexion, 211–217
Sphincter ani externus muscle, 371
Sphincter urethrae muscle, 370–371
Spinal nerve roots, 417–425
  cervical, 418–420
  lumbar, 422–424
  lumbosacral, 417–425
  thoracic, 420–422
Spinalis capitis muscle, 353
Spinalis cervicis muscle, 354
Spinalis thoracis muscle, 361
  in trunk extension, 34–37
Spine, cervical, motions of, 402
  lumbar. See also Pelvis; Trunk.
    extension of, 36, 403
    motions of, 403
  thoracic. See also Trunk.
    extension of, 35, 402–403
    motions of, 402–403

Splenius capitis muscle, 353
  in capital extension, 12–14
  in cervical rotation, 31
Splenius cervicis muscle, 354
  in cervical extension, 16–18
  in cervical rotation, 31
Sternocleidomastoid muscle, 358
  combined cervical flexion to isolate, 30
  in cervical flexion, 24–27
  in cervical rotation, 31
Sternohyoid muscle, 359
Sternothyroid muscle, 358–359
Styloglossus muscle, 290, 346–347
  in tongue posterior elevation, 295–296
  in tongue retraction, 295
Stylohyoid muscle, 356
Stylopharyngeus muscle, 304, 348–349
Subclavius muscle, 374
Subcostal nerve, 414
Subcostales muscles, 365
Suboccipital nerve, 412
Subscapular nerve, 413
Subscapularis muscle, 375
  in shoulder internal rotation, 105–107
Superficial peroneal nerve, 416
Superior gluteal nerve, 415
Superior longitudinal muscle, of tongue,
    290t, 291, 347
  in tongue channeling, 296
  in tongue curling, 297
Superior pharyngeal constrictor muscle, 304,
    348
Superior thoracic nerves, 414
Supination, elbow, in infants and children,
    259
  forearm, 118–120, 405
Supinator muscle, 378
  in forearm supination, 118–120
Suprahyoid muscles, 284t, 285, 290t,
    356–357
  in jaw opening, 286–287
Supranuclear lesions, tongue and, 293
Suprascapular nerve, 413
Supraspinatus muscle, 375
  in shoulder abduction, 90–93
  in shoulder flexion, 81–83
  in shoulder scaption, 88–89
Swallowing, 314–317
  common problems in, 315t
  elevation of larynx in, 312
  muscle actions in, 314–315
  testing of, 315–317

T
Temporalis muscle, 284, 286, 345
  in jaw closure, 287–288
Temporoparietalis muscle, 335
Tensor fasciae latae muscle, 391
  in hip abduction, 182t
    from flexed position, 186–189
  in hip flexion, 169t
  in hip internal rotation, 198–201
Tensor veli palatini muscle, 298–299, 349
  in soft palate elevation and adduction,
    302–303
Teres major muscle, 376
  in shoulder extension, 84–87
  in shoulder internal rotation, 105t
Teres minor muscle, 375–376
  in shoulder external rotation, 102–104
  in shoulder horizontal abduction, 94t
Thoracic extension, 35, 402–403
Thoracic nerves, spinal roots of, 420–422

Thoracic region, nerves of, 414
Thoracic spine. See also *Trunk*.
    extension of, 35, 402–403
    motions of, 402–403
Thoracodorsal nerve, 413
Thorax, muscles of, for respiration, 363–366
Thumb, abduction of, 156–159, 406
    adduction of, 160–162, 406
    extension of, interphalangeal, 154–155, 406
        metacarpophalangeal, 152–153, 405
    flexion of, interphalangeal (IP), 148, 150, 405
        metacarpophalangeal, 148–149, 405
    motions of, 405–406
    muscles acting on, 385–387
    opposition of, 163–165, 406
Thyroarytenoid muscle, 308t, *309*, 351
Thyrohyoid muscle, 359
Tibial nerve, 416
Tibialis anterior muscle, 395–396
    in foot dorsiflexion and inversion, 218–220
Tibialis posterior muscle, 396
    in foot inversion, 221–223
Toe(s), extension of, 232–233, 410
    flexion of, distal joint, 230–231, 409–410
        proximal joint, 227, 229, 409
    great. See *Hallux*.
    muscles acting on, 397–401
Tongue, and bilateral paresis, 293
    and supranuclear vs infranuclear lesions, 293
    channeling of, 296
    curling of, 297
    deviation of, 294
    examination of, 292
    motions of, criteria for grading, 297
    muscles of, 290–297, 346–347
        extrinsic, 290, 346–347

Tongue, muscles of *(Continued)*
        intrinsic, 290t, *291*, 347
    posterior elevation of, 295–296
    protrusion of, 294
    retraction of, 295
    unilateral weakness of, 292–293
Trace (grade 1) muscle, 5
Transverse arytenoid muscle, 308, *309*, 351
Transverse lingual muscle, 290t, *291*, 347
    in tongue channeling, 296
Transversus abdominis muscle, 367
    in forced expiration, 54–55
    in quiet inspiration, *50*
Transversus menti muscle, 342–343
Transversus perinei profundus muscle, 369
Transversus perinei superficialis muscle, 369
Transversus thoracis muscle, 365
Trapezius muscle, 371–372
    in scapular abduction, 58t
    in scapular adduction, 69–72
    in scapular depression and adduction, 73–75
    in scapular elevation, 65–68
    in shoulder scaption, 88t
Triceps brachii muscle, 377
    in elbow extension, 114–117
    in shoulder extension, 84t
    in shoulder horizontal abduction, 96
Trigeminal (V cranial) nerve, 411
Trochlear (IV cranial) nerve, 411
Trunk, extension of, 34–37
    in infants and children, 243–244
    lumbar, 36, 403
    lumbar and thoracic, 36–37
    thoracic, 35, 402–403
    flexion of, 41–44, 403
    in infants and children, 240–241
    muscles of, 360–371
    rotation of, 45–49, 403
    in infants and children, 242

U
Ulnar nerve, 414
Upper extremity. See also specific structure.
    motions of, 403–407
    muscles of, 371–387
        innervation of, 412–414
Upright motor control, 320–324
Uvula, muscle of. See *Musculus uvulae*.

V
Vagus (X cranial) nerve, 411
Vastus intermedius muscle, 394
    in knee extension, 207–209
Vastus lateralis muscle, 394
    in knee extension, 207–209
Vastus medialis longus muscle, 394
    in knee extension, 207–209
Vastus medialis oblique muscle, 394–395
    in knee extension, 207–209
Vertebrohumeral muscles, 373–374
Vertical lingual muscle, 290t, *291*, 347
    in tongue channeling, 296
Vocal cords. See also *Larynx*.
    abduction and adduction of, 312

W
Wrist, extension of, 128–131, 405
    flexion of, 124–127, 405
    motions of, 405
    muscles acting at, 378–380

Z
Zero (grade 0) muscle, 5
Zygomaticus major muscle, 278, 282, 342
Zygomaticus minor muscle, 342